D1067977

THE CURIOUS MISTER CATESBY

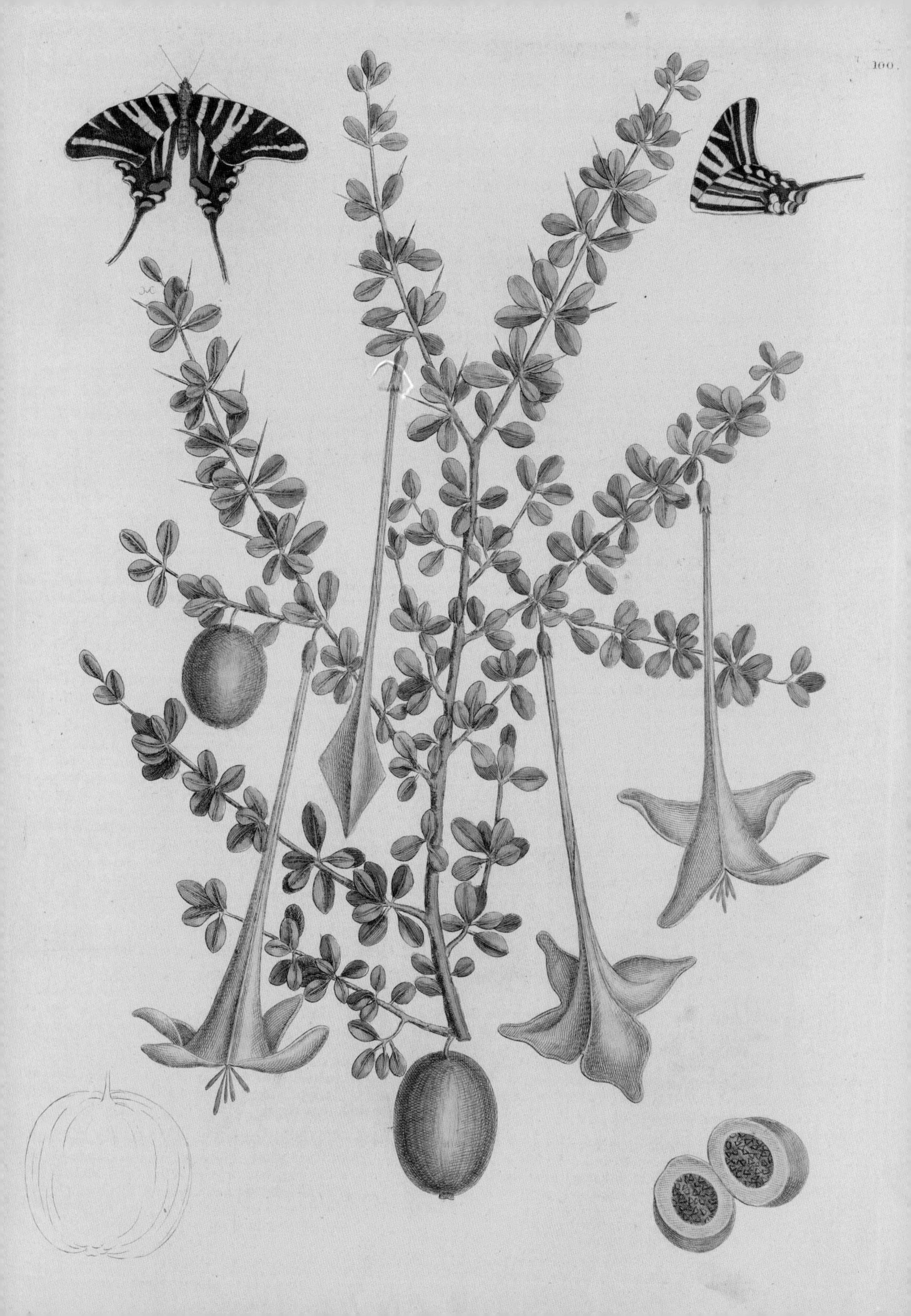
100.

THE CURIOUS
Mister Catesby

a "truly ingenious" naturalist explores new worlds

EDITED FOR THE CATESBY COMMEMORATIVE TRUST BY

E. CHARLES NELSON AND DAVID J. ELLIOTT

Foreword by Jane O. Waring

The Catesby Commemorative Trust

THE UNIVERSITY OF GEORGIA PRESS *Athens & London*

IMAGES: Frontispiece (p. ii), *Catesbaea spinosa*, lily thorn, a native of the Bahamas, named in honor of Mark Catesby (plate 100, M. Catesby, 1743, *The natural history of Carolina . . .*, volume II (digital realization of original etchings by Lucie Hey and Nigel Frith, DRPG England; courtesy of the Royal Society ©); copyright page (p. iv), Mark Catesby's autograph on the verso of a sketch of the American lotus (*Nelumbo lutea*) in Oxford University Herbaria (see figures 13-2 and 13-3, p. 178); p. vi, Southern catalpa, plate 49, M. Catesby, 1730, *The natural history of Carolina . . .*, volume I (digital realization of original etchings by Lucie Hey and Nigel Frith, DRPG England; courtesy of the Royal Society ©).

Published by the University of Georgia Press
Athens, Georgia 30602
www.ugapress.org

Designed by Erin Kirk New
Set in Adobe Caslon Pro
Printed and bound by Pacom Korea

The paper in this book meets the guidelines for permanence and durability of the Committee on Production Guidelines for Book Longevity of the Council on Library Resources.

Most University of Georgia Press titles are available from popular e-book vendors.

Printed in Korea
19 18 17 16 15 c 5 4 3 2

Library of Congress Cataloging-in-Publication Data

The curious Mister Catesby : a "truly ingenious" naturalist explores new worlds / edited for the Catesby Commemorative Trust by E. Charles Nelson and David J. Elliott ; foreword by Jane O. Waring.
pages cm. — (Wormsloe Foundation nature book)
Includes bibliographical references and index.
ISBN 978-0-8203-4726-4 (hardcover : alk. paper) 1. Catesby, Mark, 1683–1749. 2. Catesby, Mark, 1683–1749—Influence. 3. Catesby, Mark, 1683–1749 Natural history of Carolina, Florida, and the Bahama Islands. I. Nelson, E. Charles. II. Elliott, David J. (David John), 1935–
QH31.C35C87 2015
508.092—dc23 2014032431

British Library Cataloging-in-Publication Data available

ENCOURAGERS OF THIS WORK

Addison Publications
Addlestone Library,
College of Charleston
Arader Galleries
Audubon Prints and Books

Sylvia W. Bacon
Barrier Islands Foundation
Mr. and Mrs. Craig Barrow III
J. Anderson Berly III
Mr. and Mrs. Charles P. Berolzheimer II
Blackwing Foundation
Boxwood Garden Club
J. Elizabeth Bradham
Mr. and Mrs. Robert Brooke

Carlsen Design
Carolina Antiques Maps & Prints
Austin E. Catts
Ceres Foundation
Charleston Library Society
Charleston Mercury
Charleston Museum
Charleston Renaissance Gallery
Dr. Juliet Clutton-Brock
Lori Cohen
Cowardin's Jewelers

Edward Daniell, CPA
Charles P. Darby III
Discovery Editions
Gina Douglas
Charles H. P. Duell

Mr. and Mrs. O. Ralph Edwards
Sallie Lou Elliott
Professor & Mrs. W. Hardy Eshbaugh

Garden Club of America
Garden Club of Virginia
Gibbes Museum of Art

Mr. and Mrs. Christopher Hammond
Lou Hammond and Associates
Mary E. S. Hanahan
Elizabeth Wakeman Henderson
Foundation
Herzman-Fishman Foundation
Shauneen Hutchinson

Island Arts

Dylan Jones

Mr. and Mrs. Peter Kellogg
Town of Kiawah Island
Kiawah Island Golf Resort
Kurt R. Klaus, Esq.
Brantley Carter Bolling Knowles
Dr. John C. Kotz

Hugh C. Lane Jr.
Douglas B. Lee
Lloyd Library & Museum
Cartter Lupton

Osborne Mackie
Elizabeth P. McLean
Craig H. Metz
Middleton Place
Mills Bee Lane Foundation

National Geographic Society
National Museum of Natural History
National Society of Colonial Dames
of America in the Commonwealth
of Virginia
Dr. Sue Nelson
Cheryl Newby Gallery

Old Salem Museum and Gardens
Mr. & Mrs. James Carlisle Oxner Jr.

Mr. and Mrs. Scott Parker
Pathfinder Fund
Mr. & Mrs. Henry M. Paulson Jr.

Preservation Society of
Charleston
Publications Group, United
Kingdom

Professor John H. Rashford
His Grace The Duke of
Richmond and Gordon
John H. Rivers Jr.
Colonel J. G. Richards Roddey

Tina R. Schell
Thomas Schreck
Ginger Scully
Elfrida Barrow Sinkler
Smithsonian Libraries
Society for the History of
Natural History

Mrs. Leigh Thompson
Three Chopt Garden Club
Touchpoint Communications,
LLC
Laura Towers
Tuckahoe Garden Club of
Westhampton
Kirstie Tucker
Willard Sherman Turell
Herbarium, Miami University

United States Botanical Garden

Dr. Eveline Waring
Jane O. Waring
(Mrs. Charles W. Waring)
Washington Fine Properties
Wilton House Museum
Wormsloe Foundation Inc.

T. 49
MC
Icterus minor. Mas & Femina.
The basterd Baltimore Bird.
Bignonia Urucu folijs &c.
The Catalpah Tree.

CONTENTS

FOREWORD

When Mark Catesby set sail from England for Virginia in the early months of 1712 he began a journey that concluded, in one sense, when the eleventh part of "the most magnificent Work I know of, since the Art of Printing has been discover'd" was delivered to his "Encouragers." He was not quite twenty-nine years old and a bachelor when he started that voyage in the company of his eldest sister, Mrs. Elizabeth Cocke. Thirty-five years would elapse before he completed his monumental *The natural history of Carolina, Florida and the Bahama islands*, by which time he had a son and a daughter. In October 1747, three months after finishing his book, he married Elizabeth Rowland, who had been his partner for at least the past seventeen years.

Virginia must have enthralled Mark Catesby, yet we have only very scant records of what he saw or did during his seven years there. In 1722, he set sail for North America again, this time under the auspices of the Royal Society of London, then under the leadership of Sir Isaac Newton. Catesby landed at the port of Charleston, determined and prepared. The Carolinas and the Bahamian islands offered extraordinary opportunities for his exploration and documentation of their natural history. When he returned to England after another three years in the New World he brought back with him records of the plants and animals, and indeed the people, he had encountered.

The place he returned to, London, was a bustling, sprawling city that had recovered from the devastating fire of 1666, which destroyed so much. It was a lively place, sparkling with new buildings, the most magnificent being Sir Christopher Wren's masterpiece, St. Paul's Cathedral. At the cathedral's dedication on 2 December 1697, the sermon was preached by the Bishop of London, the great plant collector Henry Compton, to whom Catesby would send a parcel of seeds from Virginia. Settling almost in the great cathedral's shadow in the parish of St. Giles's Cripplegate, Mark Catesby began another journey, learning the skills of the engraver, to allow him to put on paper, then the only medium for international communication, a record of what he had seen: the birds and fishes, insects and snakes, flowers and fruits of the New World. When George Frideric Handel's oratorio *Messiah* received its London premiere on 23 March 1743, Catesby was close to completing the tenth part of his book. He must have been in London when Handel's *Music for the Royal fireworks* was first performed on 21 April 1749, an event that caused a three-hour traffic jam on London Bridge.

Mark Catesby died just before Christmas 1749 at age sixty-six. His burial place at St. Luke's Church, Old Street, London, is no longer known. Instead, his memorial comprises perhaps a hundred copies of his very remarkable book:

they are the surviving copies; perhaps another hundred have been "lost." "The whole was done within my house, and by my own hands," he wrote on the last page. In simple terms, this means that Mark Catesby not only engraved each of the 220 plates but also, once these were printed, colored the prints (perhaps not all of them). The total number needed for the first edition was probably thirty thousand.

Yet, as this book demonstrates, Mark Catesby was not simply an artist and consummate craftsman. He was also a man who engaged in scholarly debates about birds and who sought to promote hardy American plants as subjects for gardens in both Europe and North America. He was keenly aware of his contemporaries and learned (and copied) from them, and his voyage to Virginia in 1712 was the catalyst.

The Catesby Commemorative Trust celebrated his masterpiece in conjunction with the three hundredth anniversary of his first voyage to the New World by holding a symposium during November 2012 in three centers: Washington, D.C.; Richmond, Virginia; and Charleston, South Carolina. The publication of *The curious Mister Catesby: a "truly ingenious" naturalist explores new worlds* is the culmination of that symposium. The Trust believes this volume to be the most comprehensive and accurate book written to date about Mark Catesby.

JANE O. WARING (MRS. CHARLES W. WARING)
President
Catesby Commemorative Trust

INTRODUCTION

The objective of this book is to review the life and work of Mark Catesby in a way that will be valuable both to the interested and informed general reader and to the scholarly community. Included is a new biography of Catesby based on research using original sources, as well as a history of his famous book, *The natural history of Carolina, Florida and the Bahama islands*. Exploration of conditions in the different places where he lived, worked, and traveled is the topic of several chapters. The book also covers other naturalists who influenced Catesby, examining their lives and works to provide a broad context of the world as he, and they, knew it. The chapters about Catesby's impact on those who followed him, whether scientists, artists, or gardeners, demonstrate that he was not an isolated figure but a man at the center of London's horticultural and learned communities.

The present authors are recognized as leading experts in their respective fields. Readers will note that two distinguished Catesby scholars, Dr. Amy Meyers and Dr. Alan Feduccia, have not written chapters. This was only due to their being unavailable; both were very helpful at earlier stages in the creation of this book. The selection and recruitment of authors were undertaken by Leslie Overstreet (Curator, Natural-History Rare Books, Smithsonian Libraries), Gina Douglas (Honorary Archivist of the Linnean Society of London), and David Elliott (Executive Director of the Catesby Commemorative Trust).

This book contains important new information about Mark Catesby, signaling major changes to the "traditional" views of his social and economic status. This is principally due to the very significant increase in the accessibility of original sources mainly through the Internet.

We have, with one exception, emphasized factual and objective conclusions based on sound evidence and original sources (which are cited in the notes and bibliography). The exception is based on the fact that there is remarkably little information about Mark Catesby as an individual, and no portrait of him is known. William Byrd II wrote, when Catesby was in Virginia, that "he was so merry he sang," but the other pen-portraits were written when he was an old man. One, by his friend Emanuel Mendez da Costa, described Mark as "tall, meagre, hard favoured, and [with] a sullen look." The other was penned by Linnaeus's "disciple" Pehr Kalm, who found a short-sighted Catesby willing to discuss the relative qualities of colonial punches.

Therefore, a conjectural written portrait developed from a solid technical and professional base seems to be useful, and this is provided here by Cynthia P. Neal, who was engaged to produce the successful public broadcasting documentary *The Curious Mister Catesby* on the recommendation of officials at the

U.S. Fish and Wildlife Service who considered her to be the best producer of wildlife conservation films working in the United States. Her first step was to hold a pre-script conference in the Lowcountry with Henrietta McBurney, Alan Feduccia, Suzanne Linder Hurley, Leslie Overstreet, and scriptwriter Mike Purswell. Unanimous agreement was reached on the direction of the film. All participants, plus Dr. Amy Meyers, reviewed both the script and the completed documentary before its public exhibition. The premiere was at the Royal Society in London, where the documentary was complimented by Sir David Attenborough and received a favorable review in *The Times*.

Overall, we believe that we have advanced knowledge of Mark Catesby and his explorations, collections, artwork, and publications and that we have laid a new basis for assessing his importance in documenting the natural history of the New World. However, we do not believe we have had the last word. Additional sources will be found, and established sources will be reexamined. All of us involved in writing this book are pleased to have made our contribution, and we look forward to what comes next.

E. CHARLES NELSON   DAVID J. ELLIOTT

CATESBY CHRONOLOGY

	Mark Catesby and his family	*Relevant concurrent events*
1683	Mark Catesby born 24 March; baptized 30 March in Castle Hedingham	William Dampier started his first circumnavigation
1684		Coldest winter in living memory in Britain (December 1683–March 1684)
1685		King Charles II died 6 February; succeeded by his brother James as King James II of England and Ireland and VII of Scotland; George Frideric Handel born 23 February; Johann Sebastian Bach born 21 March
1686	John Catesby (eldest brother) admitted to Clifford's Inn, 16 February	New York granted city charter
1687	John Catesby (eldest brother) enrolled as pensioner in Queens' College Cambridge, 2 November	Isaac Newton's *Philosophiæ naturalis principia mathematica* published
1688	Ann Catesby (second sister) born 20 March, baptized 3 April	King James II and VII deposed and flees to France; approximate year rice production begins in South Carolina; "Carolina Gold" was to make it wealthiest colony in America
1689		King William III (Prince of Orange) and Queen Mary II crowned; Joseph Goupy born in France
1690	John Catesby (eldest brother) died, buried 5 December at Castle Hedingham	Battle of the Boyne ends King James II's effort to regain throne; John Carteret born 20 March (sponsor of Mark Catesby at Court in 1729)
1691		William Dampier returns to England via Cape of Good Hope; Thomas Knowlton born (gardener, friend of Mark Catesby)
1692	Jekyll Catesby (elder brother) admitted to Inner Temple, 23 January	John Banister accidentally shot in Virginia
1693		Samuel Dale published *Pharmacologia*; George Edwards born 3 April (naturalist, friend of Mark Catesby); Salem, Massachusetts, witch trials ended with twenty executions
1694		Richard Boyle (Earl of Burlington) born 25 April (was to subscribe for three copies of *The natural history of Carolina . . .*)

	Mark Catesby and his family	*Relevant concurrent events*
1695		Henry Purcell died 21 November; music he composed for Queen Mary's funeral, held in Westminster Abbey on 5 March, was also performed at his own funeral
1696		James Oglethorpe (founder of state of Georgia) born 22 December
1697	John Catesby (youngest brother) born, baptized 30 June	Dampier published *A new voyage round the world*; Christopher Wren's new St. Paul's Cathedral in London dedicated
1698		First successful steam pump patented by Thomas Savery
1699	Elizabeth Catesby (sister) married William Cocke 4 September in Sudbury without her father's consent	Maria Sibylla Merian sails to Surinam; Dampier, commanding HMS *Roebuck*, sails to New Holland (Australia); John Bartram born 23 March
1700	John Catesby (father) made his will 18 November, naming three sons (Jekyll, Mark, John) and two daughters (Anne and "disobedient" Elizabeth)	John Lawson explored from Charleston into North Carolina; Gregorian calendar adopted in western Europe but not in Britain and Ireland
1701		HMS *Roebuck* wrecked on Ascension Island; malaria forced Maria Sibylla Merian to return to the Netherlands
1702		King William III died; succeeded by his sister-in-law, Anne Stuart
1703	John Catesby (father) died, buried 12 November in Castle Hedingham	War of the Spanish Succession (Queen Anne's War) started; Isaac Newton elected President of the Royal Society; Dampier published part 1 of *A voyage to New Holland*; William Sherard appointed Consul in Smyrna; Great Storm kills many in southern England
1704	Probate granted on John Catesby's estate	British and Dutch forces capture Gibraltar; Isaac Newton published *Opticks*
1705		Reverend John Ray (naturalist) died 17 January; Merian published *Metamorphosis insectorum Surinamensium*; John Lawson settled near Pamlico River, North Carolina
1706		Leonard Plukenet (botanist and gardener to Queen Mary II) died 6 July
1707		Carl Linnaeus born 23 May in Råshult, Småland, Sweden

	Mark Catesby and his family	*Relevant concurrent events*
1708	Elizabeth Catesby (née Jekyll; mother) died, buried 7 September in Castle Hedingham	Dampier started on third circumnavigation; Georg Dionysus Ehret born 30 January in Heidelberg, Germany; St. Paul's Cathedral completed
1709		John Lawson published *A new voyage to Carolina* in London; Dampier published second part of *A voyage to New Holland*
1710	Mark, Anne, and Jekyll Catesby jointly sign indenture to sell property in Sudbury; Dr. William Cocke arrives in Virginia	The world's first copyright law, the Statute of Anne, came into effect; Royal Society of Sciences founded in Uppsala, Sweden
1711		Tuscarora Indian War starts; John Lawson killed by Tuscaroras in North Carolina
1712	Mark Catesby arrived in Virginia, 22 April, with his sister, Mrs. Elizabeth Cocke; William Byrd II was his principal host	First working steam engine built by Thomas Newcomen
1713	Mark Catesby sent plants from Virginia to Samuel Dale and Henry Compton (Bishop of London)	Treaty of Utrecht ended Queen Anne's War; Henry Compton, Bishop of London and avid gardener, died 7 July
1714	Mark Catesby made round trip to Jamaica, also visited Bermuda	King George I (Elector of Hanover) succeeded Queen Anne
1715	First printed record of Mark Catesby and plants associated with him in *Philosophical transactions of the Royal Society*	Treaty signed ending Tuscarora Indian War; Yamassee Indian War started and almost destroyed South Carolina; William Dampier died (will dated 29 November 1714, proved 23 March)
1716		William Sherard returned to England from Smyrna
1717	Jekyll Catesby (brother) died, buried 21 September in Castle Hedingham	Yamassee Indian War ended (Cherokees abandoned their alliance); Maria Sibylla Merian died 13 January
1718		Blackbeard (pirate) blockaded Charlestown, South Carolina, later (22 November) killed in battle at Ocracoke Inlet, North Carolina; War of the Quadruple Alliance (Spain versus Great Britain, France, and the Holy Roman Empire) started
1719	Mark Catesby returned to England before 15 October	Lords Proprietors overthrown; South Carolina became Royal colony; Royal Navy stationed ships in Charleston to protect against piracy

	Mark Catesby and his family	*Relevant concurrent events*
1720	Mark Catesby introduced to William Sherard by Samuel Dale in letter dated 11 May	Treaty of The Hague ends War of the Quadruple Alliance
1721	Royal Society endorsed Mark Catesby's venture to South Carolina: sponsors included new Royal Governor Francis Nicholson, Sir Hans Sloane, William Sherard, the Duke of Chandos, Dr. Richard Mead, and seven others	Colonel Francis Nicholson appointed first Royal Governor of South Carolina
1722	Mark Catesby sailed to South Carolina, arrived Charleston 3 May; letters to William Sherard dated 5 May, 20 June, and 9 and 10 December extant	Thomas Fairchild published *The city gardener*
1723	Mark Catesby explored South Carolina; letters to William Sherard, Peter Collinson, and Hans Sloane extant	Antonio Vivaldi composed *The four seasons*; poaching becomes a capital offense in Great Britain
1724	Mark Catesby's proposal for trip to Mexico not approved by sponsors; letters to William Sherard and Hans Sloane extant	Carl Linnaeus entered Växjö's cathedral school
1725	Mark Catesby informed Sherard and Sloane (only three letters, dated 5 and 10 January, extant) that he was departing for the Bahamas; in Bahamas on Christmas Day 1725	Francis Nicholson returned to London; succeeded as Royal Governor by Arthur Middleton
1726	Mark Catesby left Bahamas, returned to England	Landgrave Thomas Smith attempts overthrow of Middleton; arrested by Lieutenant George Anson, later First Lord of the Admiralty
1727		King George I died; succeeded by King George II and Queen Caroline; Anglo-Spanish War over Gibraltar and Panama started; Sir Hans Sloane elected President of the Royal Society
1728		Last raid by Yamassee tribe in South Carolina; William Sherard died
1729	Mark Catesby presented at Court in late May and gave Queen Caroline part 1 of *The natural history of Carolina, Florida and the Bahama islands*; copy presented to the Royal Society on 22 May	Queen Caroline appointed Queen Regent; Treaty of Seville concluded Anglo-Spanish War; last interests of Lords Proprietors in South Carolina liquidated; Thomas Fairchild died 10 October
1730	Parts 2 and 3 of *The natural history of Carolina . . .* presented to the Royal Society on 8 January and 19 November, respectively	*The natural history of Carolina . . .* subscriber the Earl of Wilmington appointed President of the Council
1731	Mark Catesby, son of Mark Catesby and Elizabeth Rowland, born 15 April, baptized 20 April; part 4 of *The natural history of Carolina . . .* presented to the Royal Society on 4 November	Coast guards in Cuba cut off Robert Jenkins's ear, incident led to "War of Jenkins's Ear" in 1739; Daniel Defoe, author of *Robinson Crusoe*, died 24 April

	Mark Catesby and his family	Relevant concurrent events
1732	John Catesby, son of Mark Catesby and Elizabeth Rowland, born 6 March, baptized 30 March, buried 25 August; part 5 of *The natural history of Carolina* . . . presented to the Royal Society on 23 November (completing text and plates for the first volume)	Royal charter granted to found Georgia Colony; J. J. Dillenius published *Hortus Elthamensis*
1733	Caroline Catesby, daughter of Mark Catesby and Elizabeth Rowland, baptized 27 May, buried 9 August; Mark Catesby is elected Fellow of the Royal Society and visited Holland	First performance of Handel's opera *Orlando* in London; flying shuttle patented by John Kay transformed weaving
1734	Part 6 of *The natural history of Carolina* . . . presented to the Royal Society on 4 April	Hogarth's Act passed to protect engravers against piratical copying
1735		Linnaeus moved to the Netherlands and met Johan Gronovius, published first edition of *Systema naturae*
1736	Part 7 of *The natural history of Carolina* . . . presented to the Royal Society on 15 January	Crown Prince Frederick marries Augusta of Saxe Gotha; Linnaeus visited London to meet Sir Hans Sloane, also met J. J. Dillenius in Oxford
1737	Ann Catesby, second daughter of Mark Catesby and Elizabeth Rowland, baptized 27 December; part 8 of *The natural history of Carolina* . . . presented to the Royal Society on 7 April	Queen Caroline died 20 November; *Magnolia grandiflora* in Sir Charles Wager's garden observed and painted by Georg Dionysus Ehret, August
1738		Linnaeus returned to Sweden from the Netherlands
1739	Part 9 of *The natural history of Carolina* . . . presented to the Royal Society on 7 June	Samuel Dale died 6 June; War of Jenkins's Ear (United Kingdom versus Spain in West Indies and Florida) started
1740		Wilmington, North Carolina, named for Prime Minister and Catesby "Encourager" the Earl of Wilmington; start of the War of Austrian Succession
1741		Linnaeus appointed professor of medicine at Uppsala University; Sir Hans Sloane resigned as President of the Royal Society of London, 16 November
1742		Handel's oratorio *Messiah* first performed in Dublin, 13 April; Battle of Bloody Marsh on 7 July ended Spanish effort to retain Georgia

	Mark Catesby and his family	*Relevant concurrent events*
1743	Part 10 of *The natural history of Carolina . . .* presented to the Royal Society on 15 December, and "An Account" published (completing text and plates of volume II, dedicated to Augusta, Princess of Wales); Mark Catesby and George Edwards elected members of Society of Gentlemen of Spalding, 29 December	Spencer Compton (Earl of Wilmington) died 2 July
1744		William Byrd II and the Duke of Chandos died; French invasion of England thwarted by a storm; twenty-two-year-old Eliza Lucas reintroduced indigo production in South Carolina, resulting in fabulous wealth for the colony
1745		Initial successes of Jacobite (Stuart) uprising created panic in London; first singing of British national anthem "God Save the King"
1746		Linnaeus published "Stewartia"; Battle of Culloden ended Jacobites' efforts to regain British throne
1747	Mark Catesby and Elizabeth Rowland married October; Mark Catesby read "Of birds of passage" to Royal Society on 5 March; "Appendix" to *The natural history of Carolina . . .* presented to the Royal Society on 2 July	Johann Jacob Dillenius died 2 April
1748	Pehr Kalm visited Mark Catesby, May	End of War of Jenkins's Ear; major fire in City of London not far from Mark Catesby's home
1749	Mark Catesby died 23 December, buried in St. Luke's Churchyard, Old Street, London	The first official performance of Handel's *Music for the Royal fireworks*
1753	Elizabeth Catesby made will 4 January; died and was buried 18 February, probate granted 29 August	

THE CURIOUS MISTER CATESBY

1

"The truly honest, ingenious, and modest Mr Mark Catesby, F.R.S.": documenting his life (1682/83–1749)

E. CHARLES NELSON

Most of the individuals who grew up in rural England during the last decades of the seventeenth century have left no records of their lives except perhaps for the briefest entries in a parish register of their baptism, marriage, or burial. The same is true for the vast majority of the inhabitants of the burgeoning metropolis of London, where Mark Catesby lived for at least two decades before 1750.

Mark Catesby, honest, ingenious, modest,[1] and, by his own admission, curious, is a little different and has left traces. While no portrait of him is known, Emanuel Mendez da Costa (1717–1791) stated that, when he knew him late in life, Catesby had a "tall, meagre, hard favoured, and sullen look, and was extremely grave or sedate, and of a silent disposition; but when he contracted a friendship was communicative, and affable."[2] He was "well known to, and much esteemed by, the curious of this and other nations, and died much lamented by his friends," to quote the brief obituary published in January 1750 in *The gentleman's magazine*.[3] Yet, we still do not know much about Mark Catesby's life, although what is known amounts to a great deal more than for countless of his fellow Englishmen who lived between 1683 and 1749.

The register of the parish of Castle Hedingham, in the county of Essex, is unusually expansive about the baptisms of the sons of John Catesby. The entry for Mark (figure 1-1), probably John's fifth son, is precise, giving not just the date of his baptism but also the date of his birth, a fact noted for only a small proportion of the children baptized in St. Nicholas's Parish Church (figure 1-2) during the late 1600s: "Mark Catesby son of John Catesby gent and Elizabeth his wife Baptiz March 30th Natus March 24th 1682."[4] The register is also revealing about the social status of John Catesby—he was a gentleman, and the only other person described by that term at this period was Nicholas Jekyll, who was Mark's maternal grandfather. Thus the first glimpse of Mark Catesby is as a baby a few days after the start of the year 1683, for he had been born on the eve of a new year, on the eve of the Feast of the Annunciation, which was then, under the Julian calendar, New Year's Day.[5]

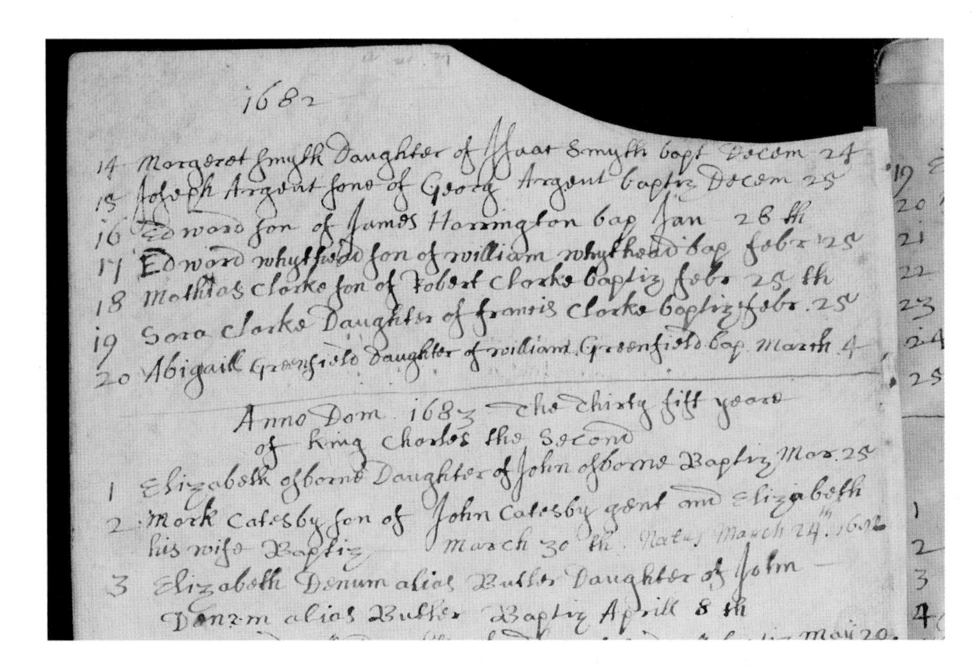

1682

14 Margaret Smyth Daughter of Isaac Smyth bapt Decem 24
15 Joseph Argent sone of Georg Argent baptiz Decem 25
16 Edward son of James Harrington bap Jan 28 th
17 Edward whythead son of william whythead bap febr 25
18 Mathias Clarke son of Robert Clarke baptiz febr 25 th
19 Sara Clarke Daughter of francis Clarke baptiz febr 25
20 Abigail Greenfield daughter of william Greenfield bap March 4

Anno Dom 1683 the thirty fifth yeare of king Charles the Second

1 Elizabeth osborne Daughter of John osborne Baptiz Mar 25
2 Mark Catesby son of John Catesby gent and Elizabeth his wife Baptiz March 30 th. Natus March 24th 1682
3 Elizabeth Denum alias Butler Daughter of John Denum alias Butler Baptiz Aprill 8 th

FIGURE 1-1. The entry recording Mark Catesby's baptism in the register of St. Nicholas's Parish Church, Castle Hedingham, Essex. (Reproduced by courtesy of Essex Record Office, Chelmsford.)

FIGURE 1-2. St. Nicholas's Parish Church, Castle Hedingham, Essex. (© E. C. Nelson 2012.)

There are no documents extant from the next two decades in which Mark Catesby's name appears. We know nothing about his boyhood or education. We must wait until after his father's death to find more written evidence of his life. Stepping back three decades, John Catesby had married on or, more probably, after 16 May 1670, the date of the marriage license issued by the Dean and Chapter of Westminster Abbey. In this, John was described as "of Sudbury, Suffolk, Gent[leman]., Bach[elo]^{r}, ab[ou]^{t} 28." His bride, Elizabeth Jekyll, age about eighteen, "of Hedingham Castle, Essex, Sp[inste]^{r},"[6] was marrying with her father's consent. The license permitted the marriage to be solemnized in St. Andrew's Holborn, or Gray's Inn, or Charterhouse Chapel, all in London. There are other documents that link John Catesby to the town of Sudbury in Suffolk: although it is in a different county, Sudbury is only about eight miles, or about two hours on foot, to the northeast of Castle Hedingham. In 1669 John was appointed the town clerk of Sudbury, and in 1673 he became mayor for the first time. He held the office on six occasions and was named as such in the Letters Patent of King James II, dated 26 March 1685, granting a new charter and making Sudbury a "free incorporated borough."[7]

Mark was surely present on 12 November 1703 when his father was laid to rest at St. Nicholas's Church in the same graveyard as Mark's brothers John (buried in 1690), Samuel (buried in 1676), and Henry (buried in 1678). Probate was granted on 2 January 1704/5, and in the probate records there is a transcript of John Catesby's will, which had been witnessed and sealed on 18 November 1700.[8] Three sons and two daughters are named: Jekyll, Mark, John, Ann, and "my disobedient Daughter Elizabeth" (figure 1-3). Jekyll, Mark's older brother, was bequeathed properties in St. Dunstan's parish in London. Their young brother, John,[9] who was a boy of only about seven, would inherit the family home and surrounding gardens, a farm called the Holgate, at Sudbury. Mark's inheritance was substantial:

> I give and devise to my son Mark his Heirs Executors and Administrators All my Houses and buildings with the Yards Gardens and Orchards to every of them belonging situate and being in Sudbury in the County of Suff and in the parish of St Bridgett ats Brides London—and all other my Houses in London not before bequeathed (except my House or Messuage hereafter mentioned and building thereto belonging called the Holgate) I give him also my seaven acre piece of Land lying at the upper end of Gallow ffields in Sudbury aforesaid and my piece of Land called the Harppiece or otherwise And my piece of Lands lying behind my House in Mr Gainsborowes occupation in Sudbury aforesaid And all my Lands in Chilton and Great Cornard in the County of Suffolk

There were other bequests that indicate John Catesby senior was a well-to-do landed gentleman with extensive properties in the very heart of London, including a "House called or known by the name and sign of the Man in the Moon," as well as in Sudbury, and the advowson and right of patronage of the

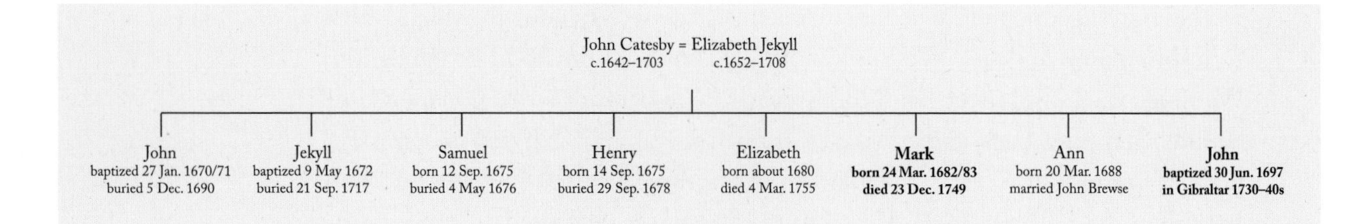

FIGURE I-3. Catesby family tree, showing Mark Catesby, his parents, and his siblings.

rectory of Pentlow in Essex. On the other hand, there is no evidence that his profession was as a lawyer or solicitor, as often stated.

Thus when he was just twenty-one years old, Mark Catesby came into possession of houses in Sudbury,[10] with their gardens and orchards, and several plots of land in the nearby villages of Chilton and Great Cornard, and also houses in London. Some of his property was in the vicinity of London's Fleet Street, a short distance from Sir Christopher Wren's new St. Paul's Cathedral, then still being built following the destruction of the previous building in the Great Fire of London in 1666. St. Bride's Parish Church, another of Wren's post-conflagration masterpieces, with its magnificent wedding-cake steeple, had been completed in 1703.

An indenture signed on 25 October 1710 by Mark, Ann, and Jekyll is the first occasion on which our subject is known literally to have made his own mark (figure I-4).[11] This document detailed property that the three wished to sell, including houses and land in Sudbury that Mark had inherited a few years earlier.

Another year passes before Mark is again sighted, and this time there is a contemporary record of him by an independent person. On 23 April 1712, according to the preface of *The natural history of Carolina, Florida and the Bahama islands*, he had arrived in the colony of Virginia on the east coast of North America. A week later, accompanied by his brother-in-law, he went to meet one of the colony's most prominent citizens, William Byrd II (1674–1744), who lived outside Williamsburg, at Westover on the James River. Byrd kept a cryptic diary,[12] and as Mark Catesby's twentieth-century biographers, George Frederick Frick (1925–1991) and Raymond Phineas Stearns (1904–1970), wrote: "We are fortunate that . . . the few pages of the diary of William Byrd which were crossed by Catesby provide us with the most intimate and detailed account available for any period of his life."[13] On 22 April Byrd noted that "the fleet has come in" from England, including those ships "that had my goods, thank God."[14] His informant, Captain B-r-k-l-t, let Byrd know that "Mrs Cocke also was come in the Harrison and two of her children." As Mrs. Cocke was Mark's eldest sister, Elizabeth, there can be little doubt that he had sailed in the same ship.[15] In other words, Mark traveled to Virginia probably as chaperone and guardian, responsibilities necessary at a period when it would have been unwise, even for an evidently married lady with a young family, to travel without a

FIGURE 1-4. Signature of Mark Catesby on an indenture. (Reproduced by courtesy of Suffolk Record Office, Bury St. Edmunds.)

suitable male escort. Not being married himself, and presumably not having over-riding responsibilities in England, Mark was able to escort Elizabeth and to sail across the Atlantic.

Dr. William Cocke (1672–1720) had arrived in Virginia in 1710 without his wife. Elizabeth Catesby had married him, without her father's consent, on 4 September 1699 in St. Peter's Parish Church, Sudbury. When John Catesby made his will fourteen months later, his anger was still evident, although he was unwilling to leave Elizabeth and her children destitute.[16] Following their mother's death—Elizabeth Catesby, widow, was buried beside her husband on 7 September 1708 at St. Nicholas's, Castle Hedingham—the imposition of paying Elizabeth's annuity fell to John, her youngest brother, who was still a minor.

Exact information about Mark Catesby's life in Virginia is frustratingly sparse beyond the time he spent with William Byrd. One small glimpse is

Sudbury in 1714 VALERIE HERBERT

The wealth that led to the building of Sudbury's three fifteenth-century churches was history when Mark Catesby sailed for Virginia in 1712. Writer and journalist Daniel Defoe (c. 1660–1731), the author of *Robinson Crusoe*, painted a bleak picture of the place in his *Tour through the eastern counties of England*, published in 1722: "I know nothing for which this town is remarkable except for being very populous and very poor. . . . the number of the poor is almost ready to eat up the rich." At this time the town mainly sustained itself by weaving cloth for mourning clothes and for flags—mostly destined for the Royal Navy.

A map of the town, drawn in 1714 probably by Cornelius Brewer (of Borley, Essex), names streets and shows buildings and structures in perspective. These include the Workhouse, where some of Defoe's poor were destined to end their days; the Butter Cross, an open-sided building where farmers' wives from the countryside sold their produce; and the Cage or lock-up, which held minor miscreants. The prison or, rather, the House of Correction was elsewhere in the town. Most of the houses shown in perspective belonged to the rich and/or powerful, many to the borough's aldermen, town elders chosen by their peers.

Only one large house is drawn in perspective in North Street. It is numbered (indistinctly, on the roof) 22, which corresponds to "Mr Catesby's House" in the key. It appears that this house with its range of dormer windows was the Catesby property, clearly not that of Mark's father, John, as he had died a decade before; the most likely candidate is Mark's brother Jekyll, Mark having sold his property in North Street in order to fund his trip to Virginia.

The house still exists. It is at the location indicated on Cornelius Brewer's map and is identifiable with the cartographer's sketch. A photograph dating from the beginning of the twentieth century shows the building with some of the dormer windows of 1714 intact and in the correct proportions.

In recent times, this property has had a number of retail uses, including being partly incorporated into a store in the defunct F. W. Woolworth chain. But the interior of the remaining part has carved and chamfered timbers that clearly predate the Catesby family's activities in Sudbury.

So far it has been impossible to prove who lived in "Mr Catesby's House," but it might well have been part of the property wealth that financed Mark Catesby on his first quest.

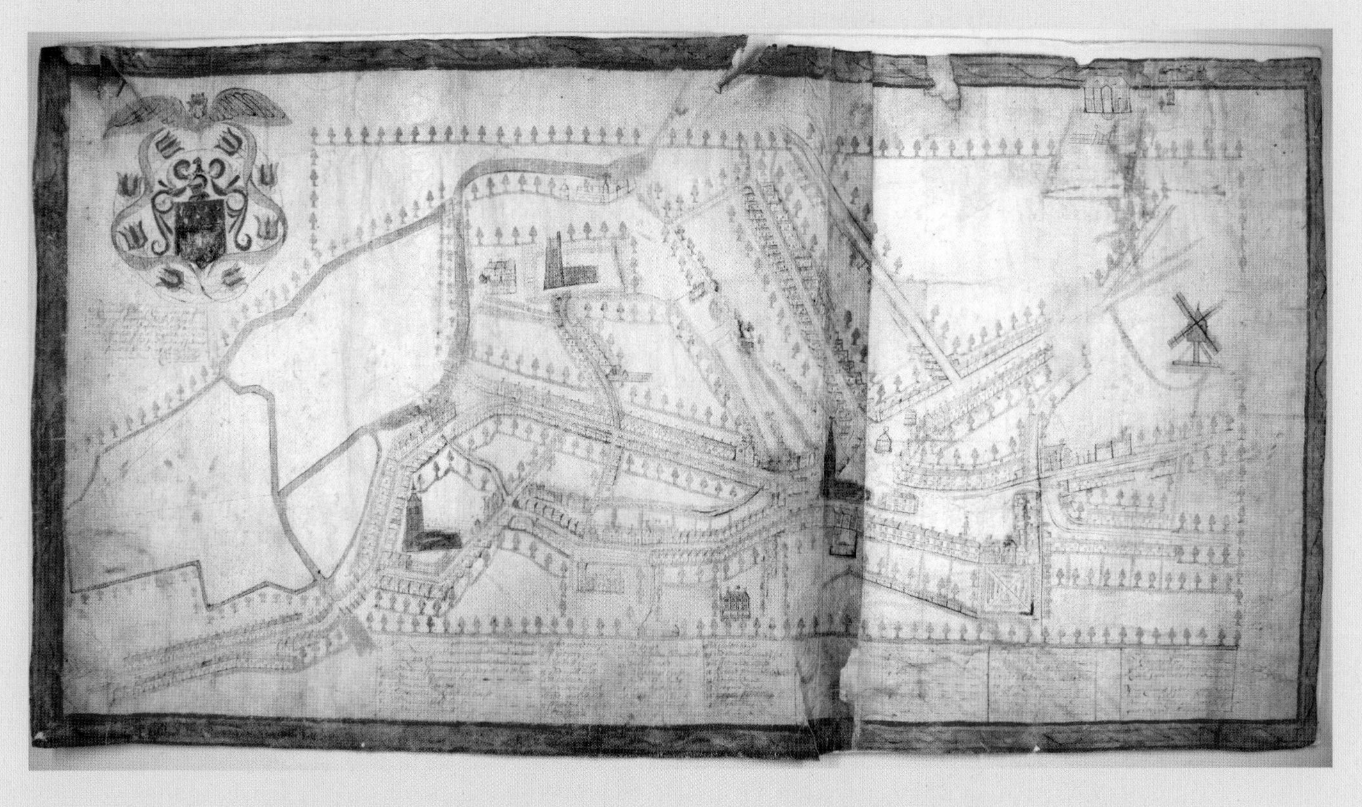

FIGURE 1-5. Plan of Sudbury, Suffolk, made in 1714. (Reproduced by permission of Sudbury Town Council; images by courtesy of Suffolk Record Office, Bury St. Edmunds.)

given in a letter his uncle, Nicholas Jekyll, wrote to the Reverend William Holman on 20 September 1712. "Last post my nephew Mark wrote me a letter dated 1st of August; they were all well . . . ," Jekyll related, adding: "Part of his time he uses in making a large collection of Plants &c For Mr Dale, having the assistance of Colonel Bird [*sic*], a vast rich man there. . . ."[17] Thus, soon after his arrival in Virginia, Mark had begun collecting. He pressed and dried specimens of the Virginian flora, making herbarium specimens, some of which survive, and labeling them carefully with finely inscribed tickets. Among the extant specimens is an example of the yellow passionflower (*Passiflora lutea*) "gathered out of a pomgranate hedge" in William Byrd's garden.[18] Catesby gathered the specimens and seeds principally for Samuel Dale (c. 1659–1739), an apothecary and physician of Braintree, Essex, who had a keen interest in botany. A list of "Seeds from Virginia sent by Mr Catesby to Mr Dale" exists among the papers of James Petiver (c. 1665–1718): more than one hundred different plants are named.[19] Dale was acquainted with the Catesby family; he had visited Nicholas Jekyll at Castle Hedingham in 1711.[20] Through the Lieutenant Governor of Virginia, Alexander Spotswood (1676–1740), Catesby also sent seeds to the Bishop of London, Dr. Henry Compton (1632–1713), who had a passion for exotic plants.[21]

In *The natural history of Carolina, Florida and the Bahama islands*, which Catesby did not start writing until more than a decade after this visit, he provided a few more fragments. With some unnamed companions he traveled from the James River to the Appalachians in 1714.[22] That year he also had time to take passage on a boat carrying cargo, including sheep, to Jamaica, and he must have stayed there for some time, because he saw cocoa plantations[23] and managed to collect specimens of the Jamaican flora for Dale.[24] There is ample evidence, including a pressed specimen of the Bermudan blue-eyed grass (*Sisyrinchium bermudiana*),[25] that he also visited Bermuda at this time. In *The natural history of Carolina* . . . he reported that he had shot tropicbirds during their nesting season "at *Bermudas* . . . from the high rocks that environ those Islands," although he could not climb the cliffs to see the nests or eggs.[26] Catesby observed the "Plat Palmetto" there:

> In Bermudas its Leaves were [f]ormerly manufactured, and made into Hats, Bonnets, &c. and of the Berries were made Buttons. This is the slowest grower of all other Trees, if Credit may be given to the generality of the Inhabitants of *Bermudas*, many of the principal of whom affirm'd to me, that with their nicest Observations, they could not perceive them to grow an Inch in height, nor even to make the least Progress in fifty Years, yet in the Year 1714, I observ'd all these Islands abounding with infinite Numbers of them of all Sizes.[27]

Furthermore, in a letter written more than two decades later, he discussed the endemic juniper of Bermuda, and the context also indicates that on the return voyage from Jamaica he visited Hispaniola and Puerto Rico too.[28]

By 1715 plants raised from seeds sent by Catesby from Virginia were thriving in Thomas Fairchild's nursery at Hoxton on the northeastern outskirts of London: their quaint names, as published in the *Philosophical transactions of the Royal Society* by James Petiver, were "*Fairchild*'s broad *Bobart*" (raised from seed received from that "*curious* Botanist Mr. *Mark Catesby* of Virginia"); "Herman's *Virginia* yellow Basil"; "Munting's *yellow* Maracoc"; and "White *Virginia* Bindweed, with a blackish bottomed *Flower*" (also from "the inquisitive Mr. *Catesby*").[29] Other seedlings, including at least one plant of rabbit-tobacco (*Pseudognaphalium obtusifolium*), were blooming in Chelsea Physic Garden.[30] Meanwhile, on the other side of the Atlantic, Catesby had established a garden of his own at Williamsburg and was cultivating and assessing the value of some of the native plants: on the label attached to a pressed specimen of "Bushing brake" (eastern baccharis, *Baccharis halimifolia*) he commented that he had "removed some plants of it into my garden—Where 'tis very ornamental producing fulle of white downy flowers in October. . . ."[31]

Samuel Dale referred to plants received from Catesby in a letter dated 17 August 1718, and during the early months of 1719 he dispatched several consignments of American plants to William Sherard (1659–1728). It was in another letter to Sherard, dated 15 October 1719, that Dale announced:

> Mr Catesby is come from Virginia. . . . He intends againe to return, and will take an oppertunity to waite upon you, with some paintings of Birds etc., which he hath drawn. Its pitty some incouragement can't be found for him, he may be very usefull for the perfecting of Natural History.[32]

We have no explicit information about why Mark Catesby decided to leave Virginia. His older brother, Jekyll, who continued to live in England, probably acted on Mark's behalf in financial and property matters, but he was buried in the graveyard of St. Nicholas's Church, Castle Hedingham, on 21 September 1717. Thus, Mark assumed the role of head of the family. His youngest brother, John, was only just twenty years old and so, legally, was still a minor. On learning of his brother's death, Mark may have decided that he had to return to England to ensure that his property and finances were properly managed, as well as to support John. When Mark was back in England he visited Sudbury and Castle Hedingham, as well as Braintree and London because he acted as a carrier of letters between Dale and Sherard. However, he remains elusive in documents during this period, although his acquaintances evidently were discussing the idea mooted by Dale that Catesby intended to go back to North America. At a meeting of the Council of the Royal Society of London on 20 October 1720,[33]

> Colonel Francis Nicholson going Governor to South Carolina was pleased to declare that he would allow Mr. Catesby, recommended to him as a very proper person to Observe the Rarities of the Country for the uses and purposes of the

Society the Pension of Twenty Pounds per Annum during his Government there, and at the Same time to give him Ten pounds by way of advance for the first half Years payment and so for the future a Years pay beforehand.

Several weeks later, William Sherard, writing to Dr. Richard Richardson,[34] confirmed the arrangement but also indicated that more funds would be required to ensure success:

> Mr. Catesby, a Gentleman of small fortune, who liv'd some years in Virginia with a relation, pretty well skill'd in Natural history who designs and paints in water colours to perfection, is going over with General Nicholson, Governor of Carolina; that Gent. allows him 20£ a year and we are indeavouring to get subscriptions for him, viz. Sir Hans [Sloane], Mr. Dubois, and myself, who are all that have yet subscribed to him but I'm in hopes to get the Duke of Chandos, which will be a great help.

"A gentleman of small fortune" suggests that Mark Catesby had some personal income from his property, enough to maintain his status as a gentleman, but probably inadequate to meet the costs of a lengthy overseas journey. A handsome, illustrated edition of *Aesop's fables* bearing his signature and the date 1720 also indicates he still had enough money to spend on luxuries (figure 1-6).[35]

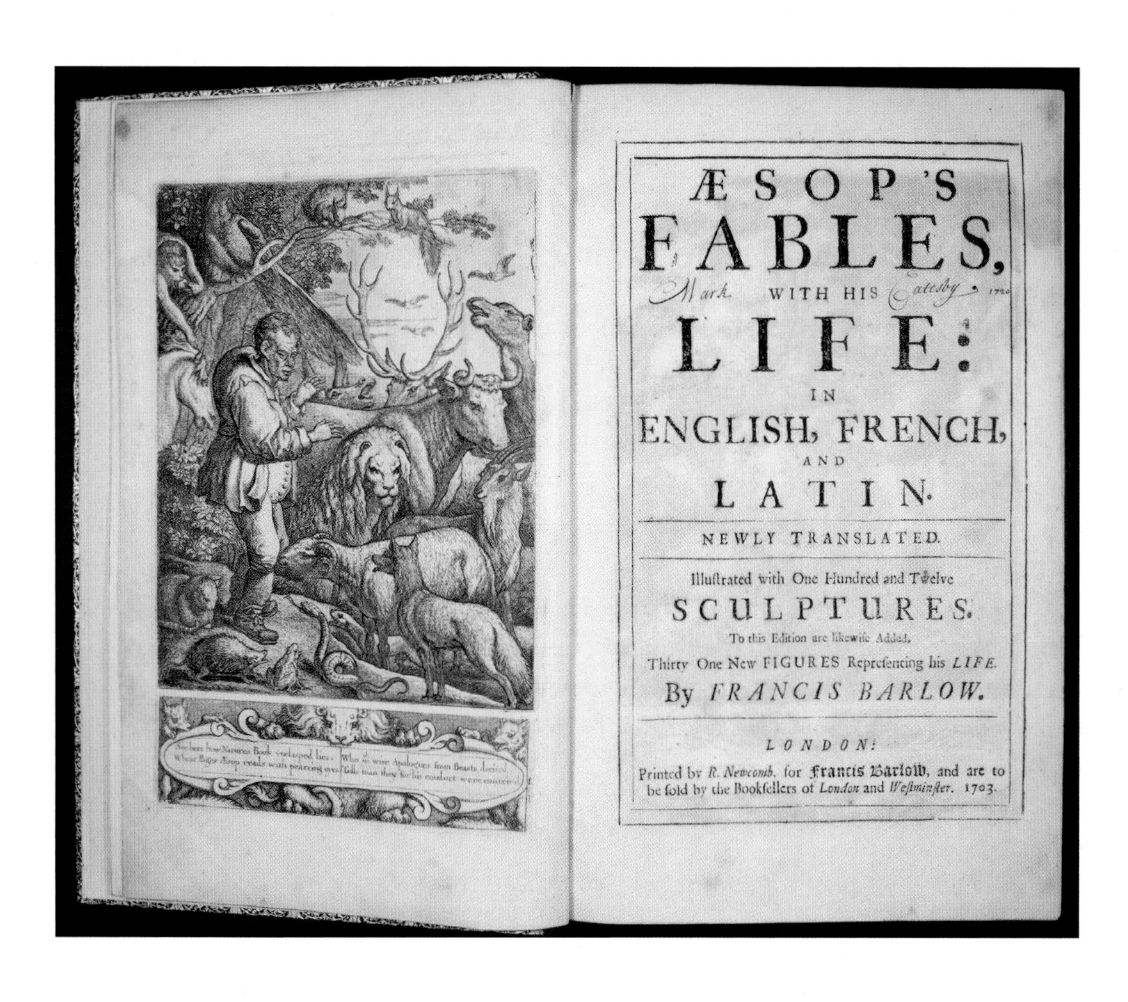

FIGURE 1-6. Mark Catesby's copy of Francis Barlow's illustrated edition of *Aesop's fables*. (Photograph by James Dewrance, reproduced by courtesy of Tom Schreck.)

Meanwhile, Francis Nicholson (1655–1726/27) left England to take up the post of Royal Governor of South Carolina in the spring of 1721 and, having paid ten pounds as agreed, complained that Catesby was not accompanying him. Catesby's departure was eventually announced by Sherard in a letter to Richardson dated 27 January 1721/22: "Mr Catesby goes next week for Carolina. He has put off his going till the last ship."[36] And, according to a letter Catesby himself wrote soon after he reached Charleston, "I came from London the beginning of Feb. last and left all well in Essex and Suff[olk]."[37] An interesting twist on this is contained in one of Nicholas Jekyll's letters to William Holman: "My Nephew Mark wrote word the Ship he endevoured to . . . was lost in going, but the truth was thus. In the ocean that ship foundered and sank, and as many as could got into [a] Boat, the rest (who were the most) perrishd. . . ."[38]

In fact, Mark Catesby spent his twenty-ninth birthday afloat in the North Atlantic. He whiled away the time watching the birds and porpoises and catching fish:[39]

> In Lat 25 and 300 leagues from land (the Canary Iles being ye nearest land) We had an owl and other Birds hovering about our ship. how an Owl [illegible] . . . so incapable of long flights that they [illegible] . . . down by Boys on shore should hold it so long on the wing is to me surprizing On the Coasts of America we had several other Birds come on Board us which gave me an opertunity of discribing some of them as well as some fish. Amongst the Birds the Turn stone or Sea plover of Mr Willoughby [figure 1-7] was one which by comparing with his discription of it agrees exactly. I had not the same opertunity at Sea of meeting any plants except the Gulf weed with which the Ocean at places was covered.

He landed at Charles Town on 3 May and two days later wrote to William Sherard: this is the earliest extant letter in Catesby's hand, the first autobiographical manuscript that survives.[40]

> Honrd Sr
> As you commanded me I take ye first opertunity of letting you know of my arrival here 3d May which day fortnight before we were in Soundings, and within a few leagues of our port, but a violent and contrary wind set us back into the Current of the Mexican Gulph, otherwise we have had a short and easy voyage The Governor received me with much kindness to which I am Satisfyed Your Letter contributed not a little. . . .[41]

(Mark's memory slipped a little, for when he composed the preface for his book, probably in 1733 before the fifth part of the first volume was issued, he recalled that "I arrived in *Carolina* 23d. of *May* 1722, after a pleasant tho' not a short Passage.")[42]

From this time onward, Catesby's fairly frequent letters to William Sherard, Sir Hans Sloane (1660–1753), Peter Collinson (1693/94–1768) and others in England, and a few letters to his family in Virginia provide dated reference

Morinellus marinus *Frutex &c.*

points for his work and travels and a few glimpses of his personal life. For example, from the seal on one of his letters it is evident that he possessed a signet ring with a Catesby crest on it: two lions passant (figure 1-8).[43]

FIGURE 1-7. (*Opposite*) Ruddy turnstone, plate 72, M. Catesby, 1731, *The natural history of Carolina . . .* , volume I. (Digital realization of original etchings by Lucie Hey and Nigel Frith, DRPG England; courtesy of the Royal Society ©.)

Catesby used Charleston as a base, traveling inland to collect. While a trip to Mexico was mooted in the late summer of 1724, this was abandoned, and early in January 1725 Catesby prepared to depart for the Bahamas. He had first suggested the idea of spending a year in the Bahamas during the previous January in a letter to Sherard and evidently continued to nurse the plan, writing to Sherard about it again in August 1724: "I am determined with my selfe tho' at my own expence to continue here another Year or at the Bahama Ilands but Your Sentiments concerning it shall be my guide which I beg you'l please to favour me with when you write next."[44] In early January 1725 he announced his imminent departure from Charleston in separate letters to Sloane and Sherard. To Sloane on 5 January he wrote:

> I am S^r preparing to goe to the Bahama Islands to make a further progress in what I am about. This will add another Year to my continuance in America. And tho' I doe not expect a continuance of my full Subscriptions Yet I hope partly by your interest a[n]d continuance of Your Favours, I may expect the greater part of it. . . . I promise my self great variety of shells and animals not to be found here Whatever commands Sr You'l pleas to Honour me with, please to direct to Carolina and I shall have them conveyed to me.[45]

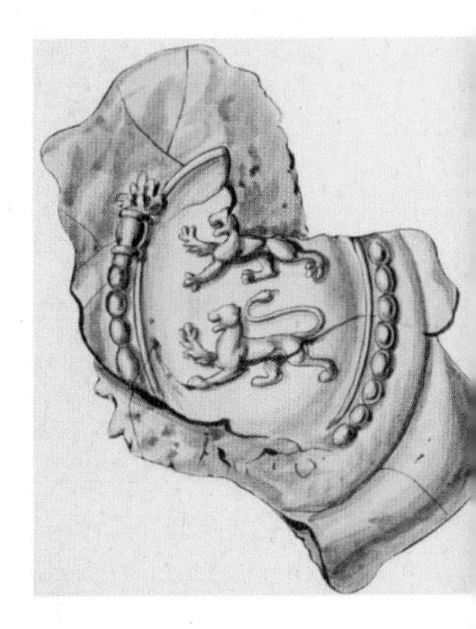

FIGURE 1-8. Seal impression used by Mark Catesby. (From L. H. Jones, *Captain Roger Jones of London and Virginia* [1891].)

Thus Catesby spent most of 1725 in the Bahamas, departing for home sometime after December, during which there were a couple of days "so cold, that we were necessitated to make a Fire in the Governors Kitchen to warm us."[46]

Back in England sometime around the spring of 1726, Mark Catesby must have started work on the production of *The natural history of Carolina, Florida and the Bahama islands*. On his own authority he changed his plan of having the hundreds of images etched by plate-makers on the Continent and instead, "At length by the kind Advice and Instruction of that inimitable Painter Mr. *Joseph Goupy*, I undertook and was initiated in the way of Etching them myself."[47] So, once again, Mark Catesby becomes elusive, learning how to engrave and then laboriously engraving the plates for his book. He also wrote the English text that accompanied each plate. In a parallel column on each page, this text was rendered in French: "As to the *French* Translation I am oblig'd to a very ingenious Gentleman, a Doctor of Physick, and a *French-man* born, whose modesty will not permit me to mention his Name."[48] (Catesby was evidently able to read French.)[49]

A letter from George Rutherford, who was married to Rachel Cocke (Mark's niece), confirms Catesby's presence in London in the summer of 1728, but he was still elusive. Writing on 27 June 1728, Rutherford told another of Mark's nieces that "your Uncle M^r. Mark Catesby is now in London but I can't tell you where he lodges."[50] Meanwhile, the first twenty plates of *The natural history of Carolina* . . . were printed with the accompanying letterpress, and Mark colored them, one by one, himself. In the middle of May 1729, the first part of the work was ready, and towards the end of that month several British newspapers published this highly significant announcement:[51]

> **LONDON.**
> **The laſt Week Mr. Catesby (introduced by the Right Honourable the Lord Cartaret) preſented to her Majeſty the firſt Volume of his Natural Hiſtory of Florida, Carolina, and the Bahama Iſlands; containing a great Variety of the Animal and Vegetable Productions of thoſe Countries, which the Author hath been ſeveral Years collecting and delineating from the Life.**

One newspaper added: "This is said to be a Work superior to any Thing of the Kind."[52]

Queen Caroline had just been appointed the Queen Regent, as King George was absent in his German dominions. Moreover, Lord Carteret was still holding on to his position as one of the Lords Proprietors of the Carolina Colony and was the only one who had refused to give up his share when the Lords Proprietors' administration was overthrown in 1719. The precise date of the audience with the Queen has not been discovered, but it was most probably before Catesby attended the Royal Society's meeting on 22 May. He also presented the first part to the Society, commencing the process of publication of *The natural history of Carolina* . . . , which would continue for eighteen years until the final part, the Appendix, was completed: "Catesby's Noble Work is finished . . ." Peter Collinson told the Swedish botanist Carl Linnaeus (1707–1778) in a letter written on 15 April 1747.[53] A copy of the Appendix was presented to the Royal Society on 2 July 1747.[54]

During the two decades between his return from the Americas and the conclusion of his publication, Catesby settled in London. He needed space so he could engrave the plates, and space to lay out the sheets for coloring, as well as space to store his original paintings and the copper plates onto which he engraved their images. On 1 March 1729/30, writing to his niece in Virginia, he headed his letter Hoxton, which was also one of the addresses that he advertised to anyone interested in becoming a subscriber:[55]

> GENTLEMENS Names will be enter'd
> *By* W. Innys, at the West End of St Paul's; John Brindley, Book-Binder to Her Majesty and to his Royal Highness the Prince of Wales, at the Kings Arms, New Bond-Street; and by the Author at Mr. Fairchild's, in Hoxton, where may be seen the original paintings.

Thomas Fairchild (c. 1667–1729), plantsman and nursery proprietor of Hoxton, was certainly well acquainted with Mark Catesby. The clearest indication of their closeness is the fact that Catesby was one of the three individuals present to witness Fairchild's will on 20 February 1728/29 and that one of the bequests Fairchild included was "Item I give and bequeath . . . To my friend M^e. Cateby [*sic*] one guinea for a Ring. . . ."[56] Fairchild died seven months later on 10 October 1729, and his nephew, Stephen Bacon, took over the business, so that when the title-page of the first volume was printed in 1732, Bacon's name was on it: "Printed at the Expence of the Author; and Sold . . . by the Author at Mr. Bacon's in Hoxton."[57]

After 1729 information about Catesby's activities in London becomes more abundant. Before July 1730 he had met Elizabeth Rowland and set up home with her in the parish of St. Giles-without-Cripplegate,[58] which lies on the eastern side of the City of London, a short walk south of Hoxton. The parochial register contains records of the birth and baptism of three children of Elizabeth and Mark between April 1731 and May 1733, as well as the burials of the two youngest: Mark (born 15 April 1731), John (born 6 March 1731/32, buried 25 August 1732), and Caroline (baptized 27 May 1733, buried August 1733).[59] Their second daughter, Ann, was baptized in St. Luke's Church, Old Street, on 27 December 1737[60]—it is significant that Mark Catesby died in "his House behind St Luke's Church" twelve years after Ann's christening, suggesting that the family had not moved. It is noteworthy too that Mark did not marry Elizabeth Rowland until another decade had passed, on 8 October 1747,[61] in St. George's Chapel, Hyde Park Corner, on the western side of London. Little is known about Mrs. Elizabeth Catesby except that she had another daughter named in her will as Elizabeth Rowland.

Having often been a guest at meetings of the Royal Society, on 1 February 1732/33 Mark Catesby was nominated for election as a member, which process was completed on 26 April 1733 (figure 1-9) and afterward he was able to add the initials F.R.S.—Fellow of the Royal Society—after his name. That is how his name appeared on the black-and-red title-page of the first volume of *The natural history of Carolina . . .*, indicating that that page was not printed until late April 1733 (despite being dated MDCCXXXI). In the summer of that year, he is known to have crossed the English Channel and visited Holland, offering to "be of any sarvice" while there to his friend, the gardener Thomas Knowlton (1691–1781), then working at Everingham in Yorkshire.[62]

Work on producing *The natural history of Carolina . . .* went a lot more slowly than Mark had advertised. In the *Proposals* (see figure 12-1, p. 157) he stated that he "intended to publish every Four Months Twenty Plates and their Descriptions"—sixty plates, or three parts, a year. On 14 August 1730, more than a year after the first part had been presented to Queen Caroline, Nicholas Jekyll reported that "Nep[hew] Marks 3d Part is done, all but Collurring. . . ."[63] Jekyll also indicated that eighty plates had been completed. Meanwhile, Catesby also undertook other work, the most remarkable being the engraving of a broadsheet catalog for Christopher Gray (1694–1764), whose nursery was in Fulham.[64] The engraving incorporated the superb portrait of *Magnolia*

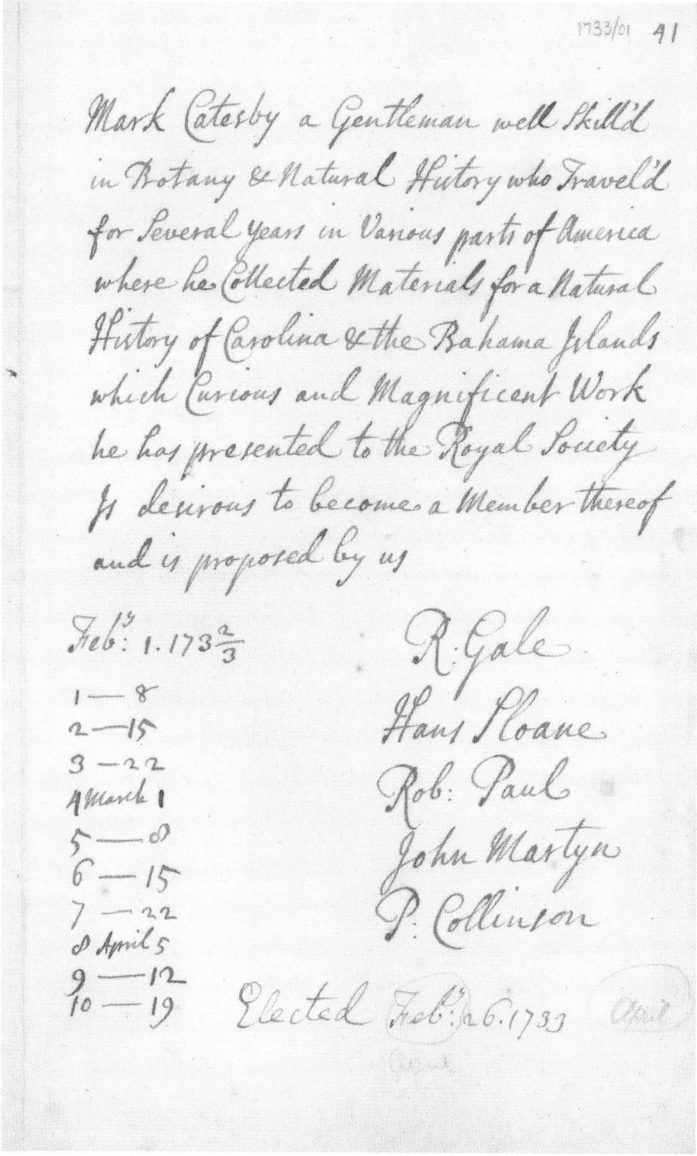

1733/01 41

Mark Catesby a Gentleman well Skill'd
in Botany & Natural History who Travel'd
for Several years in Various parts of America
where he Collected Materials for a Natural
History of Carolina & the Bahama Islands
which Curious and Magnificent Work
he has presented to the Royal Society
Is desirous to become a Member thereof
and is proposed by us

Feb: 1. 173 2/3

R. Gale
Hans Sloane
Rob: Paul
John Martyn
P. Collinson

1—8
2—15
3—22
4 March 1
5—8
6—15
7—22
8 April 5
9—12
10—19

Elected Feb: 26. 1733 April

FIGURE 1-9.
Mark Catesby's nomination paper as Fellow of the Royal Society of London; he was elected on 26 April 1733, not February as indicated. (© Royal Society.)

grandiflora that Georg Ehret had produced.[65] This broadsheet (see figure 7-6, p. 94) is a *tour de force* of the art of engraving—bear in mind that every letter and every word had to be engraved as a mirror image.

Only twice in *The natural history of Carolina . . .*, in the last part and in the Appendix, did Catesby mention plants growing in his own garden. Referring to *Kalmia latifolia* (mountain laurel), he stated that "some Bunches of Blossoms were produced in July 1740, and in 1741, in my Garden at Fulham," and to *Stewartia malacodendron* (silky camellia) he acknowledged that "for this elegant Plant I am obliged to my good friend Mr. [John] Clayton, who sent it me from Virginia, and three months after its arrival it blossom'd in my garden at Fulham in May 1742."[66] Fulham lies on the western side of London, about seven miles from St. Luke's, Old Street. No contemporary source has been found to verify these references.[67] The most reasonable inference is that because there was no garden attached to his house in east London, Catesby made "a garden of his own within the grounds of Fulham Nursery,"[68] which was managed, if not owned, by Christopher Gray.

On 29 December 1743 Mark Catesby and his friend George Edwards were made members of the Society of Gentlemen of Spalding, in Lincolnshire, although they almost certainly were not present at the meeting (figure 1-10).[69] In the mid-1740s, now past sixty years of age, Mark remained productive. He engraved twenty more plates for an Appendix to *The natural history of Carolina* . . ., a strange potpourri of animals and plants, some of which were not even from the Americas. He prepared, but did not publish, his *Hortus Britanno-Americanus*, a selection of eighty-three trees and shrubs suitable for growing in gardens in Britain and Ireland. This also required engraving copper plates, but there is no evidence that the work was printed and hand-colored by Catesby himself. He remained active in the Royal Society, presenting his celebrated paper about bird migration at the meeting on 5 March 1747.[70]

When he married Elizabeth Rowland in October 1747, Mark Catesby was sixty-four years old. He had just completed his magnificent book, presenting the final part to the Royal Society on 2 July 1747:[71]

> The whole was done within my house, and by my own hands; for as my honour and credit were alone concerned, I was resolved not to hazard them by committing any part of the Work to another person, besides, should any of my original Paintings have been lost, they would have been irretrievable to me, without making another voyage to America, since a perpetual inspection of them was so necessary towards the exhibition of truth and accuracy in my descriptions.[72]

During the spring of 1748 Catesby met the young Finnish naturalist Pehr Kalm, who would soon travel to North America. Kalm spent "nearly the whole afternoon" of 23 May 1748

> at the house of *Mr. Catesby* . . . [who] seemed to be a man of nearly sixty years, and somewhat short-sighted. He now devoted his time to reading, and to further elaborating the *Natural History* . . . which consisted to two large Volumes in Regal Folio . . . and both together now cost in England twenty-two to twenty-four guineas, therefore not for a poor man to buy.[73]

Catesby's health was certainly in decline; a much-quoted comment from a letter written by Thomas Knowlton to Dr. Richard Richardson on 18 July 1749 was that "poor M^{r}. Catesby's Legs swell & he looks badly, D^{rrs}. Mead & Stack said there were little hopes for him long on this side the Grave. . . ."[74] Announcements of Catesby's death, which occurred on 23 December 1749, quickly appeared in *The gentleman's magazine* and even in newspapers such as the Scottish broadsheet *The Caledonian mercury* on Tuesday, 2 January 1750:[75]

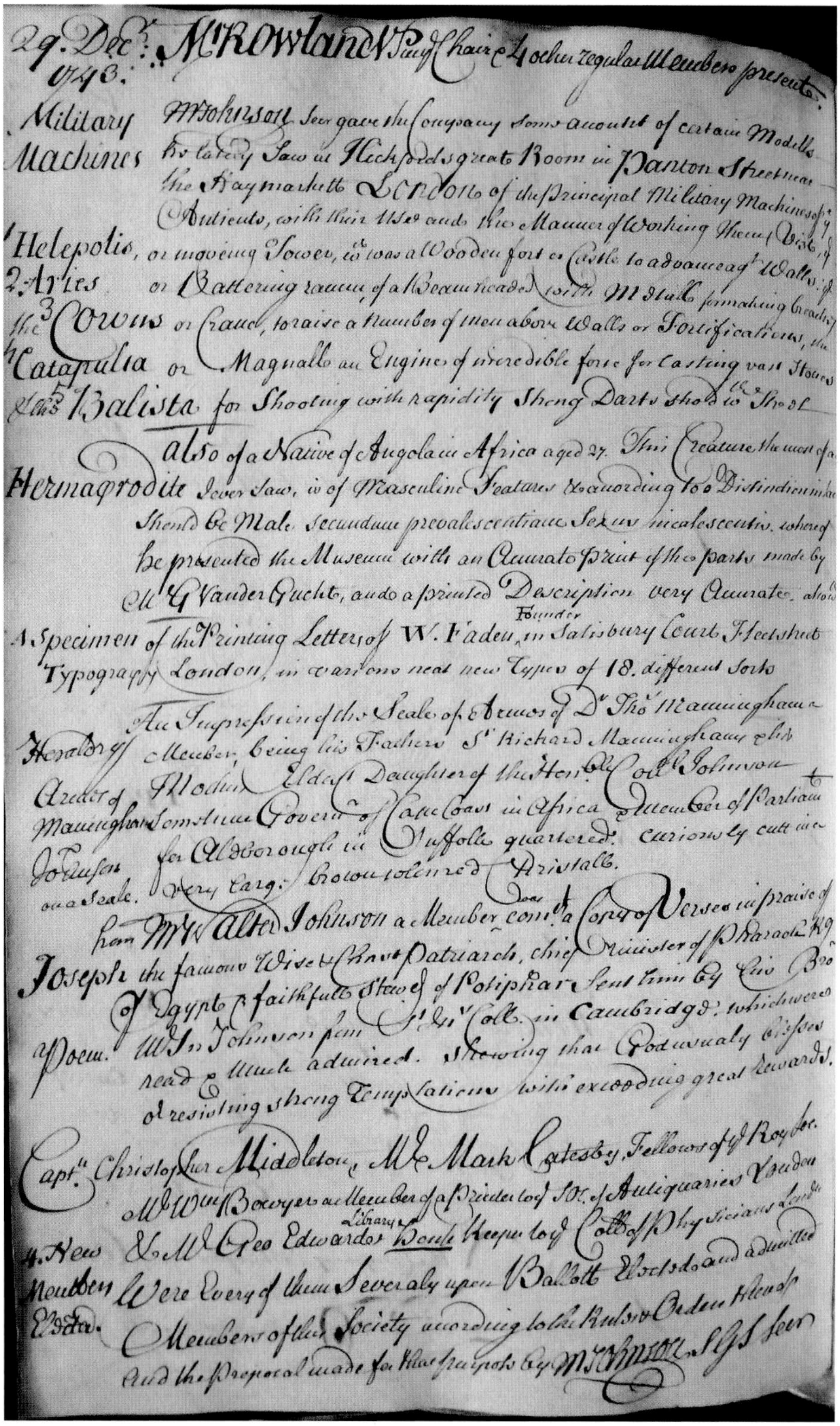

29. Decr: 1743: Mr Rowland V. Pres. Chair & 4 other regular Members present.

Military Machines — Mr Johnson Secr gave the Company some account of certain Modells he lately saw in Hickford's great Room in Panton Street near the Haymarkett London of the Principal Military Machines of the Antients, with their Uses and the Manner of Working them Viz. 1 Helepolis, or moveing Tower, wch was a Wooden fort or Castle to advance agt Walls. 2 Aries or Battering ramm of a Beam headed with Metall for making Breaches. the 3 Corvus or Crane, to raise a Number of Men above Walls or Fortifications, the 4 Catapulta or Magnall an Engine of incredible force for Casting vast Stones & the 5 Balista for Shooting with rapidity Strong Darts shod wth Steel

Hermaphrodite — also of a Native of Angola in Africa aged 27. This Creature the most of a Hermaphrodite I ever saw, is of Masculine Features & according to a Distinction [illegible] Should be Male secundum prevalescentiam Sexus incalescentis. whereof he presented the Museum with an Accurate Print of the parts made by Mr G Vander Gucht, and a printed Description very Accurate. also

Typography — A Specimen of the Printing Letters of W. Faden, Founder in Salisbury Court Fleetstreet London, in various neat new Types of 18. different Sorts

Heraldry of Arms of Manningham Johnson on a Seale. — An Impression of the Seale of Armes of Dr Thos Manningham a Member, being his Fathers Sr Richard Manningham & his Mother Eldest Daughter of the Honble Coll Johnson sometime Governr of Cape Coast in Africa & Member of Parliamt for Aldborough in Suffolk quartered. curiously cutt in a very large Brown coloured Cristall.

Poem. — from Mr Walter Johnson a Member comd a Copy of Verses in praise of Joseph the famous Wise & Chast Patriarch, chief Minister of Pharaoh Kg of Egypt & faithfull Steward of Potiphar sent him by his Bror Mr Jn Johnson fem St Jn Coll. in Cambridge: which were read & much admired. Showing that God usualy blesses [the] resisting strong Temptations with exceeding great Rewards.

4. New Members Elected. — Capt. Christopher Middleton, Mr Mark Catesby, Fellows of ye Roy. Soc. Mr Wm Bowyer a Member of & Printer to ye Soc. of Antiquaries London & Mr Geo Edwards Library Keeper to ye Coll of Physicians Lond. Were Every of them Severaly upon Ballott Elected and admitted Members of this Society according to the Rules & Orders thereof And the Proposal made for that purpose by Mr Johnson Secr

FIGURE 1-10. Minute book of the Society of Gentlemen of Spalding showing the election of Mark Catesby and George Edwards. (Reproduced by courtesy of the Society of Gentlemen of Spalding.)

LONDON.

On Saturday Morning died at his Houſe behind St. Leke's Church, aged ſeventy, the ingenious Mr. Mark Catesby, who was Author of a Work, entituled, *A Natural Hiſtory of Carolina.*

An account of Mark Catesby's final weeks, written by his friend George Edwards a dozen years after the events described, is the closest to a contemporary record that has survived, but it is demonstrably inaccurate. In a letter to Thomas Pennant dated 5 December 1761,[76] Edwards wrote:

> Whither his [Catesby's] death was natural or accidental, it is hard to determin. In Crossing the way in holbore, he fell and was taken up Senceless and So continued 2 or 3 days, when he dyed. He receiv'd in his fall a bruse in his head which might caus his death, or the fall might be from an appoplectic fit. He had his son with him, a Boy of 8[77] years old, who could give not Satisfactory account. He was put in to a Coach and carried home to his wife in that Condition.

Edwards attended Catesby's funeral. He "was buried in the Church yard of St Luke in Old Street, London," and his name in the parochial register was spelled "Marke Sketesby." The cause of his death was given as "Age" (figure 1-11).[78] He was survived by his wife and their eldest son and youngest daughter, Mark and Ann, as well as by his stepdaughter, Elizabeth.

No will is extant and there is no record of probate. George Edwards recalled that Mark Catesby's "whole fortune: household Stuff, Coppy of his history, original drawings, and Copper plates did not amount in the whole to 700£."[79] This is confirmed by *The gentleman's magazine*'s obituary, which was written by Peter Collinson.[80] However, Catesby did not die without making some provisions for his family and friends. One of his prized possessions was a copy

FIGURE 1-11. The entry recording the burial of Mark Catesby as "Marke Sketesby" in the register of St. Luke's Parish, Old Street, London. (Reproduced by permission of London Metropolitan Archives.)

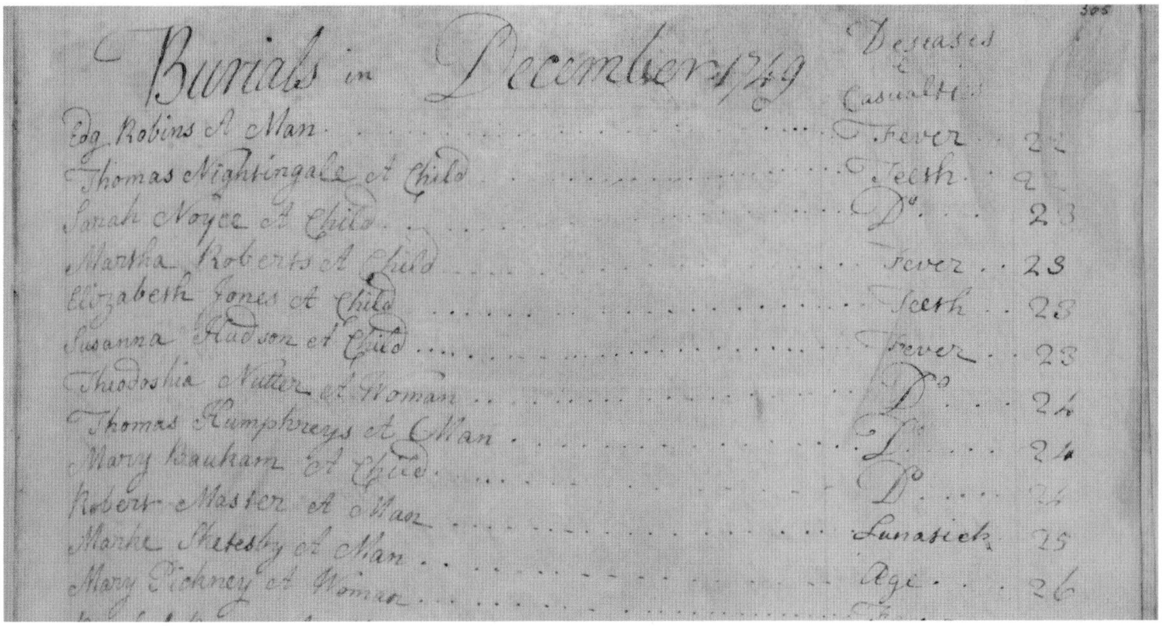

of *Hortus Cliffortianus*, with plant descriptions written by Carl Linnaeus and illustrations mainly based on drawings by Georg Ehret: "George Clifford gave this book to Mark Catesby, and he at his death will'd it to Thos. Knowlton."[81]

Elizabeth Catesby made her will, witnessed by Peter Collinson and Mary Artha, on 4 January 1753.[82] She died of consumption (figure 1-12) six weeks later and was buried in the churchyard of St. Luke's on 18 February 1753 (figure 1-13).[83] Mark and Elizabeth lay at rest together for more than two centuries until the entire churchyard was excavated late in 2000. Their burial place was not identified, but all the exhumed remains were reburied.[84]

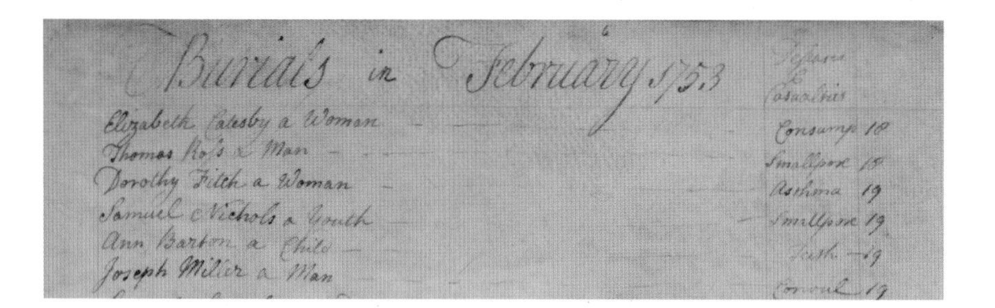
Burials in February 1753

Elizabeth Catesby a Woman — Consump 18
Thomas Ro's a Man — Smallpox 18
Dorothy Filth a Woman — Asthma 19
Samuel Nichols a Youth — Smallpox 19
Ann Barton a Child — Teeth 19
Joseph Miller a Man — Convul 19

FIGURE 1-12. The entry recording the burial of Mrs. Elizabeth Catesby in the register of St. Luke's Parish, Old Street, London. (Reproduced by permission of London Metropolitan Archives.)

FIGURE 1-13. St. Luke's Church, Old Street, London. (© E. C. Nelson 2013.)

2

Behind the scenes: Catesby the man, viewed through the lens of a camera

CYNTHIA P. NEAL

I do not recall ever having heard the name Mark Catesby until about ten years ago. Then I was contracted to create a film about this remarkable man. From there Catesby grew to be a close companion, and I discovered that we shared many similar interests and experiences. These are the interests and experiences that I want to share with the hope that his art and his science—his genius—will be more meaningful.

As the producer and director of the film *The Curious Mister Catesby*, I was challenged to attempt to understand and put into words and pictures Mark Catesby's heart and passion—to find the man behind the art and the science. There are precious few references to the personal nature of the man either in his writing or in the writings of his contemporaries.

My first responsibility as the film's director was to listen to and read and study the materials that have been meticulously researched and published by academics, art historians, and scientists. Then my work was to take their work and give it wings, to make it fly off the screen and be remembered by the many people who have no idea who the "curious Mister Catesby" might have been and, further, probably don't much care.

Early in the film's production planning, something about this Englishman and his adventures became clear to me. Like Catesby, I came to Virginia early in my life with a good name and a decent education but not much money or experience in southern high society. Still, I had lots of curiosity and tons of youthful energy. I loved my years in Virginia. I even married one of Mr. Jefferson's men. I forged the foundations for lifelong friendships and a personal direction that would become not just a career but also a passionate lifestyle. Catesby also enjoyed his time in Virginia. I don't know if he had a grand romance, but I hope he did. He left, as I did, inspired to know more about wild places and the wondrous things that live in them.

Twenty years after I left Virginia, with a successful film-making career covering four continents, some very nice awards, and many weary miles behind

FIGURE 2-1. *The Curious Mister Catesby.* Producer/director Cynthia Neal with the Executive Producer, David Elliott. (© Pamela Cohen.)

me, I was contracted for my first job in the Carolina Lowcountry. Never before had I seen such a place. The people were a breed unto themselves; the coast was different, along with the food, the weather, the trees, the plants, the swamps and rivers, the insects. Nearly three hundred years before me, Mark Catesby also had been fascinated by this remarkable place.

For the next ten years, my camera crew and I spent several months of each year shooting all across the Lowcountry. Each of the films that I created required learning more and more about the natural history and the conservation challenges of the many ecosystems that comprise the area. The body of work—the sheer volume of footage that was shot and edited—represents one of the most productive and exciting periods of my career. Like Catesby before me, in the struggle to collect those images, I fell in love with this strange and wonderful place.

FIGURE 2-2. *The Curious Mister Catesby.* Film crew moving camera and sound recording equipment into the marsh. (© Pamela Cohen.)

It is a fluke of history that Mark Catesby and I traveled the same paths and rivers, watched many of the same birds, visited many of the same plantations, suffered many of the same challenges, framed many of the same pictures, and did it all for so many of the same reasons—three hundred years apart. That is what brings me to this point with what is certainly an extremely personal and well-informed authority concerning who was Mark Catesby the man. Indeed, I have walked in his shoes and swatted swarms of mosquitoes.

FIGURE 2-3. *The Curious Mister Catesby*. Producer/ Director Cynthia Neal on location with the Director of photography, Chris Conder, and the Grip, Patrick Wilson. (© Pamela Cohen.)

I believe Mark Catesby was filled with personal charm, had well-developed social skills, and was an attractive, interesting personality. Others have said that Catesby must have been a shy or introverted man. In one production he was cast as a flighty, silly mooch. This speculation, I believe, is based on a lack of information and a small measure of intuition. There are bits and pieces of evidence that provide some insight into the man, for example, when he was observed drinking, singing, and laughing while visiting with the Byrds at Westover in Virginia. It is not a stretch to think he must have been socially adroit, if not downright charming. Think of the numbers of homes that were opened to him throughout his time in the colonies. He was a guest in someone's home for the better part of ten years—conceivably more than three thousand nights. Regardless of how often the linens were changed in the eighteenth century, that still represents a lot of clean sheets and towels. I am assuming there also would have been a few meals, an ale or two, perhaps a horse or carriage. In order to enjoy this degree of hospitality, Mark Catesby must have been a welcomed and charming man. Likewise, when he was in London, Catesby gained the confidence and business of the best and the brightest Englishmen, from Fellows of the Royal Society to the finest artists and craftsmen of the time. He was able to sell approximately 180 copies of one of the most expensive books of the time. Those are feats accomplished by a gregarious man with a confident, outgoing personality.

What do we know about his physical presence and his stamina? I believe he was a strong, athletic person who was remarkably fit. Over the years some have observed that many of his specimens were collected and a large number of his drawings were done from the comfort of shaded verandas and the fine homes where he lodged. In fact, some of his paintings have been discounted as being common garden vines and yard birds of little value. The implication is that he lived an easy life at the expense of his hosts and that he was, if not lazy, certainly less than adventuresome and perhaps a little prissy. It is easy to dispel the criticism about garden vines and yard birds. His English sponsors had never seen them before and raved for more and more. It was immaterial to Catesby's mission where the specimen was identified so long as it was from this side of the Atlantic and new to England. Apart from his garden and yard work, Catesby took to the wilderness with the enthusiasm of a world-class explorer. During every season, in all sorts of weather, Mark Catesby walked for miles in rough, forested, semitropical terrain; he paddled deep rivers and slogged through murky swamps. He scrapped his way through tangles that would have brought many men to end the journey. Granted, he did befriend and come to depend on Indian guides, but still, knowing the wilderness of the Carolinas as I do, I can attest to this certainty: Catesby had to have been possessed of not only great curiosity and passion but also the constitution of an iron man to have stayed on task day after day, month after month, as he did in this hostile environment.

Those who have spent any time with Catesby's creations soon discover his wit and humor. They are important and attractive parts of his personality and help bring into better focus who was Catesby the man. Time and again we are

FIGURE 2-4. Alligators lurk almost everywhere there is water in the Lowcountry. (© Pamela Cohen.)

struck by his deliberate humor. There is a southern leopard frog standing on its head (*NH* II: plate 70), a bluejay sticking out its tongue (*NH* I: plate 15), a northern curly-tailed lizard clinging to both ends of the fragmented stem of a night-scented orchid (*NH* II: plate 68), and the commentary of an apparently dead American robin beside a supposed medicinal Virginia snakeroot (*NH* I: plate 29), but some may not have taken much notice of the crafty ways Catesby's signature was disguised in some of his etchings. One particularly whimsical example is what at first appears to be a spider or an insect dangling on the end of a long piece of spun web. On close inspection, however, it actually turns out to be Catesby's initials about to become a tufted titmouse's lunch (*NH* I: plate 57). I believe these delightful and quirky signs show us the sense of humor of a clever man, not one who was loud or crude but one whose keen, clear eye saw subtle things in nature and then grinned and chuckled. I think Mark Catesby had a nice, easy smile.

FIGURE 2-5. Great egret. (© Pamela Cohen.)

Aside from all else that I admire about Mark Catesby, the one characteristic that stands above all the rest is his phenomenal powers of observation and recall. In this arena, he surely was a genius. I suspect he had a photographic memory. How else could he capture the remarkably detailed and accurate representations of birds? I can tell you from serious firsthand experience that these birds will not take direction and sit still for one minute more while I get the picture framed just exactly right. Most everyone has had the experience of trying to identify a bird and, when asked the color of its beak or legs, having no recollection of the bird even having legs, much less color. Clearly, we must acknowledge that, for several reasons, Catesby killed his share of animals and plants. He did, however, declare his intention to not kill subjects in order to create his images. If we were to split the difference and guess that he sketched *plein air* even half of his subjects, think of the number of birds, fishes, reptiles, and mammals that Catesby painted correctly down to the slightest tiny color shading and fin or feather texture. Think of carrying his eighteenth-century art box of tints, inks, pencils, charcoals, and powder plus the porous handmade papers into the swamp during the heat and humidity of summer. How did he accomplish this remarkable work—this man who was not a trained artist? I believe Catesby's ability to observe and share his impressions of the natural world is a stroke of true genius. Otherwise, how can we explain his ability to recall and create the paintings and drawings that are so compelling that they continue to enthrall us after three centuries?

In order to understand the depth and scope of Catesby's work and how he accomplished it, one must go to the wilderness and experience the conditions that confronted Catesby without the benefit of sunscreen, insect repellent, a GPS, or motorized vehicles.

FIGURE 2-6.
Freshwater swamp in the South Carolina Lowcountry.
(© Pamela Cohen.)

I submit that Mark Catesby, first and foremost, was a remarkably intelligent man with a genius for observation and recall. He obviously possessed a strong natural talent for drawing. These abilities were advanced by a charming, witty, and energetic personality. He described himself as being passionately curious. The joy he found in nature is illustrated throughout his work. He was a strong, healthy man who lived a long, highly productive life filled with creative and intelligent associates. His was a life well lived. I am very sure I would have enjoyed being his friend. It has been quite an adventure getting to know him in the twenty-first century.

3

Mark Catesby's botanical forerunners in Virginia

KAREN REEDS

During the century before Mark Catesby's birth, a handful of English adventurers and naturalists had eagerly sought the plants that grew wild in "Virginia," as the southeastern corner of North America was long known. None of them, however, was able to achieve what Catesby did in *The natural history of Carolina, Florida and the Bahama islands*: a substantial, firsthand, beautifully illustrated account of the flora and fauna of this territory. Our appreciation of Catesby's accomplishment grows when we look at some of the reasons for his predecessors' limited results.

In each of the three generations before Catesby, his botanical forerunners form two overlapping groups: the collectors and the describers, or, to put it slightly differently, those who saw America firsthand and those who wrote about it.

Three generations of botanical explorers and writers

In the spring of 1585, the mathematician-navigator Thomas Harriot (c. 1560–1621) and the artist John White (c. 1540–c. 1593) took part in Walter Raleigh's second expedition to Roanoke Island, off the coast of what was later named North Carolina. Their task was to investigate the natural resources that could support a colony. Harriot wrote up his observations for Raleigh's investors in *A Briefe and True Report of the New Found Land of Virginia* (1588). The 1590 edition, enhanced with large engravings by Theodor de Bry after John White's paintings of America's "naturall inhabitants," became a European best seller.[1]

An album containing thirty-six watercolors by White of American plants and animals circulated in Elizabethan London, as did some specimens.[2] John Gerard (1545–1612), a surgeon and skilled gardener, illustrated one Virginian plant in his *Herball, or, Generall Historie of Plantes* (1598): the leaves and pods of the common milkweed (*Asclepias syriaca*) labeled "Indian Swallow woort, *Vincetoxicum Indianum*" (figure 3-1).[3] The specially prepared woodcut was

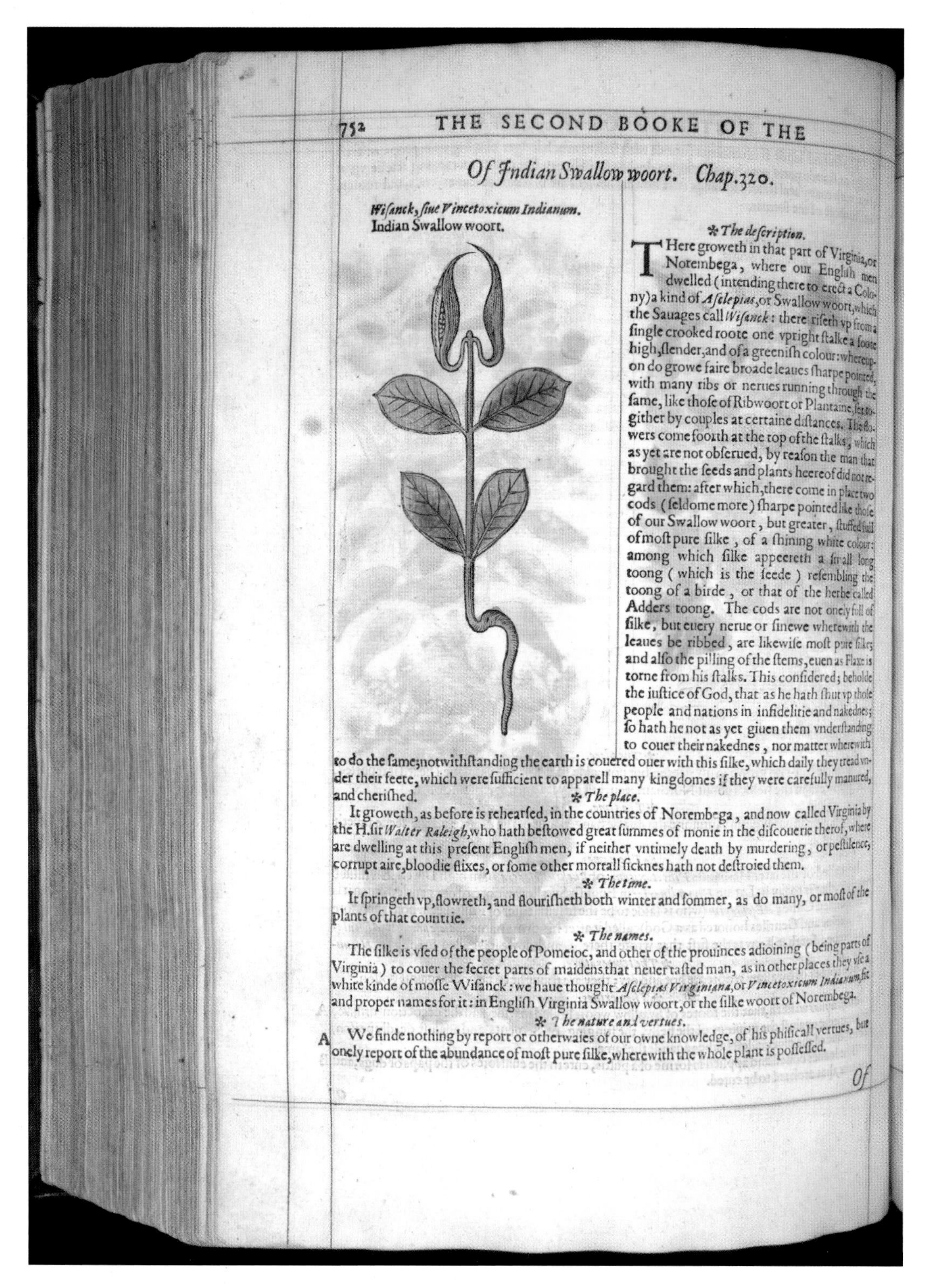

752 THE SECOND BOOKE OF THE

Of Indian Swallow woort. Chap.320.

Wisanck, siue Vincetoxicum Indianum.
Indian Swallow woort.

The description.

THere groweth in that part of Virginia, or Norembega, where our English men dwelled (intending there to erect a Colony) a kind of *Asclepias*, or Swallow woort, which the Sauages call *Wisanck*: there riseth vp from a single crooked roote one vpright stalke a foote high, slender, and of a greenish colour: whereupon do growe faire broade leaues sharpe pointed, with many ribs or nerues running through the same, like those of Ribwoort or Plantaine, set togither by couples at certaine distances. The flowers come foorth at the top of the stalks, which as yet are not obserued, by reason the man that brought the seeds and plants heereof did not regard them: after which, there come in place two cods (seldome more) sharpe pointed like those of our Swallow woort, but greater, stuffed full of most pure silke, of a shining white colour: among which silke appeereth a small long toong (which is the seede) resembling the toong of a birde, or that of the herbe called Adders toong. The cods are not onely full of silke, but euery nerue or sinewe wherewith the leaues be ribbed, are likewise most pure silke; and also the pilling of the stems, euen as Flaxe is torne from his stalks. This considered; beholde the iustice of God, that as he hath shut vp those people and nations in infidelitie and nakednes; so hath he not as yet giuen them vnderstanding to couer their nakednes, nor matter wherewith to do the same; notwithstanding the earth is couered ouer with this silke, which daily they tread vnder their feete, which were sufficient to apparell many kingdomes if they were carefully manured, and cherished.

The place.

It groweth, as before is rehearsed, in the countries of Norembega, and now called Virginia by the H. sir *Walter Raleigh*, who hath bestowed great summes of monie in the discouerie therof, where are dwelling at this present English men, if neither vntimely death by murdering, or pestilence, corrupt aire, bloodie flixes, or some other mortall sicknes hath not destroied them.

The time.

It springeth vp, flowreth, and flourisheth both winter and sommer, as do many, or most of the plants of that countrie.

The names.

The silke is vsed of the people of Pomeioc, and other of the prouinces adioining (being parts of Virginia) to couer the secret parts of maidens that neuer tasted man, as in other places they vse a white kinde of mosse Wisanck: we haue thought *Asclepias Virginiana*, or *Vincetoxicum Indianum*, fit and proper names for it: in English Virginia Swallow woort, or the silke woort of Norembega.

The nature and vertues.

A We finde nothing by report or otherwaies of our owne knowledge, of his phisicall vertues, but onely report of the abundance of most pure silke, wherewith the whole plant is possessed.

Of

FIGURE 3-1. Common milkweed (*Asclepias syriaca*): stalk, leaves, and "cods"; woodcut (after watercolor by John White, c. 1585), p. 752, J. Gerard, *The herball, or generall historie of plantes* (1597). (Courtesy of the History of Science Collections, University of Oklahoma Libraries.)

clearly based on John White's painting. Gerard described the Algonquians' use of the milkweed's silken fluff "to couer the secret parts of maidens," information that could also have come from White.[4]

In 1633 Gerard's *Herball* was substantially enlarged and corrected by Thomas Johnson (c. 1600–1644), a botanically learned apothecary.[5] To the revised "Indian Swallow-woort" chapter, Johnson gave a fuller description and added a flower to the woodcut. Gerard had only seen the milkweed "cods," but Johnson had seen "the growing and flouring plant"—evidence of the continuing transmission of seeds from North America during the first three decades of the seventeenth century (figure 3-2).[6] Thanks to Hans Sloane who commissioned a copy of the album of White's natural history paintings, these Elizabethan watercolors directly influenced Mark Catesby.[7] Catesby adapted several for *The natural history of Carolina* . . . and, for his tiger swallowtail butterfly image, explicitly acknowledged the manuscript of Walter Raleigh belonging to Sloane.[8]

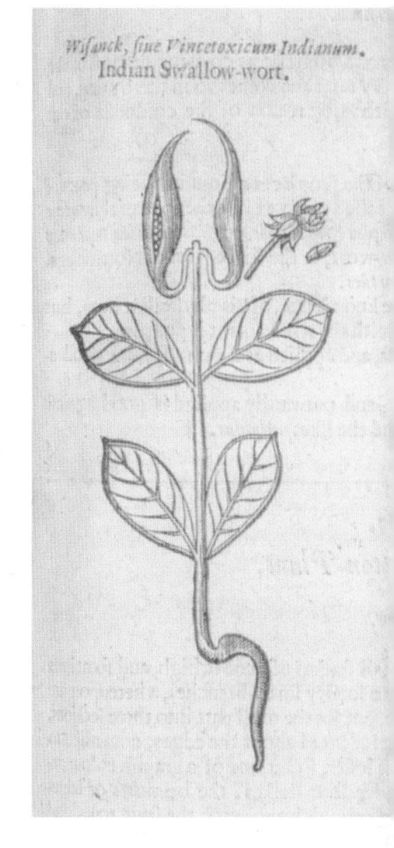

FIGURE 3-2. Common milkweed (*Asclepias syriaca*) with detail of flower added to original 1597 image; woodcut, p. 899, Thomas Johnson's revision (1636 reprint) of Gerard's *Herball* (1598). (Courtesy of the History of Science Collections, University of Oklahoma Libraries.)

A generation after Harriot, White, and Gerard, John Tradescant the Younger (1608–1662) voyaged to Virginia at least three times, in 1637–1638, in 1642, and in 1653–1654. On the first trip, his assignment—possibly from King Charles I—was to "gather up all rarityes of Flowers, plants, shells &c."[9] In his profession, adventurous spirit, love of curiosities, and intense interest in the New World, the younger Tradescant was following his father's example. John Tradescant the Elder (1570–1638), who had served several noblemen and Charles I as gardener, had brought back exotic plants from travels as far north as Russia and as far south as Morocco. A friend of Captain John Smith (1580–1631), he had invested heavily in the Jamestown enterprise. Through the father's success in cultivating a plant "brought . . . out of Virginia [and] deliuered to Iohn Tradescant," the Tradescant name achieved botanical immortality: Carl Linnaeus bestowed the name *Tradescantia virginiana* on "the soon fading Spider-wort of Virginia, or Tradescant his Spider-wort."[10] The Tradescants themselves published little, however.[11] For full descriptions and pictures of their plant discoveries, contemporaries had to turn to the books of the Tradescants' friends: John Parkinson (1567–1650) and Thomas Johnson.

Parkinson, an apothecary and a fellow gardener, recorded the elder Tradescant's detailed observations of the Virginia spiderwort in 1629 in *Paradisi in sole paradisus terrestris*.[12] In 1632, while revising Gerard's *Herball*, Johnson also visited Tradescant's garden, where he saw, among other plants, the "much admired Snakeweed of Virginia" (Virginia snakeroot, *Aristolochia serpentaria*) in bloom (figure 3-3).[13] Soon after the younger Tradescant's return from Virginia, Parkinson published accounts of the newly discovered American plants in his immense herbal, *Theatrum botanicum, the theater of plants*, issued in 1640, which became England's new authority on plants. Where Parkinson's first book on garden plants, *Paradisi in sole*, had identified barely a dozen plants as "Virginian," his *Theatrum botanicum* quadrupled that number.[14]

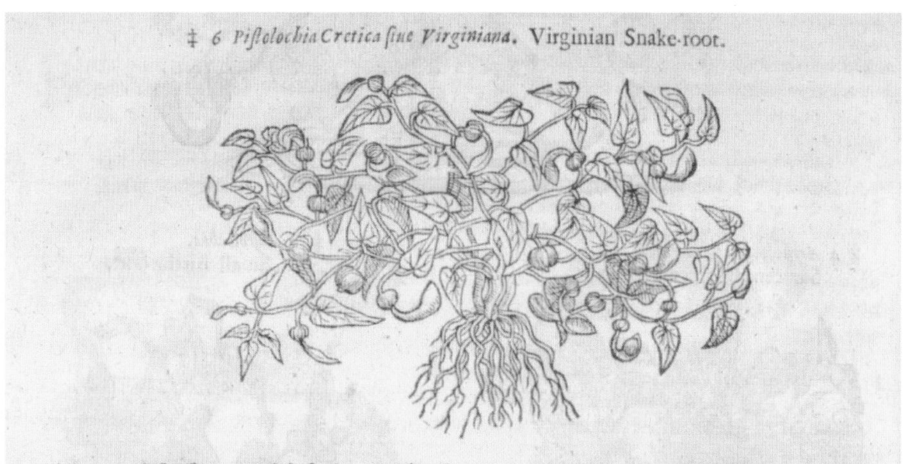

‡ 6 *Piſtolochia Cretica ſiue Virginiana.* Virginian Snake-root.

‡ 6 *Cluſius* figures and deſcribes another ſmal *Piſtolochia* by the name of *Piſtolochia Cretica*, to which I thought good to adde the Epithite *Virginiana* alſo, for that the much admired Snakeweed of Virginia ſeems no otherwiſe to differ from it than an inhabitant of Candy from one of the Virginians, which none I thinke will ſay to differ in *ſpecie*. I will firſt giue *Cluſius* his deſcription, and then expreſſe the little varietie that I haue obſerued in the plants that were brought from Virginia, and grew here with vs: it ſends forth many ſlender ſtalks a foot long, more or leſſe, and theſe are cornered or indented, creſted, branched, tough, and bending towards the ground, or ſpred thereon,

FIGURE 3-3. “Pistolochia Cretica, siue Virginiana. Virginian Snakeroot . . . The much admired Snakeweed of Virginia” (*Aristolochia serpentaria*) seen in the garden of John Tradescant the Elder, 1634: woodcut, p. 848, Thomas Johnson’s revision of Gerard’s *Herball* (1633, 1636). (Courtesy of the History of Science Collections, University of Oklahoma Libraries.)

Mark Catesby’s most immediate botanical forerunner was John Banister (1650–1692), an Anglican minister who had gone to Virginia in 1678 in the double capacity of missionary and naturalist and lived there until his early death. Although Catesby did not say so, he may well have felt that he was carrying on Banister’s unfinished work.[15] In their curiosity and taste for adventure, the two had much in common. Neither had become a practicing apothecary or physician, the traditional professions for educated young men with a bent for natural history. They thus exemplified the turn of botany toward the broader scientific aims of the recently founded Royal Society of London.

Although Banister and Catesby belonged to different generations, their intellectual circles included several influential English naturalists whose lives intersected with both of theirs: John Ray (1627–1705), Europe’s pre-eminent botanist; Henry Compton (1632–1713), Bishop of London; Martin Lister (1639–1712), a physician and conchologist; Leonard Plukenet (1642–1706), an ardent botanist and gardener to Queen Mary; James Petiver, an apothecary and self-publisher of natural history works; Jacob Bobart the Younger (1641–1719), Keeper of the University of Oxford’s botanic garden; as well as William Sherard and Hans Sloane.[16] In Virginia, Banister and Catesby enjoyed the patronage, hospitality, and library of the Byrds of Westover, Banister with Colonel William Byrd I (1652–1704), and Catesby with his son. William Byrd II was the most direct link between Banister and Catesby: before departing for school in England at the age of seven, he had almost certainly met Banister at Westover.[17]

As an undergraduate at Oxford University and then as a chaplain, Banister had the benefit of studying under Oxford's first Professor of Botany, Robert Morison (1620–1683). Morison's lectures, herbarium, books, and demonstrations in the university's botanic garden gave Banister his first encounters with American plants.[18] Banister's natural history expertise commended him to Compton, who was in charge of the Church of England's overseas missions and was also "a great lover of . . . Exotic Plants."[19] To the bishop, the Reverend John Banister must have seemed a heaven-sent candidate, equally competent to spread the Gospel and gather seeds.

Four months after reaching Virginia in December 1678, Banister thanked Morison for his support. The letter and accompanying specimens recorded Banister's first observations and his desire to explore inland: "I long to see what Naturals . . . the great Ridge of Mountains that Runs across the Continent does produce [and to go] among ye Indians . . . to inform myself of ye Names but especially the Virtues of Plants, which Nature has taught them to a Miracle."[20] A decade later, Banister's zest for exploration was undiminished: "Had I an Estate would bear out my Expense, there is no part of this, or any other Country that would afford new matter, though under ye Scorching Line, or frozen Poles my genius would not incline me to visit."[21]

Banister's enthusiasm was buoyed by praise from the greatest botanist of the age. Calling him a "most consummate botanist," John Ray had published Banister's Virginia discoveries in the magisterial *Historia plantarum*.[22] Such recognition encouraged Banister to pursue his own special project: an illustrated natural history of Virginia. Sadly, the project died with him.

Obstacles to publication

Of Catesby's forerunners, the four stay-at-homes—Gerard, Johnson, Parkinson, and Ray—all published on a massive scale, each trying to encompass everything that was known in their day about the world's plants. Their books remain key sources about pre-Linnaean knowledge of plants from North America. The four travelers to Virginia—Harriot, White, the younger Tradescant, and Banister—were, however, only able to publish tantalizing snatches of what they had seen in the New World. What prevented the adventurers from publishing their own discoveries?

Beyond the sheer labor of writing and the expenses of seeing a book through the press, the obstacles were manifold: hazards of exploration, difficult personal circumstances, lack of training or inclination, turbulent politics, and limited resources.

For Thomas Harriot, the chief obstacle to publishing a fuller account was a disaster as he left Virginia.[23] "Boysterous" weather had put the ship in such peril of running aground that the sailors cast overboard all his maps, "Bookes and writings."[24] It is remarkable that, working from memory, he could give as much information as he did about the names and uses of plants in his report—and all the more so because his scientific strengths were mathematics, astronomy, physics, and navigation, not natural history. After 1588 Harriot's only continuing interest in Virginian plants was tobacco for his pipe.[25]

The storm that laid waste to Harriot's manuscripts may also have destroyed John White's eyewitness paintings of New World plants. Dazzled by Virginia's possibilities, White made two brief return visits not to paint but to try to establish a new Roanoke settlement. The tragic venture of the Lost Colony preoccupied White for the rest of his life.[26] Through de Bry's engravings, White's pictures of American Indians became iconic images, but White himself put nothing into print.

John Tradescant the Younger, by contrast, was skilled in everything to do with plants, and his luck was far better. He and his Virginia collections survived the trips back to England, and many of the plants grew successfully in his garden in London alongside American rarities that his father had received over the years.

However, the Tradescants' publishing ambitions met with much less success. Only one copy of the elder Tradescant's printed catalog of his garden is known to survive.[27] In *Musaeum Tradescantium*, published in 1656, the younger Tradescant cataloged the extraordinary array of curiosities he and his father had assembled and exhibited. But Tradescant's hopes for a sumptuous illustrated volume were frustrated by personal misfortunes. Amid the turmoil of the Commonwealth under Oliver Cromwell, the Tradescants' earlier association with King Charles I may also have been problematic. As published, *Musaeum Tradescantium* had no pictures. Terse phrases described the outlandish artifacts and natural specimens; a bare list of Latin and English plant names commemorated the Tradescant garden. Consequently, the best evidence of the Tradescants' knowledge of American plants comes through John Parkinson's *Theatrum botanicum* rather than from the two gardeners themselves.[28]

Of these early investigators of Virginian plants, John Banister had the best hope of getting his discoveries into print. He was easily the best suited to the task by personal inclination and by training. To his own delight in plants, he added stamina, perseverance, intelligence, ambition, and (as John Ray declared) "his knowledge of all the pertinent scientific literature."[29] Moreover, through Oxford, the Royal Society, and the London botanists, he had the right connections to help him turn his manuscript and drawings into a book.

After surviving the voyage to Virginia, incursions by Indians, a fall from a horse, and a dozen years of plant hunting in the wilderness, Banister's luck ran out. On a trip with William Byrd I to the "Land of Eden" on the lower Roanoke River, Banister was accidentally shot by a fellow member of the party who had perhaps mistaken Banister for a wild animal.[30]

Banister had transmitted lists, notes, drawings, and specimens to his English colleagues at every opportunity. After his death, most of his papers, natural history specimens, and drawings were forwarded to Bishop Compton. Over the next two decades, English naturalists continued to mine Banister's materials. In the Royal Society's *Philosophical transactions*, Martin Lister published extracts from Banister's letters about insects and shells. Banister's pressed plants found their way to Oxford, where Jacob Bobart the Younger was editing the third part of Robert Morison's *Plantarum historia universalis Oxoniensis*, published in 1699. Plukenet's *Phytographia*, issued in parts between 1691 and 1696, mentioned Banister frequently and incorporated engravings made from Banister's drawings (sometimes uncredited); and these were further analyzed in the third volume of John Ray's *Historia plantarum*, published in 1704.[31] Banister's personal working draft of his catalog of Virginian plants went to James Petiver, and after Petiver's death, it went into Hans Sloane's manuscript collection.[32] Later still, Banister's observations were used by Carl Linnaeus and by Frederick Pursh (1774–1820) in *Flora Americae septentrionalis*, the first thorough conspectus of North American plants, published in 1814.

When Byrd I's son-in-law, Robert Beverley (c. 1673–1722), wrote his *History and present state of Virginia*, issued in 1705, he acknowledged John Banister's "curious collections for a *Natural History* of Virginia" but did not disclose just how heavily he had relied on Banister's manuscript. Beverley's section "the Works of Nature" reproduced Banister's draft virtually verbatim.[33] Through this range of sources, Mark Catesby could have encountered Banister's work before or during his Virginia years or in preparing his own *The natural history of Carolina . . .* later on.

"for want of Books and other helps"

John Banister put his finger on another impediment to writing an account of the plants of the New World: the lack of adequate reference books. Although he had studied the exotic plants available in the gardens and herbaria of Oxford and London, Banister was staggered by his first encounters with Virginia's flora. He wrote to Robert Morison in April 1679, "I have met with a great Number of Trees, Shrubs, and herbs that I know not what to make of for want of Books and other helps & Assistances pertinent to him that undertakes & intends to go through with such a Work."[34]

The problem had arisen from the start of the English exploration of Virginia. Of the various sixteenth-century herbals and explorers' accounts that described American plants, Harriot and White apparently only knew one: *Ioyfull nevves ovt of the newe founde worlde*. Harriot cited this panegyric on American materia medica in his comments on "Winauk," sassafras (*Sassafras albidum*).[35] The original author, a Spanish physician, Nicolás Monardes (1493–1588), had himself only known these wonder drugs from reports and samples brought back to Seville. The books that John White carried on his 1587 voyage to Roanoke did not prevent a harrowing encounter with the poisonous manchineel tree (*Hippomane* sp.) on the island of St. Croix.[36] Although White's story was quoted, along with many like it, in Hans Sloane's 1696 catalog of Jamaican plants, Catesby did not heed the warnings and endured his own painful experience with "Mancaneel" in the Bahamas.[37]

By the early seventeenth century, botanical information from around the world was accumulating rapidly and making its way into print in herbals and voyage narratives. The younger Tradescant could consult many new books for his own Virginia travels, whether or not he carried them along. His *Musaeum* bore witness to his knowledge of the botanical literature available by mid-century.[38]

During Banister's years in Virginia, the "want of Books" was eased to some degree. William Byrd I used his visits to London to stock his own shelves and acquire books on Banister's behalf. Notably, in 1688 Byrd brought back the second volume of Ray's *Historia plantarum*, with its praise of Banister, right after its publication. A gift from Plukenet, the copy seems to have been in the

FIGURE 3-4. John Banister's inscription, "Ex Biblijs J. Banister in Virginiâ. 1688 [From the books of J. Banister, in Virginia, 1688]," on the title page of his copy of John Parkinson, *Theatrum botanicum* (1640). (Courtesy of Burlington County Lyceum of History and Natural Sciences. Photograph by Chris Lippa.)

nature of a consolation prize to Banister in lieu of funding for his natural history of Virginia.[39] On Banister's death, his books were incorporated into the Byrd library, which William Byrd II enthusiastically enlarged. In the 1780s the whole collection—the largest private library in England's North American colonies—was dispersed by auction, John Banister's books among them.[40]

"Conscious of my own Inability": Catesby and his forerunners

John Ray's *Historia plantarum* was not the only important botanical book to reach John Banister in 1688. A used copy of John Parkinson's *Theatrum botanicum* also arrived that year, as Banister's inscription on the title-page attests (figure 3-4).[41] In the *hortus siccus* Banister had compiled for himself at Oxford, he had labeled many pressed plants with Parkinson's names and descriptive phrases.[42] That herbarium had accompanied him to Virginia, but it could not substitute for having his own copy of *Theatrum botanicum*. The well-worn herbal makes it easy for the reader to imagine John Banister at work. Out of the 250-plus American plants named in his working draft catalog,[43] nearly three dozen invoked Parkinson—more than any other author (Ray's *Historia plantarum* only appears once).[44] For every plant Banister encountered in Virginia, native or introduced, he would have consulted *Theatrum botanicum*, enlarging his fund of botanical knowledge with every search.

The Parkinson volume was far too bulky to carry into the backwoods. Banister probably kept his books, specimens, and papers at the Byrd plantation. There he could take advantage of the library and share his discoveries with his interested patron—and with the curious child, William Byrd II, who grew up to contribute his own natural history observations to the Royal Society and to befriend Mark Catesby.[45]

Byrd II, following "a strong inclination to promote naturall history," tried at least three times between 1697 and 1710 to recruit a "man of skill in the works of nature" to continue Banister's project.[46] Catesby's arrival in Virginia on 23 April 1712 meant that Byrd's wish had been unexpectedly granted. Within a month Catesby began his first extended visit to Westover. On 27 May Byrd recorded in his diary: "In the afternoon, we went into the library to see some prints. We spent the afternoon in conversation and in the evening we took a walk."[47]

So the door to the Byrd library was open to Catesby, and the Parkinson herbal and dozens of pertinent books on natural history, horticulture, art, and travel were there to be consulted.[48] Judging from *The natural history of Carolina . . .*, however, Catesby did not go in to read them. During their two years of close association, Catesby and Byrd II apparently preferred to spend their time together in the open air, planning improvements to the Westover garden, walking through the plantation, or traveling.[49]

Catesby was aware that he had let an opportunity pass by. *The natural history of Carolina . . .* began with a disclaimer: Catesby had been deprived "of all Opportunities and Examples to excite me to a stronger Pursuit after those Things to which I was naturally bent." Even when fulfilling his "passionate Desire" to see exotic plants in their native lands, he admitted—"I am ashamed to own it"—that he had not made good use of his seven years in Virginia.[50] Although he had sent some pressed and living plants to Samuel Dale in Braintree and to some other friends at their request, he had not made systematic collections, notes, or drawings for himself.

Catesby's self-deprecating rhetoric—"Tho conscious of my own Inability"—was customary in authors' prefaces of the period; and from the accounts of others who knew Catesby in later years, he was indeed modest by temperament. There is, however, a certain disingenuousness in Catesby's excuse: "my residing too remote from London the Center of all Science."[51] In rural Suffolk and Essex, where Catesby probably spent his youth, there were two men deeply interested in plants: John Ray and Samuel Dale.[52] Catesby's uncle, Nicholas Jekyll of Castle Hedingham (1655–after 1725?),[53] was acquainted with both Dale and Ray and could have provided introductions. Catesby's home in Sudbury lay about sixteen miles from Dale in Braintree and four more from Ray in Black Notley. There is, however, no firm evidence that Catesby took advantage of this proximity. Ray may have served the young naturalist as an exemplar of a life devoted to natural history, but it seems unlikely that Ray actively mentored Catesby—Ray was too burdened by ill health and anxious to complete large projects during the decade before his death in 1705.[54] Nor is there any sign that Catesby used the auction catalog of Ray's extensive library to guide his own early study of natural history.[55]

Long before he had any thought of going to Virginia, Catesby could have learned something about John Banister from Dale. In Dale's *Pharmacologia*, which was published in 1693, a year after Banister's death, Banister's manuscript catalog served as Dale's authority for two different American snakeweeds (figure 3-5)—a telling indication of how quickly Banister's Virginia papers circulated posthumously and how closely connected Dale was to the London naturalists acknowledged in his preface.[56] Catesby could also have seen two brief accounts of Banister's Virginian plants, published by Ray in 1698 and Petiver in 1707.[57]

Thus, in principle, both before and during his stay in Virginia, Mark Catesby had greater access to existing accounts of American botany than any naturalist before him and plenty of reasons to consult them. Yet, when he came to writing his own account, he paid little attention to the botanical literature. He had a good excuse, of course. What he said about New World fish also held for the plants, "there being not any, or a very few of them described by any Author."[58] However, that had been even truer for Banister, who strove hard to give chapter and verse whenever he could.

By considering *The natural history of Carolina, Florida and the Bahama islands* in relation to earlier herbals and plant lists, we can see that Catesby broke with

but of a darke or ſullen yellow colour, and ſomewhat leſſe alſo; after which come very ſmall heades with ſeede, ſomewhat like to the *Piſtolochia* but leſſer: the rootes are a number of very ſmall blackiſh gray fibres or threds, as ſmall almoſt as haires, which have both an aromaticall and reſinous ſmell, when they are drie, more than when they are greene, and of an aromaticall reſinous aſtringent taſte, without any great or manifeſt heate.

Polyrhizos Virginiana. The rattle Snakeweede of *Virginia*.

The Place.

It groweth very frequent in the upper parts of our *Virginian* plantation, in the fields and champion countries, where under the graſſe and herbes, that venemous rattle Snake lurketh and abideth, ready to bite whomſoever ſhall come neare unto it.

The Time.

It flowreth with us in *Iune* and *Iuly*.

The Names.

This may very well be referred to the *Piſtolochia* I ſaid, but I have ſevered it being ſo notable an Antidote; but by what Latine name it might beſt be called, either *Piſtolochia Virginenſis*, or as I doe *Polyrhizos Virginiana*; I leave it to every man to doe as he will, or untill a fitter Latine title may be given it, if it be thought expedient: our people in *Virginia* doe there call it the Snakeweede, or Snakeroote, and thereupon may be called *Colubrina*

FIGURE 3-5. "Polyrhizos Virginiana: The rattle Snakeweede of Virginia . . . so certain a remedy against the biting of that venemous rattle Snake" (*Aristolochia serpentaria*), pp. 420–421, in John Banister's copy of John Parkinson, *Theatrum botanicum* (1640). The specimen pressed in the herbal cannot be dated. (Courtesy of Burlington County Lyceum of History and Natural Sciences. Photograph by Chris Lippa.)

a practice that went back to antiquity. While these books traditionally devoted a substantial part of each entry to disentangling a plant's assorted names and citing many authorities, as a rule *The natural history of Carolina* . . . rarely cited other authors in the text and confined nomenclature to the chapter headings. These headings typically consisted of a single long descriptive phrase-name, sometimes briefly referencing one or two authors, and sometimes noting "the English and Indian Names they are known by in their Countries."[59] For the Latin names, Catesby declared himself "beholden to the . . . Learned and accurate Botanist Dr. Sherard."[60] This was both polite and politic: since Sherard had died in 1728, the formulation side-stepped the choices Catesby made in the matter of names. There is no certainty which of them should be credited with *The natural history of Carolina* . . .'s minimal references to the earlier writers on Virginian plants: three references to Parkinson's *Theatrum botanicum* and four chapter headings citing John Banister.[61]

We do not know whether Catesby's decision not to cite his predecessors arose from his own "intellectual diffidence,"[62] an impatience with the confusion of names, his belief that "Figures convey the strongest Ideas,"[63] a desire to emphasize other kinds of relationships in nature, or simply a pragmatic recognition that fewer words saved money. Ultimately, Catesby had no need to look back on his stay in Virginia with regrets. It was both his apprenticeship and opportunity for reflection. That is where he learned how to conduct "Enquiry in so remote parts."[64] That is where he trained his eye and hand. That is where he worked out his own "perfect understanding" to "Illuminating Natural History."[65] Those seven years did not go to waste.

4

Maria Sibylla Merian (1647–1717): pioneering naturalist, artist, and inspiration for Catesby

KAY ETHERIDGE AND
FLORENCE F. J. M. PIETERS

In 1699 a fifty-two-year-old woman and her younger daughter embarked in Amsterdam on a two-month voyage to the Dutch colony of Surinam in South America. Maria Sibylla Merian undertook this journey specifically to study insects in the tropical jungle and to document their metamorphoses and food plants, a study that today would be considered ecology. By this time in her life Merian was a renowned naturalist and had published two books on European caterpillars, moths, and butterflies. The book that would result from this extraordinary undertaking, *Metamorphosis insectorum Surinamensium*,[1] was the first to show New World plants and animals together in colorful images (figure 4-1) and would make a strong impression on Mark Catesby, as it did on other naturalists who followed her. Along with her volumes on European caterpillars, Merian's book on New World organisms, like Catesby's, changed the way in which the natural world was perceived and portrayed.

Maria Sibylla Merian's background and artistic training made her uniquely suited to write and illustrate the first books to portray the interactions of animals and plants. Born in the German city of Frankfurt am Main in 1647, Maria Sibylla grew up among artists and publishers (figure 4-2). During the second part of her life she lived in Amsterdam, where her work is still celebrated today. Her name, "M. S. MERIAN," appears on the façade of the Artis Library (built in 1867) along with those of thirty-five famous men of science, including Aristotle, Pliny, and Linnaeus.[2] The name Artis is derived from the motto "Natura artis magistra" of the Zoological Society of Amsterdam, which founded a zoo, a zoological museum, and a library in Amsterdam in 1838.[3] This motto seems an appropriate starting point for a discussion of Merian and Catesby, as it means: Nature is the teacher of the arts and sciences. Interestingly, in her native Germany, Maria Sibylla Merian seems to be regarded as the earliest female German artist of renown,[4] whereas in the Netherlands she is generally considered to be the first important female scientist.

FIGURE 4-1. Banana (*Musa × paradisiaca*) with moth and larva of the bullseye moth (*Automeris liberia*); plate 12, M. S. Merian, 1719, *Over de voortteeling en wonderbaerlyke veranderingen der Surinaemsche insecten.* (Courtesy of Artis Library, University of Amsterdam.)

Maria Sibylla's father, the prominent engraver, painter, and publisher Matthäus Merian (1593–1650), died when she was only three years old. Her mother, Merian's second wife, Johanna Sibylla Heim (c. 1620–1690), then married the versatile painter Jacob Marrell (1613–1681). Within these households, Maria Sibylla was taught and influenced artistically by her stepfather and her half-brothers. Her favorite half-brother, Caspar Merian, was probably her teacher in the arts of etching and engraving on copper, and printing, but it was her stepfather to whom she owed much of her excellent artistic education. Jacob Marrell has long been celebrated for his splendid flower paintings, which in the tradition of that period included insects and other small animals to enliven the composition and to denote important symbolism. Animals such as a butterfly or a lizard could symbolize resurrection, and a snail could indicate caution or laziness; floral symbolism was complex as well, with withering flowers indicating mortality, columbine signifying the Holy Spirit, or others,

FIGURE 4-2. Maria Sibylla Merian; anonymous, 1679, oil on canvas. (Reproduced by permission of Kunstmuseum Basel.)

such as lily-of-the-valley, representing humility (figure 4-3).[5] Apart from masterworks in oil, Marrell also produced highly detailed and life-sized portraits of tulips in watercolor on parchment for catalogs used by the tulip trade in Holland.[6] At least four of his "tulip books," sometimes with prices of the bulbs indicated, are extant.[7] It is striking that he added small animals to these images even though they were generally not of use in such a trade catalog (figure 4-4). His stepdaughter also produced beautiful and detailed images of tulips and decorated her early flower paintings with little animals, but she would later include insects with a new and clearly scientific purpose (figure 4-5).

Maria Sibylla's artistic training in the relatively liberal Marrell household was fortuitous. At that time in Germany, girls were barred from painters' guilds, and in some towns it was forbidden for women to sell their own oil paintings.[8] In addition, Maria Sibylla grew up among printing presses and a vast array of books, a promising start in life for someone who would go on to publish several natural history books of her own. The Merian family firm published natural history books such as John Jonston's *Historia animalium*, an early zoological encyclopedia for which the plates were engraved by her half-brothers. Jonston's volume on insects, which included large plates with many creatures in rows, may have particularly caught her eye (figure 4-6).[9] Such childhood influences appear to have piqued her curiosity, for at the age of thirteen, Maria Sibylla observed the metamorphosis of silkworms and documented

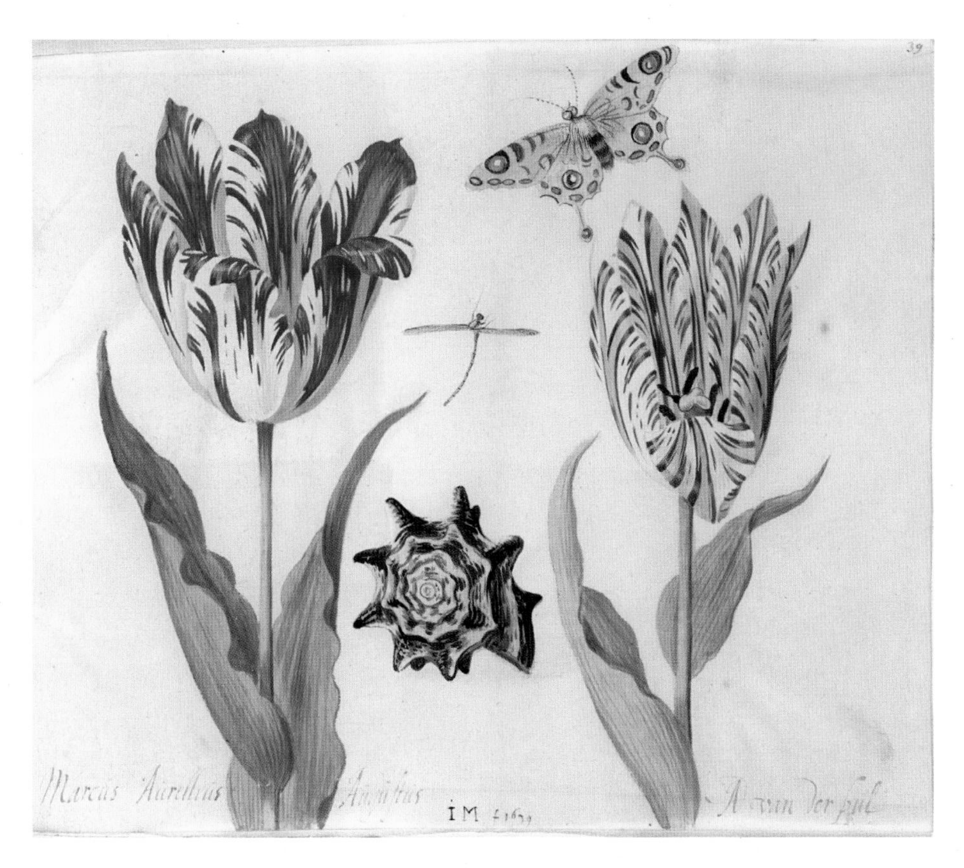

FIGURE 4-4.
Watercolor on vellum, 1639, in a tulip book by Jacob Marrell with butterfly, shell, and dragonfly added. (Reproduced by permission of Rijksmuseum, Amsterdam.)

FIGURE 4-5.
Metamorphosis of a moth (possibly the purple clay moth, *Diarsia brunnea*) on a tulip; plate 2, M. S. Merian, 1679, *Der Raupen wunderbare Verwandelung und sonderbare Blumen-nahrung*. (Courtesy of Netherlands Entomological Society [NEV], Amsterdam.)

FIGURE 4-3.
Flower still-life painting by Jacob Marrell, with snail and lily-of-the-valley in the foreground; undated, oil on canvas. (Courtesy of Art Gallery P. de Boer, Amsterdam; private collection.)

FIGURE 4-6. Plates 4 and 21 from J. Jonston's book on insects were originally engraved by Matthaeus Merian jr and Caspar Merian for the Latin edition (1653). These images are from a re-engraved and hand-colored Dutch edition (1660). (Courtesy of Artis Library, University of Amsterdam.)

their transformation from egg to adult. The same patience and skill needed to depict living organisms accurately stood her in good stead as she began to raise insects and study their life-cycles. Thus began a lifetime of pioneering empirical research that breathed the spirit of the early Enlightenment.

When she was eighteen, Maria Sibylla married Johann Andreas Graff, who had been a pupil of her stepfather. They had two daughters: Johanna Helena was born in 1668, and Dorothea Maria ten years later. From 1668 the family lived in Nuremberg, where Maria Sibylla gave lessons in painting and embroidery to young women and published her *Neues Blumenbuch* in three installments. This flower book often included insects in its plates.[10] More strikingly, however,

Maria Sibylla had expanded her investigations of the life-cycles of moths and butterflies, and in 1679 she published the first volume of her studies of the metamorphoses of European moths and butterflies, *Der Raupen wunderbare Verwandelung und sonderbare Blumen-nahrung* (The wondrous transformation of caterpillars and their curious diet of flowers) (figure 4-7).[11] After the death of her stepfather in 1681 she moved back to Frankfurt to assist her mother, and there she published the second volume of *Raupen*.[12] About four years later she moved with her two daughters and her mother to Wieuwerd in Friesland (part of the Dutch Republic) and joined her half-brother Caspar in a Protestant community called the Labadists. During her stay at Wieuwerd, she continued her investigations of the life-cycles of insects. Studying natural history was considered an appropriate occupation for a devout person because it was believed that one could come nearer to God by examining his creations.[13]

Around this time Maria Sibylla began to arrange the loose pages of her detailed paintings on vellum and notes into a large journal containing her observations on metamorphosis, insect behavior, and what we now think of as

FIGURE 4-7. Mulberry wreath (*Morus* sp.) with early life stages of the silkworm (*Bombyx mori*) and several unidentified insects; frontispiece, M. S. Merian, 1679, *Der Raupen wunderbare Verwandelung und sonderbare Blumen-nahrung*. Of particular interest is the depiction of larvae hatching from eggs, published at a time when the idea of spontaneous generation of insects was still entrenched. (Courtesy of the University Library Johann Christian Senckenberg, Frankfurt am Main.)

the ecology of insects. She continued to add to this journal for three decades. For Maria Sibylla as naturalist and artist, this notebook was an indispensable record and an invaluable source of carefully depicted models. The small paintings documented the precise colors of the living insects, their eggs, larvae, and pupae, and in doing so recorded information that would be lost in preserved specimens. Scholars consider this an essential tool for their research on Maria Sibylla Merian, both because her original hand-painted works depict so many details and because some of Merian's notes were not included in her published works. For instance, in 1686 she described the development of frog's eggs and the metamorphosis of tadpoles long before the observations of the same phenomenon by the Dutch microscopist Antony van Leeuwenhoek (1632–1723) were published in 1699.[14] As Maria Sibylla was continuing her work in Wieuwerd, her husband tried unsuccessfully to join the Labadist community as well, and one result of this family drama was their separation.[15] Maria Sibylla left the community and moved with her daughters to Amsterdam in 1691; shortly afterward her marriage was formally dissolved, and she published thereafter under the name Maria Sibylla Merian.

In Amsterdam, Merian made a living selling paintings of flowers and insects and trading in specimens. Like Frankfurt and Nuremberg, Amsterdam was an important hub of the European book trade, and, moreover, it was an epicenter of international trade. By this time, Merian was well respected among naturalists, scholars, and collectors interested in natural history due to her *Raupen* books, and this gave her entry into the drawing rooms, libraries, and natural history collections of Amsterdam, including that of the Director of the East India Company, Nicolaas Witsen.[16] She wrote about this in the introduction to *Metamorphosis*:

> In Holland I marveled to see what beautiful creatures were brought in from the East and West Indies . . . in which collections I found these and countless other insects, but without their origins and generation; that is, how they change from caterpillars to pupae and so forth. This prompted me to undertake a long and expensive journey and to travel to Surinam in America . . . to continue my observations there. . . .[17]

Thus Merian made the astounding decision to travel to Surinam to study tropical insects first hand. In June 1699, after having drawn up her will, she embarked for Surinam with her daughter Dorothea. Maria Sibylla Merian became the first naturalist to undertake a voyage specifically to study metamorphoses of New World organisms and one of the few to undertake such a journey without the backing of a wealthy patron or in service to a government. Merian was forced by reasons of health to return to Amsterdam after just two years in Surinam, but in that short time she observed, collected, and documented almost two hundred species of tropical plants and animals. Soon after her return to Amsterdam she began the work necessary to publish her observations, and in 1705 *Metamorphosis insectorum Surinamensium* was published.

For the remainder of her life Merian worked on various projects, including a third volume of her caterpillar book; this volume was edited by Dorothea and published soon after Merian died at the age of sixty-nine, on 13 January 1717.[18] Apparently she had been living in rather poor circumstances, and it is interesting that around the time of her death, Merian's research journal was sold to Robert Erskine (1677–1718), a Scot who was court physician to the Tsar of Russia, Peter the Great.[19] Erskine also bought for the Tsar almost three hundred of Merian's watercolors on vellum at the cost of three thousand guilders, roughly the price of an average house in 1717 Amsterdam. The next year, Dorothea moved with her second husband, Georg Gsell, to Saint Petersburg, where they were appointed art advisors to the Tsar and teachers at the newly founded Russian Academy of Arts and Sciences.[20] Meanwhile, *Metamorphosis* firmly established Maria Sibylla Merian's fame, and it went through several editions during the eighteenth century.

Merian's books

During her lifetime Merian published the life-cycles and habits of more than 250 species of insects, and by elucidating their relationship to plants, she can be called one of the first ecologists. Merian's published images were the first to combine biologically linked plants and insects on the same page; in her *Raupen* books, Merian arranged the life stages of each insect, usually a moth or butterfly, around a plant that served as food for their respective caterpillars (figure 4-8).[21] She described the reproduction and metamorphoses of butterflies and moths and was one of the earliest naturalists to write about the insects' defensive behavior and locomotion, as well as myriad other biological details of larvae and adults. She also recorded factors that are part and parcel of contemporary ecological science, including descriptions of environmental effects on insect development and abundance and observations on food choice and feeding behavior.[22] Through her observations of insects in nature and by raising them through all life stages, Merian demonstrated that they reproduce by mating and egg production, which was notable at a time when spontaneous generation of insects was still accepted by many scholars. By the time Merian published her first "caterpillar book" in 1679, she was able to state conclusively that caterpillars hatched from eggs laid after male and female butterflies and moths mated.[23]

Before Merian's innovative compositions were published, plants and animals were illustrated separately for the most part, a model unchanged since medieval encyclopedias and herbals. An entomological example can be seen in Jonston's encyclopedia, in which insects were not only separated from plants, but adults and larvae were depicted on separate plates (see figure 4-6).[24] Even Merian's near contemporary, Hans Sloane, had his illustrators follow this traditional approach in his volumes on Jamaica (figure 4-9),[25] so that similar organisms were arranged on a page with no regard to the other plants or animals that

FIGURE 4-8. One of Merian's earliest studies of metamorphosis, of a small tortoiseshell butterfly (*Aglais urticae*) on the small nettle (*Urtica urens*); plate 44, M. S. Merian, 1679, *Der Raupen wunderbare Verwandelung und sonderbare Blumen-nahrung*. (Courtesy of Netherlands Entomological Society [NEV], Amsterdam.)

might appear in the same habitats. But Merian continued her more naturalistic approach in constructing her images for *Metamorphosis*. As she did for almost every entry on moths and butterflies, she also described the caterpillar she had found feeding on the leaves and the dates of its pupation and of its emergence as an adult. *Metamorphosis*, like her earlier *Raupen* books, focused on insects but included much more information on plants, perhaps because the species pictured would have been unfamiliar to her European audience. For example, in the text accompanying a plate depicting the inflorescence of a banana (see figure 4-1, p. 40), she described the growth-form of the plant and the taste and texture of the fruit.[26] Much of the information on the uses of plants was obtained from slaves on the sugar plantations where she had stayed or from indigenous people who had helped her collect specimens. In *Metamorphosis* she often acknowledged the role of slaves and "her Indian," writing in one entry that a plant she wished to study had been "dug up by the roots by my Indian and brought back to my house and planted."[27] Merian often noted potential medical or other uses of the plants, such as the trunk of the fan palm, which when cooked "tastes better than artichoke hearts."[28]

As well as adding more botanical information in *Metamorphosis*, Merian broadened her coverage beyond moths and butterflies to include some spiders, reptiles, and amphibians in the hopes of stimulating interest in a further volume (figure 4-10).[29] Her fascination with reproduction and development features in the text and plates of frogs; she was one of the first to represent amphibian reproduction accurately (figure 4-11).[30] In addition to life-cycles, Merian also depicted food chains and portrayed both adult insects and their larvae while feeding. In *Metamorphosis* she included the predation of insects upon frogs (figure 4-11), reptiles upon insects (figure 4-10),[31] and the startling image of a giant spider preying upon a hummingbird while other spiders and ants feed and forage (figure 4-12).[32] The text that accompanied the latter image was more than twice the length of most, and in addition to material on the spiders and the hummingbird, it included a description of the migratory foraging raids of army ants, the ability of leaf-cutter ants to defoliate a tree overnight, and even the fact that the leaf-cutters took their harvest back to deep underground nests.[33] Between her own observations and what she gleaned from her local sources, Merian built up a vivid picture of the teeming life of the Surinam jungle.

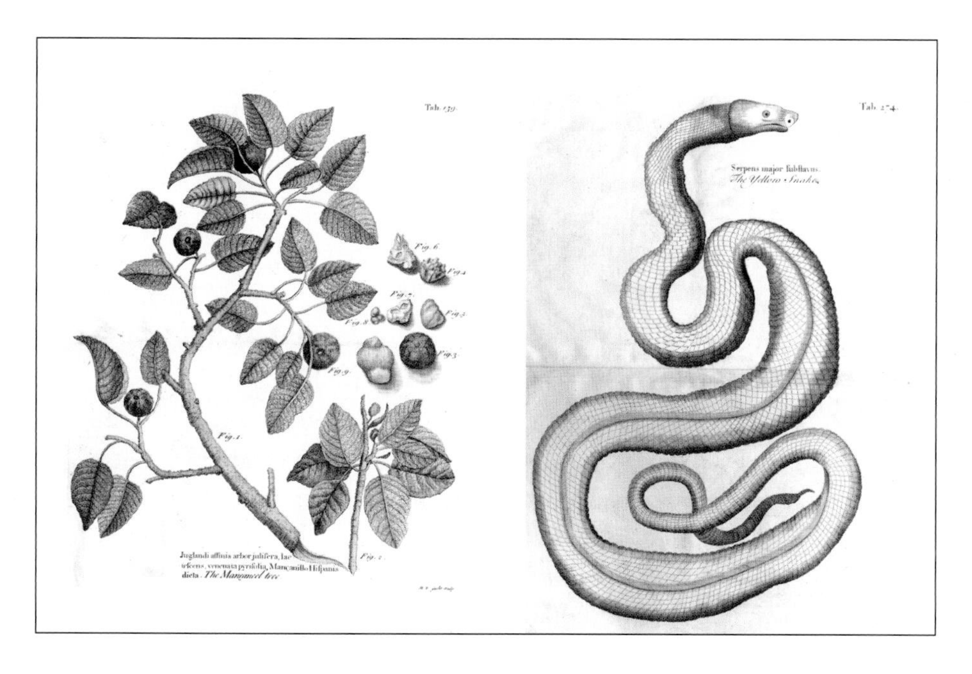

FIGURE 4-9. The poisonous manchineel tree (*Hippomane mancinella*) and the Jamaican boa or yellow snake (*Epicrates subflavus*); plates 159 and 274, respectively, Hans Sloane, 1725, *A voyage to the islands Madera, Barbados, Nieves, S. Christophers and Jamaica. . . .* (Courtesy of Missouri Botanical Garden, St. Louis.)

FIGURE 4-10. Metamorphosis of the white peacock butterfly (*Anartia jatrophae*) on cassava (*Manihot esculenta*) with tegu lizard (*Tupinambis merianae*; named after Merian in 1839 by Duméril & Bibron) and unidentified ant; hand-painted version of plate 4, M. S. Merian, 1705, *Metamorphosis insectorum Surinamensium*; pen and ink with watercolor and bodycolor on vellum. (© Trustees of The British Museum, London.)

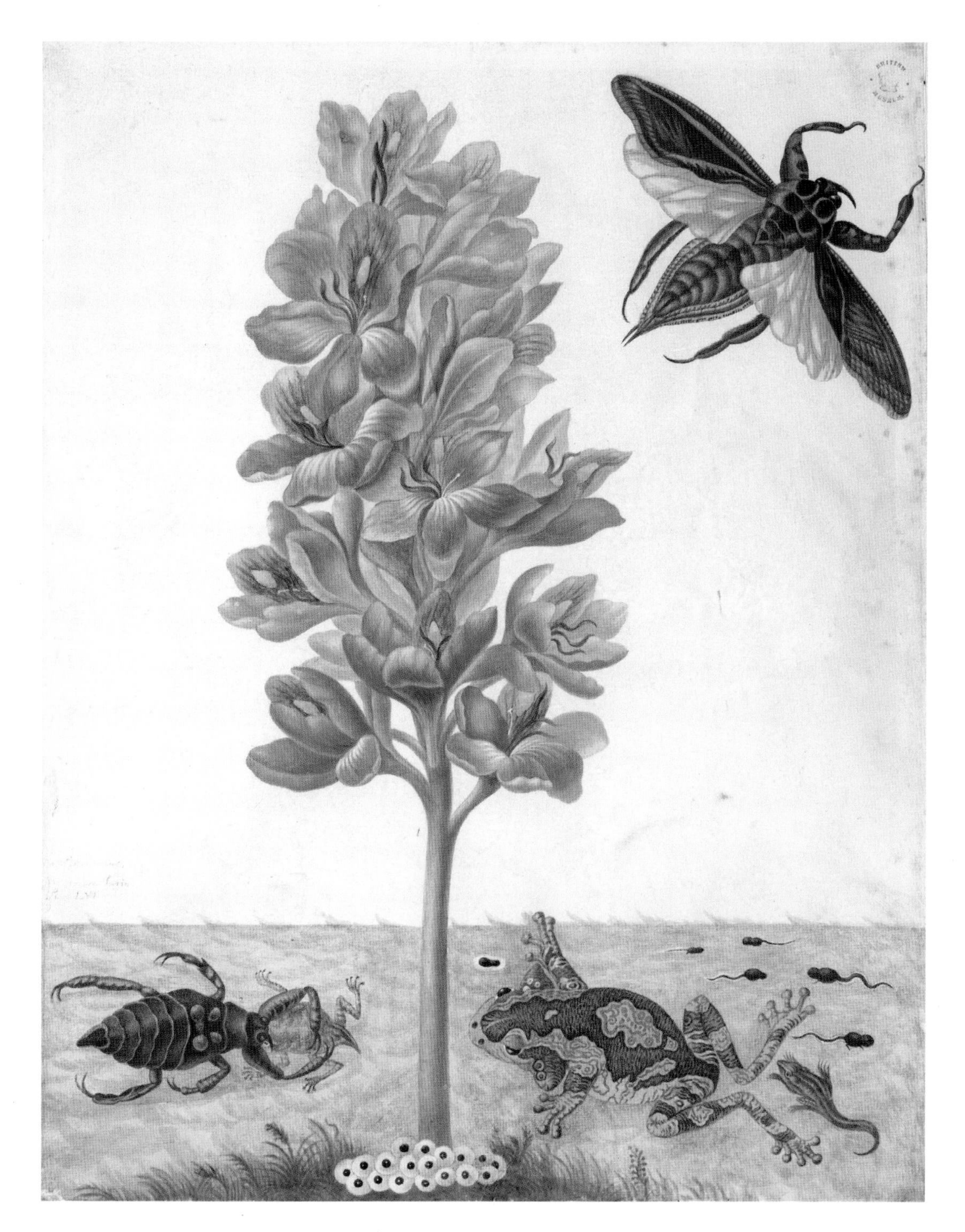

FIGURE 4-11. Marbled tree frog (*Phrynohyas venulosa*) with tadpoles and eggs, water hyacinth (*Eichhornia crassipes*), and water bug juvenile and adult (*Lethocerus grandis*); hand-painted version of plate 56, M. S. Merian, 1705, *Metamorphosis insectorum Surinamensium*; pen and ink with watercolor and bodycolor on vellum. (© Trustees of The British Museum, London.)

FIGURE 4-12. The pink-toed tarantula (*Avicularia avicularia*), brown huntsman spider (*Heteropoda venatoria*), an unidentified orb weaver of the Araneid family, a roach (*Blattaria* sp.), army ants (*Eciton* sp.), leaf cutter ants (*Atta cephalotes*), and a guava tree (*Psidium guineense*). The hummingbird does not precisely match any known species. Hand-painted version of plate 18, M. S. Merian, 1705, *Metamorphosis insectorum Surinamensium*; pen and ink with watercolor and bodycolor on vellum. (© Trustees of The British Museum, London.)

Merian's influence

The *Raupen* books on European insects and *Metamorphosis* provided new models for representing nature that would be echoed in the work of other artists and naturalists in the eighteenth century and beyond.[34] *Metamorphosis* in particular was to become one of the most influential natural history books ever published. It was striking both in its size, roughly 22 inches (54 centimeters) in height, and in its content. The large pages allowed Merian to portray her plant and insect subjects as life-sized, adding further to the biological information contained in the groundbreaking work. Merian offered the option of hand-colored copies, and those books were among the first publications to show New World organisms in colorful splendor. Not only the Tsar but many other influential collectors and naturalists sought out both Merian's paintings on vellum and her books. The London pharmacist James Petiver sent specimens to her and corresponded with her about the possibility of an English edition of *Metamorphosis*. Hans Sloane acquired early editions of Merian's books and a considerable number of her paintings, as did Dr. Richard Mead (1673–1754), whose collection of Merian's paintings and books is now in the Royal Library at Windsor Castle. Sloane and Mead also subscribed to Mark Catesby's book.

Merian's vibrant and dramatic images and her writings would have been viewed with great interest by a circle of naturalists who communicated and worked with these influential patrons, including Mark Catesby and Eleazar Albin (c. 1690–1742). Albin referred frequently to Merian's *Raupen* books in his 1720 work on moths and butterflies, and his compositions of insect life-cycles and the larval host-plants mirror her work to a large extent.[35] Catesby's *The natural history of Carolina . . .* was the most important book about New World organisms to follow *Metamorphosis* and has several similarities to Merian's. The size and layout of Catesby's *The natural history of Carolina . . .* echo *Metamorphosis*, and his images are structured in a similar way (compare, for example, figures 4-11, p. 51, and 4-13, p. 54).[36] As another protégé of Sloane, he would have had opportunity to study Merian's work,[37] and there is direct evidence that he did so, because he made two specific mentions of details from *Metamorphosis*. In the first, he criticized Merian's depiction of the cashew.[38] The second reference is in his description of the geographic range of the opossum, where he noted that Merian "has described them at Surinam."[39] Catesby's mention of Merian's opossum indicates that he must have used one of the posthumous editions of *Metamorphosis*, because the opossum plate was first inserted by the publisher of the 1719 edition.[40] It is possible that Catesby had studied the bilingual 1726 *Metamorphosis*;[41] this edition had text in Latin and in French in a format similar to that used by Catesby in his *Natural history of Carolina . . .* , although his text was in English and French.[42]

FIGURE 4-13. Canada lily and beetles; plate II, M. Catesby, 1747, *The natural history of Carolina . . .* , volume II: Appendix. Catesby depicted several insects in *The natural history of Carolina . . .* , including the dung beetle (*Canthon pilularius*) (*lower left*) and rainbow scarab beetle (*Phanaeus vindex*) (*lower right*), with the Canada lily (*Lilium canadense*). (Digital realization of original etchings by Lucie Hey and Nigel Frith, DRPG England; courtesy of the Royal Society ©.)

Merian's images of interacting organisms may have led Catesby, consciously or otherwise, to portray his subjects in a more lively way. Catesby also followed Merian by depicting plants and animals close to life-sized, as with his full-page image of a bullfrog (see figure 17-8, p. 242). Similarly, he included images such as a sea turtle with her eggs and a tree frog eyeing a spider as prey, akin to the way Merian pictured reptiles with their eggs and animals in various acts of predation. However, major differences exist between these two pivotal natural history books, and one is that Catesby's focus was clearly vertebrates, while Merian included only a few of these. The first edition (1705) of *Metamorphosis* documented two species each of frogs, lizards, and snakes,[43] and the second edition (1719) included a caiman defending its young from an egg-eating snake (figure 4-14), an adult tegu lizard, the opossums, and additional frogs.[44] The unfortunate image of the "frog-fish" added to the later editions came from the Amsterdam pharmacist and collector Albert Seba (1665–1736) and actually served to undermine Merian's reputation in later years, even though it was not her work.[45]

Whereas Merian described the food plants and ecological relationships of dozens of insects, Catesby did this primarily for birds, mammals, fish, reptiles, and amphibians. A notable exception is the Appendix of *The natural history of Carolina . . .*, in which Catesby included information about the habits of some industrious beetles (see figure 4-13).[46] Catesby also included a number of moths in the second volume, in some cases showing their pupae and in one instance a lone caterpillar.[47] However, in several cases Catesby depicted insects from the Carolinas on plants from the Bahamas, whereas Merian took care in most instances to pair insects with a host plant of their larvae.[48] In many images Catesby combined his vertebrate subjects with plants found in the same or similar habitat, although not in every case. Similarly, there is no direct association between the few vertebrates Merian depicted and the plants within the same plate other than that they were both found in Surinam.

Merian and Catesby had expertise in different kinds of animals, and Catesby's much more extensive travels are also revealed in the diversity of organisms included in his two volumes. *Metamorphosis* is an intensive look at a relatively small cross-section of terrestrial habitat, whereas Catesby's much more wide-ranging survey delved into both the terrestrial and the marine organisms of the New World. For example, Merian's image teeming with spiders, ants, and their prey depicts a small slice of a tropical jungle in an up-close view, whereas Catesby's broader perspective let him understand something about the migration of birds over long distances. But the two books are similar in several important aspects, such as portrayal and description of organisms existing and interacting in what we now think of as a habitat. Although both Merian and Catesby depicted organisms so accurately that many of their images were

FIGURE 4-14. Juvenile spectacle caiman (*Caiman crocodilus*) with the false coral snake (*Anilius scytale*); plate 69, M. S. Merian, 1719, *Over de voortteeling en wonderbaerlyke veranderingen der Surinaemsche insecten.* (Courtesy of Artis Library, University of Amsterdam.)

used by Carl Linnaeus and others to describe new species, it could be argued that their major contribution was to lay the foundation for ecological studies of interactions of organisms with their environment. Both authors showed predation and included information about reproduction; both depicted and described the interrelationships of plants and animals. Each of these pioneering naturalists learned from indigenous sources and conveyed to their readers the potential uses of New World plants. But perhaps most importantly, the vision of nature created by Merian and Catesby introduced Europeans to the American flora and fauna in a way never before seen, and the work of both served as models for future work by William Bartram (1739–1823), John James Audubon (1785–1851), and countless others. Maria Sibylla Merian and Mark Catesby viewed the living world in a new way and, by portraying what they saw through text and vivid images, changed the course of natural history.

5

William Dampier (1651–1715): the pirate of exquisite mind

DIANA PRESTON AND
MICHAEL PRESTON

Mark Catesby was born into the Age of Reason, when the study of natural philosophy, as science was then known, was becoming more widespread. King Charles II had in 1662 granted the Royal Society—or, to give it its full title, The Royal Society of London for Improving Natural Knowledge—its royal charter. It was, and remains, Britain's first and foremost scientific body. Its goal in this "Learned and Inquisitive Age" was "to overcome the mysteries of all the Works of Nature . . . for the benefit of humane life" and to undertake "an universal, constant, and impartial survey of the whole *Creation*."[1] In its work the Society emphasized building up a large body of observations not only of unusual phenomena but also of the normal and commonplace, which it rightly considered more helpful to understanding.

Among the Society's early members were the famous diarists John Evelyn (1620–1706) and Samuel Pepys (1633–1703); Christopher Wren (1632–1723), the architect of London's St. Paul's Cathedral; the astronomers Edmund Halley (1656–1742) and John Flamsteed (1646–1719); and the scientists Robert Hooke (1635–1703) and Isaac Newton (1642–1727)—the latter was the Society's President from 1703 until his death. His successor was Hans Sloane, whose collections formed the nucleus of the British Museum. Another prominent member was the well-known naturalist John Ray, who is said to have inspired Mark Catesby with a "gen[i]us for natural history."[2]

However, Catesby was also influenced by another man whose early career had taken him well outside the normal purview of the Royal Society but who had become well known to and praised by the Royal Society and its members and was the author of best-selling books studied by Catesby. His name was William Dampier (1651–1715; figure 5-1), and his influence on Catesby is demonstrated by Catesby's references to Dampier in his works and by Catesby's use of Dampier's words. For example, in *The natural history of Carolina* . . . Catesby used information from Dampier's description of the cacao tree (*Theobroma*

FIGURE 5-1.
William Dampier by Thomas Murray circa 1697. (© National Portrait Gallery, London.)

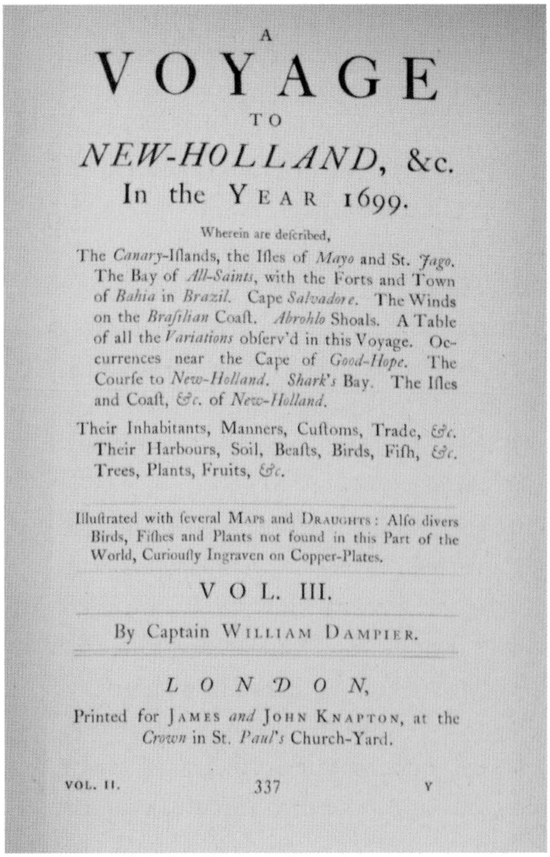

A
VOYAGE
TO
NEW-HOLLAND, &c.
In the YEAR 1699.

Wherein are deſcribed,

The *Canary*-Iſlands, the Iſles of *Mayo* and St. *Jago*. The Bay of *All-Saints*, with the Forts and Town of *Bahia* in *Brazil*. Cape *Salvadore*. The Winds on the *Braſilian* Coaſt. *Abrohlo* Shoals. A Table of all the *Variations* obſerv'd in this Voyage. Occurrences near the Cape of *Good-Hope*. The Courſe to *New-Holland*. *Shark's* Bay. The Iſles and Coaſt, *&c.* of *New-Holland*.

Their Inhabitants, Manners, Cuſtoms, Trade, *&c.* Their Harbours, Soil, Beaſts, Birds, Fiſh, *&c.* Trees, Plants, Fruits, *&c.*

Illuſtrated with ſeveral MAPS and DRAUGHTS: Alſo divers Birds, Fiſhes and Plants not found in this Part of the World, Curiouſly Ingraven on Copper-Plates.

VOL. III.

By Captain WILLIAM DAMPIER.

LONDON,
Printed for JAMES *and* JOHN KNAPTON, at the *Crown* in St. *Paul's* Church-Yard.

VOL. II. 337 Y

FIGURE 5-2.
Title-page of W. Dampier, 1703, *A voyage to New Holland in the year 1699*. (D. Preston & M. Preston.)

FIGURE 5-3. Brown booby; plate 87, M. Catesby, 1732, *The natural history of Carolina . . .*, volume I. (Digital realization of original etchings by Lucie Hey and Nigel Frith, DRPG England; courtesy of the Royal Society ©.)

cacao) telling how the "cods"—the fruits—were harvested.[3] Another passage that shows evidence of Catesby's familiarity with Dampier's works is one about vanilla. What is more, Catesby referenced Dampier's description of both the booby and the "mangrove grape tree."[4] In the case of the booby (*Sula leucogaster*) (figure 5-3), Catesby recorded that "*Dampier* says, they breed on Trees in an Island called *Bon-airy*, in the *West-Indies*, which he observes not to have seen elsewhere. While young, they are covered with a white Down, and remain so till they are almost ready to fly."[5] As for the "mangrove grape tree" (sea-grape, *Coccoloba uvifera*), Catesby reported that "*Dampier* says the Wood of this Tree makes a strong Fire, therefore used by the Privateers to harden the Steels of their Guns, when faulty."[6]

To understand Dampier's career, background, and relevance to Catesby's work, picture the following scene. In September 1683 in the Cape Verde Islands, William Dampier was lying "obscured" among the scrubby vegetation, watching birds. He had just caught his first sight of flamingos. The detail and delicacy of the description he would later write would gladden any modern ornithologist. Flamingos[7] were "much like a heron in shape," though "bigger and of a reddish colour," and were so numerous that from a distance they appeared like "a brick wall, their feathers being of the colour of new red brick." They nested in shallows "where there is much mud which they scrape together making little hillocks like small islands . . . where they leave a small, hollow pit to lay their eggs. . . . They never lay more than two eggs. . . . The young ones are, at first, of a light grey." Then, as a practical and hungry seventeenth-century sailor, Dampier noted the bird's culinary qualities: "The flesh is lean and black yet very good meat . . . their tongues [have] a large knob of fat at the root which is an excellent bit, a dish of flamingos' tongues being fit for a prince's table."[8]

Dampier also meticulously recorded in his journal the movements of tides, currents, and winds around the islands. He would later use these data to draw far-reaching conclusions about their behavior and the relationships between them. However, he would not mention that while he was deep in such observations, worthy of any natural scientist, his companions were otherwise engaged in plotting to seize a better ship for the piratical voyage to the Pacific on which they and Dampier were engaged.

These scenes highlight the contradictions in the career and character of William Dampier. A portrait of him by the artist Thomas Murray (figure 5-1), probably commissioned by Hans Sloane, shows a lean, strong-featured man with brown, shoulder-length hair wearing a plain brown coat and holding a book in his hand. A caption under this portrait styled him "*Pirate and Hydrographer*,"[9] but that tells only a small part of his story. He was a pioneering naturalist and as such a significant precursor of Mark Catesby in recording the natural world. He was also a navigator, travel writer, and explorer, but he was, indeed, quite happy in his youth to seek his fortune as a pirate.

Today, William Dampier is largely forgotten. Though celebrated in his own lifetime and for over a century after his death, he fell victim to changing times and hardening moral attitudes. The very activity that had made his achievements possible—buccaneering—was seen as a taint, overshadowing his achievements. Dampier was born in the village of East Coker in Somerset in the west of England in 1651 (figure 5-4). But when in 1907 the inhabitants proposed erecting a brass memorial to him, a local notable objected, calling him "a pirate ruffian that ought to have been hung."[10]

William Dampier started his career as a buccaneer in the Spanish Main. Yet while he marched with a buccaneer army through dense, humid, snake- and spider-infested jungles and over the Isthmus of Panama to plunder Spanish ships and cities, he began keeping notes of the natural world around him. He scratched his observations on scraps of parchment, which he stored inside a tube of bamboo stoppered with wax to protect them against the deluging tropical rains.[11]

FIGURE 5-4. William Dampier's birthplace, East Coker, Somerset, England. (© D. Preston & M. Preston.)

The more Dampier saw, the more fascinated he became, and the more determined to satisfy his compulsive curiosity. Every time the buccaneers debated where to go next, Dampier voted to sail to places he had never seen. He crossed the Pacific on a buccaneer ship, reaching the western coast of Australia in 1688. He and his companions were probably the first Britons to land on the continent, and his description of the eight weeks they spent there and his comment that "it is not yet determined whether it is an island or a main continent, but I am certain that it joins neither Asia, Africa nor America" would later catch the attention of the British Admiralty.[12]

Dampier tired of the buccaneers and after leaving Australia broke away from them to continue his travels, motivated not by thoughts of commercial gain but wishing to learn more of the world and other peoples' way of life. Returning to England in 1691 after an absence of twelve years, he began preparing his observations for publication. His books brought him from obscurity to celebrity, changed scientific perceptions, and altered the literary landscape. His work caught the attention of members of the Royal Society, who invited him to address them. His rational, logical approach to explaining natural phenomena exactly fitted the Society's philosophy. The literary world also embraced him. The diarist John Evelyn wrote, "I dined with Mr. Pepys, where was Captain Dampier, who had been a famous buccaneer."[13]

Dampier had a profound impact on later generations. James Cook (1728–1779) and Horatio Nelson (1758–1805) studied his navigational methods and used his maps. As well as influencing Mark Catesby, his work as a naturalist was well known to Alexander von Humboldt (1769–1859) and Charles Darwin (1809–1882), who used his acute observations and detailed descriptions as building blocks for their theories. Humboldt praised "the remarkable

English buccaneer," to whose works he thought "the subsequent studies of great European scholars, naturalists and travellers had added little."[14] Darwin called his work "a mine"[15] of information and felt so familiar with him that he referred to "old Dampier" in his diary.[16]

Dampier's *A Discourse of Trade-Winds, Breezes, Storms, Seasons of the Year, Tides and Currents*, published in 1699,[17] established his reputation as a hydrographer. It provided detailed information on tides and distinguished between tides and currents, showing how currents flowed farther out from shore and were much more constant in their direction of flow. Dampier also identified that "in all Places where the Trade [wind] blows, we find a Current setting with the Wind . . ."[18] and deduced that the winds were the causes of the surface currents—the first time this key connection had ever been made.

Edmund Halley, as well as Dampier, was then also investigating wind patterns. Both produced maps that became landmarks in the understanding of prevailing winds, but Dampier's were the more accurate and complete (figure 5-5). Dampier alone first created a wind map of the world, including the Pacific, a region of which Halley was ignorant. Dampier's maps, not Halley's, became the prototype for the many maps and globes picturing the trade winds that appeared throughout the eighteenth century.

Like Catesby, Dampier was intensely rational in his descriptions of natural phenomena at a time when superstitions about them abounded. For example, in their day so little was known about how the sea moved that some scholars

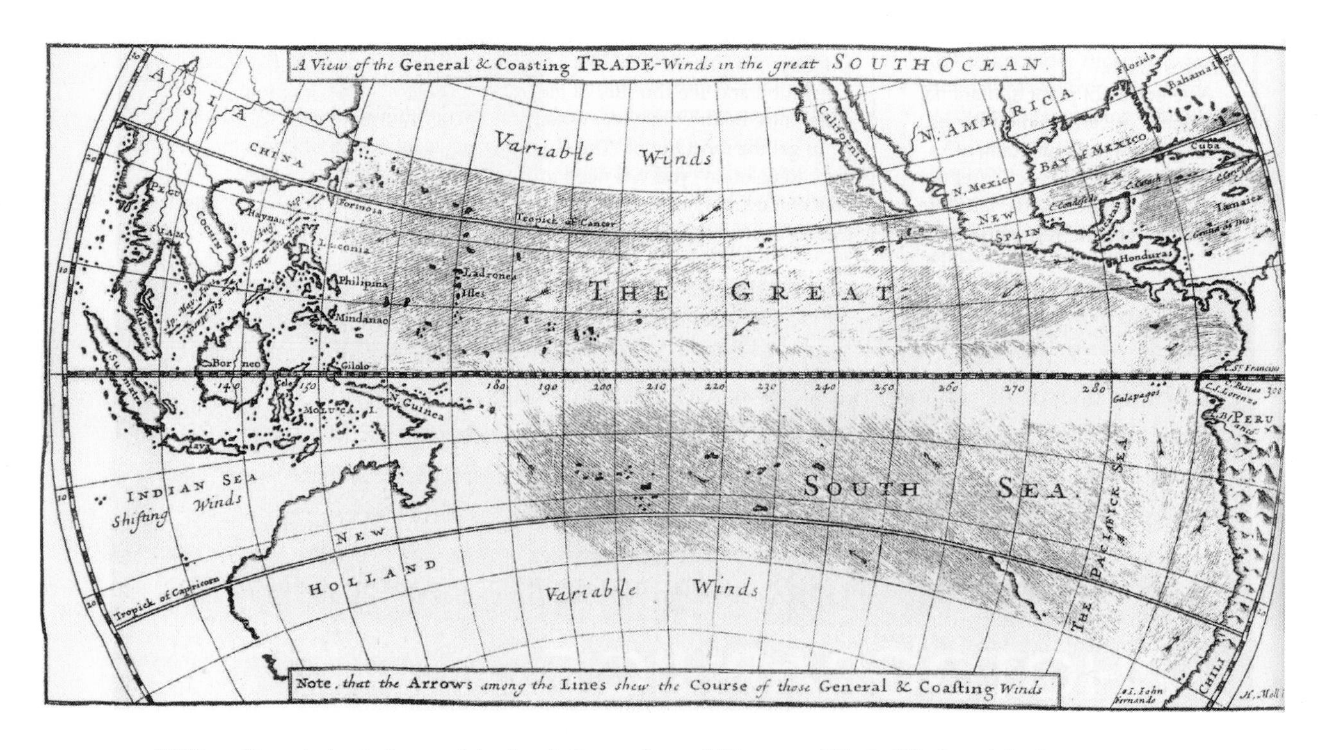

FIGURE 5-5. William Dampier's wind map of the South Ocean, from *A Discourse of Trade-Winds . . .* (1699).

suggested that the world's oceans were regulated and saved from stagnation only by the circulation of their water through the earth's center, entering at the North Pole and exiting at the South, or through gurgling subterranean passages such as under the Isthmus of Panama. As Dampier tramped the Isthmus with the buccaneers, he pondered the belief that the Isthmus was, in his words, "like an Arched Bridge, under which the Tides make their constant Courses, as duly they do under London-Bridge. . . ." He listened for the allegedly "continual and strange Noises made by those Subterranean Fluxes and Refluxes. . . ."[19] Hearing nothing of the kind, he dismissed the possibility of subterranean passages.

After a storm Dampier noticed what he called "a certain small glittering light . . . like a Star" dancing around the top of the mast. He had seen this light before, creeping over the decks like "a great Glow-worm." It was St. Elmo's fire, an electrical discharge that sometimes appears after storms at sea. Many of his contemporaries believed it was a manifestation of Christ or the Holy Spirit—it was often known as the "*Corpus Sant.*"[20] Other sailors thought the radiant light was the cavortings of hobgoblins or fairies or was "the enchanted bodies of witches."[21] The light might even be the spirit of a dead comrade warning of mortal danger. Though writing around the time of the Salem witch trials, Dampier was convinced that the light was a natural phenomenon with a rational explanation and disparaged what he called "ignorant seamen" with their "dismal Stories" about what it portended.[22]

Again, like Catesby, Dampier was an instinctive, intuitive naturalist. He pioneered what today is known as descriptive botany and zoology, in other words, the careful, detailed, and objective recording of the world's living things. When he was in South Africa, Dampier saw the skins of two zebras and later gave the first description of the zebra in English as "a very beautiful sort of wild ass," its body "curiously striped with equal lists of white and black; the stripes coming from the ridge of his back, and ending under the belly, which is white. These stripes are two or three fingers broad, running parallel with each other, and curiously intermixt. . . ."[23]

New sights, sounds, and smells consumed Dampier. He was one of the first English speakers to document the effects of marijuana, recording: "Some it makes sleepy, some merry, putting them into a Laughing fit, and others it makes mad. . . ."[24]

Another of his discoveries in the South Pacific was the breadfruit. Dampier wrote the first description in English of these trees, "as big and high as our largest apple trees," with glossy, dark-green leaves and fruits "as big as a penny loaf." He explained how the fruits were made into bread:

> The natives of this island use it for bread: they gather it when full grown, while it is green and hard; then they bake it in an oven, which scorcheth the rind and makes it black: but they scrape off the outside black crust, and there remains a tender thin crust, and the inside is soft, tender and white, like the crumb of a penny loaf.[25]

Captain William Bligh (1754–1817) quoted Dampier's glowing description in his own book of 1790, giving his side of the mutiny on the *Bounty* story as being the catalyst for Britons to see breadfruit as suitable food for their slaves and indentured servants in their American possessions and hence for the ill-fated voyage of HMS *Bounty* to Tahiti to transfer breadfruit plants to the West Indies.[26]

Dampier's description of the avocado[27] introduced the fruit to the English-speaking world. He also noted the belief that "this fruit provokes to lust, and therefore is said to be much esteemed by the Spaniards." Any Spanish settlement he had ever seen had its avocado bushes, so he decided this was probably true. Dampier described the sapodilla tree, which was "as big as a large pear tree" and with "an excellent fruit," the juice of which was "white and clammy and it will stick like glue."[28] The latex of the sapodilla today provides chicle, the basis for chewing-gum.

Dampier also examined cochineal, which at that time was thought to be made from seeds. However, he identified the real source as a small insect infesting a cactus. Dampier watched the Indians beat the plant with sticks to drive the red insects into the air so that the heat of the sun killed them and they fell onto large linen cloths spread on the ground. The Indians left the insects to dry for several days before scooping them up to make "the much esteemed scarlet."[29]

As the first naturalist to visit all five continents and to travel widely in areas largely unknown to Europeans, Dampier was able to compare and contrast animals, birds, reptiles, and plants across the globe. He was among the first party of Britons to visit the Galápagos. Here, his scientific observations of marine green turtles led him to write that they were "bastard" green turtles compared to those in the Caribbean, thus suggesting location-dependent differences within species and prefiguring Darwin. This revolutionary view was also at variance with the accepted religious doctrine.

Dampier's continuing deep interest in behavior and relationships within and between species led him into a pioneering description of migration patterns of the green turtle in the Caribbean, Atlantic, and Pacific. Perhaps even more importantly, in a study of Brazilian waterfowl during a stay in Salvador de Bahia he was the first to introduce "sub-species,"[30] both as a word and as a concept. The four types of "long-legg'd fowls" he saw wading in the swamps were, he decided, as "near a-kin to each other, as so many sub-species of the same kind." His view that some animals were more closely related than others was again formerly heretical. The accepted doctrine was that each had been created in precisely its present form and entirely independently. Darwin later regularly referred to Dampier's works during the voyage of the HMS *Beagle*, and the famous "red notebook"[31] in which Darwin first formulated his theory of natural selection quotes observations from Dampier.

Dampier was the only major well-documented British maritime explorer between Francis Drake (1550–1596) and his fellow adventurers of the sixteenth and early seventeenth centuries and James Cook and his fellow naval expeditioners in the mid- to late eighteenth century. He uniquely bridged those two eras, fusing the piratical plundering and derring-do of the former with the scientific inquiry and meticulous chart- and record-keeping of the latter. He was also the first Briton known to have circumnavigated the globe three times.

In 1699 Dampier returned to Australia, no longer a buccaneer but in command of the first British naval expedition to the continent with instructions, among other things, to bring back botanical specimens. Landing again on the west coast (figures 5-6 and 5-7), Dampier found plants (figures 5-8 and 5-9)

FIGURE 5-6. Shark Bay, Western Australia, where Dampier landed in HMS *Roebuck* in 1699 to gather botanical specimens. (© D. Preston & M. Preston.)

FIGURE 5-7. Karrakatta Bay, Western Australia, where Dampier landed with the buccaneers in 1688. (© D. Preston & M. Preston.)

FIGURE 5-8. Plants from New Holland; W. Dampier, 1703, *A voyage to New Holland in the year 1699.*

FIGURE 5-9. Plants from New Holland; W. Dampier, 1703, *A voyage to New Holland in the year 1699.*

with "either blossoms or berries"—red, white, yellow, "but mostly blue"—that smelled "very sweet and fragrant." There were also "very small flowers growing on the ground that were sweet and beautiful, and for the most part unlike any I had seen elsewhere."[32] He was correct in recognizing their uniqueness.

Although wrecked off Ascension Island on his return voyage, Dampier succeeded in bringing specimens back to England—the first botanical ones from Australia to reach Britain. Like many of Catesby's American specimens, they are preserved in the Oxford University Herbaria.

Dampier also explored New Guinea, "discovering" the large island of New Britain (figures 5-11 and 5-12). Here, as on so many occasions, he described the indigenous people for the first time in English. Some, he wrote, were "wearing ornaments of crab claws and white shells stuck through their noses." Their hair was "cut in ridges upon their heads" and, like their faces and hands, spectacularly painted. In another group the men were "finely bedecked with feathers of divers colours about their heads." The women "had no ornament about them nor anything to cover their nakedness but a bunch of small green boughs, before and behind stuck under a string which came round their waists," and they were balancing large baskets of yams on their heads. Dampier, eager to make comparisons from across his travels and to understand the structures of the lives of those he met, recalled in his book: ". . . and this I have observed, among all the wild natives I have known, that they make their women carry their burdens while the men walk ahead without any other load than their arms and ornaments."[33]

Ethnographers still quote Dampier's descriptions of foreign peoples, which are notable not only for their accuracy but also for their sympathy. When he

FIGURE 5-10. Illustrations of "dolphins" from William Dampier's *A voyage to New Holland in the year 1699* (1703). The upper figure shows the marine mammal, a relative of the whales, which we still call a dolphin or a porpoise. The lower figure shows a true fish, the dolphinfish, which had a distinctive dorsal fin that extends the entire length of the body. Also called dorado (the Spanish name) and mahi-mahi (the Hawaiian name), these fish belong to the genus *Coryphaena*.

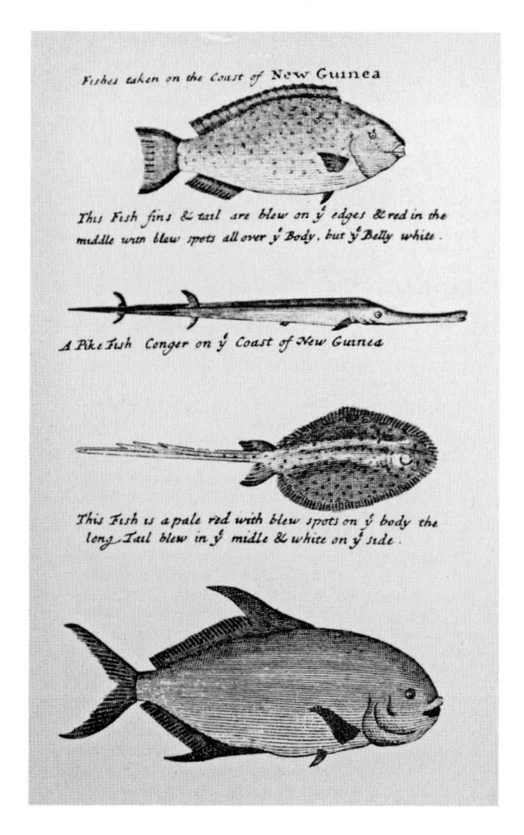

FIGURE 5-11. Fish found off the coast of New Guinea; W. Dampier, 1703, *A voyage to New Holland in the year 1699*.

FIGURE 5-12. Fish found off the coast of New Guinea; W. Dampier, 1703, *A voyage to New Holland in the year 1699*.

visited China, for example, the bound feet, or "golden lilies," as the Chinese called them, of the women shocked him. He recorded how their feet were swathed in tight bandages from infancy, in his words, "to hinder them from growing," since the Chinese esteemed little feet as "a great beauty," an "unreasonable custom" that made them "in a manner lose the use of their feet." Rather than being able to move freely, the women could "only stumble about their houses, and presently squat down . . . being, as it were, confined to sitting all the days of their lives." Dampier wondered whether it was, in his words, "a stratagem of the men's" to keep the women from "gadding about."[34]

Dampier respected and admired the Native peoples of South and Central America. His books portray them with sympathetic insight as, in his words,

> a very harmless sort of people; kind to any strangers; and even to the Spaniards, by whom they are so much kept under, that they are worse than slaves . . . This makes them very melancholy and thoughtful . . . sometimes when they are imposed on [by the Spanish] beyond their ability [to bear], they will march off whole towns, men, women and children. . . .[35]

Those who lived in their own villages were, he thought, "like gentlemen" in comparison with those who lived degraded lives on the edge of Spanish towns, forced to work by the Spanish for little or no wages.

Dampier's own inclination was to believe that the ways of his own country were best, but he did not see Europeans at the apex of a pyramid of humanity pre-ordained to rule and exploit other peoples as of divine right. He saw no reason why all races could not reach comparable levels of attainment, and he believed each could learn from the other. He also saw some virtue in the state of nature compared to living in a more sophisticated society subject to arbitrary rule.

Dampier believed in individual liberty and that dictatorship failed both ruler and ruled. He despised Spanish colonial rule and the way it degraded the local people. He preferred free trade to imperialism and believed that every man should be free to develop his potential. (Dampier reserved some of his strongest criticisms for the "lazy"—as he called them—who failed to do so.) He believed in the application of the buccaneers' democratic code and the right of everyone to have a say about the course of their lives. He even criticized his colleagues for their refusal to give a female maritime pilot a chance simply because she was a woman.

But perhaps Dampier's greatest gift was to convey in words to his fellow countrymen the frontiers opening up around them. His accounts of his voyages, the first of which was published in 1697, aroused an enthusiasm for travel writing that made it the most popular form of secular literature for the next quarter of a century and beyond. *The history of the works of the learned*, the first literary magazine in English, recommended his books to the "*Sedantary Traveller*" for the "variety of Descriptions, and the surprizingness of the

Incidents therein."[36] No gentleman's library, including Catesby's, would have been complete without copies.

As well as being a reliable recorder of facts, Dampier was sufficiently skilled to infuse and enthuse his readers with the excitement he felt. More than a century later, the poet Samuel Taylor Coleridge praised what he called Dampier's "exquisite mind"[37] and advised contemporary travel writers "to read and imitate him."[38] Dampier's writings inspired a rational wonder at what he had seen, and his simple English and homely similes connected the lives of his readers to a new and broader world. A hummingbird was, in his words, "a pretty little feathered creature, no bigger than a great, over-grown wasp." A poison blow-dart was "like a knitting needle." Dampier was the first to compare the vast expanse of the flat ocean to the surface of the millpond so familiar to English country-dwellers.

About one thousand entries in the *Oxford English dictionary* come from Dampier's publications. To take only the first three letters of the alphabet, he gave to the English language such words as avocado, barbecue, breadfruit, cashew, and chopsticks. He was the first to write in English about southeastern Asia, describing the taste and manufacture of soy sauce and of what we now know as Thai fish sauce. Anyone researching the places Dampier visited (such as logwood settlements on the Caribbean coast of Mexico, the Chinese coast, or the Pacific islands) often finds him cited as the earliest and frequently the only authority for his period.

Dampier primed the imagination of other writers. Without Dampier there might have been no Yahoos, no Robinson Crusoe, no Man Friday. Daniel Defoe studied Dampier's voyages, including his accounts of his shipwreck on the then-uninhabited Ascension Island, and his tales of his tattooed companion, the painted prince, and of a Moskito Indian named William, marooned for three years on the Pacific island of Juan Fernández. Defoe also studied the testimony of Scottish sailor Alexander Selkirk, abandoned on that same island during one of the voyages on which Dampier sailed and rescued during another. Dampier's adventures also strongly influenced Jonathan Swift's satirical book *Travels into several remote nations of the World . . . by Lemuel Gulliver*, published in 1726, eleven years after Dampier's death. Gulliver actually referred in the opening passages to "my cousin Dampier."

And what of the character of William Dampier? One of his companions called him "self-conceited,"[39] but in many ways he had reason to be. The accuracy of his writing and descriptions testifies not only to his talent but also to his remarkable dedication and perseverance. Even while dodging Spanish patrols, Dampier still found time and energy to mix ink, trim his quill pen, and scratch entries in his journal. He was also a courageous man not only in battle but in disregarding hardships to make his observations. One of his strongest self-criticisms was that he allowed an acute bout of dysentery to prevent him from noting, in detail, an eclipse of the sun. Usually such ailments meant little to him—they were merely an opportunity to try the local cures, such as yogurt, and to record the effects. He once had himself buried up to his neck in warm sand to cure a fever—it worked, although some years later a similar "cure" is said to have killed Vitus Bering.

Dampier was not necessarily an easy companion, being inclined to be critical of others. He was certainly not an easy man to be married to. He left his wife, Judith, after just a few months of marriage, telling her he was just going on a short trading voyage to Jamaica. Instead he stayed away twelve years on a round-the-world odyssey. Perhaps surprisingly, they stayed together.

On his third and last circumnavigation, which he made as a government-sponsored privateer, Dampier finally succeeded in realizing his "golden dreams" and capturing a Spanish treasure galleon. So when, after covering more than two hundred thousand sea-miles in his lifetime, he died at home in London in 1715, he was not poor. He had, however, brought greater riches to the human understanding of the world around him, and he deserves to be brought back from his voyage to oblivion.

Returning to the connections between Mark Catesby and William Dampier, they shared not only influences but also associates. Among them (in Dampier's case, after the publication of his first book) was, as already mentioned, the great naturalist the Reverend John Ray, who examined Dampier's Australian specimens and described some of them in 1704.[40] Both Dampier and Catesby were also known to William Sherard, whose collection of plant specimens was the foundation of the Oxford University Herbaria.

But what is most striking about the two men is not whom they knew in common or even that Catesby quoted Dampier but the similarity of their approach in describing the natural world—Dampier painting it primarily in words and Catesby using images as a centerpiece backed up by verbal descriptions. Both were meticulous and painstaking, wanting the general reader to appreciate and understand what they described. Just as Dampier wrote of a hummingbird as "an overgrown wasp," Catesby likened the fruit of the pawpaw (*Asimina triloba*) to a ram's scrotum.[41]

Both were intensely rational when other intelligent men were influenced by superstition and myth. A good example is that when they were describing migration patterns (in Dampier's case, those of turtles, and in that of Catesby, birds), others, such as the non-Conformist minister Charles Morton (1627–1698), Daniel Defoe's schoolmaster, seriously claimed that swallows migrated to the moon in winter.[42] Eighty years later the great lexicographer Samuel Johnson (1709–1784) maintained the still-common thesis that swallows wintered under ice in ponds.

Both Catesby and Dampier were true pillars of the Age of Reason.

6

John Lawson's *A new voyage to Carolina* and his "Compleat History": the Mark Catesby connection

MARCUS B. SIMPSON JR.

John Lawson's *A new voyage to Carolina* presents the most comprehensive account of the flora and fauna of the Carolina area prior to Mark Catesby's *The natural history of Carolina, Florida and the Bahama islands*. First published in 1709, *A new voyage* was re-issued in 1714 and 1718 with the title *The history of Carolina*, consisting of the same 258 numbered pages and unaltered sheets left over from the original 1709 printing. Lawson's book includes material on zoology, botany, anthropology, geography, and meteorology admixed with travel narratives and promotional propaganda (figures 6-1 and 6-2).[1]

In his *The natural history of Carolina . . .* , published between 1729 and 1747, Catesby acknowledged a debt to Lawson's *New voyage* for information on American Indian culture, but only when Lawson's reports confirmed Catesby's direct observations and experience.[2] Catesby did not identify other Lawson contributions, but text comparisons indicate that *A new voyage* provided content for more than thirty passages in *The natural history of Carolina . . .* , involving plants, birds, mammals, reptiles, and fish.

The importance of Lawson's life on Catesby's place in American natural science is not limited to Catesby's use of observations from *A new voyage*. At the time of his death, Lawson was engaged in an ambitious project for what he called a "Compleat History" of Carolina.[3] If Lawson had not died in the Tuscarora war of 1711, his labors might have significantly affected Catesby's contribution to colonial natural sciences, as well as the history of American botany and zoology.

Details of Lawson's early life and education are sparse, but he probably received training as a surveyor before departing from London in 1700. His interest in natural history was apparent, however, as early as 1701, more than a decade before Catesby's first sojourn in America.[4] A native of England, Lawson was advised that "Carolina was the best Country I could go to" in pursuing his "Intention . . . to travel." He sailed from London in April 1700 to Charleston (then called Charles Town), South Carolina, finally arriving by September.[5]

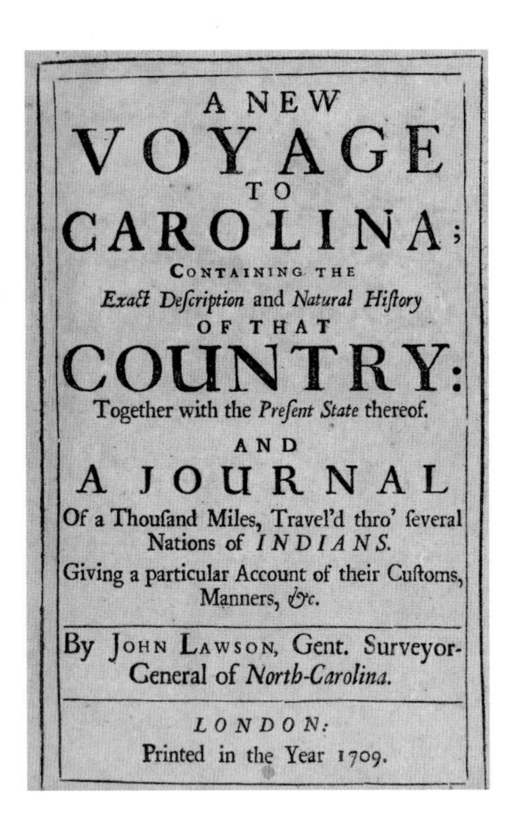

A NEW
VOYAGE
TO
CAROLINA;
CONTAINING THE
Exact Deſcription and *Natural Hiſtory*
OF THAT
COUNTRY:
Together with the *Preſent State* thereof.
AND
A JOURNAL
Of a Thouſand Miles, Travel'd thro' ſeveral
Nations of *INDIANS.*
Giving a particular Account of their Cuſtoms,
Manners, &c.

By JOHN LAWSON, Gent. Surveyor-
General of *North-Carolina.*

LONDON:
Printed in the Year 1709.

FIGURE 6-1. Title-page of John Lawson, 1709, *A new voyage to Carolina*. Published in four parts ("fascicles"), this was the second book in a multi-volume anthology entitled *A new collection of voyages and travels* (1708–1710). Lawson's book was bound within the *New collection* or as its own single-title, free-standing work. Catesby was among a number of writers who subsequently relied on Lawson's *New voyage* for information. (Courtesy of the North Carolina Collection, University of North Carolina Library at Chapel Hill.)

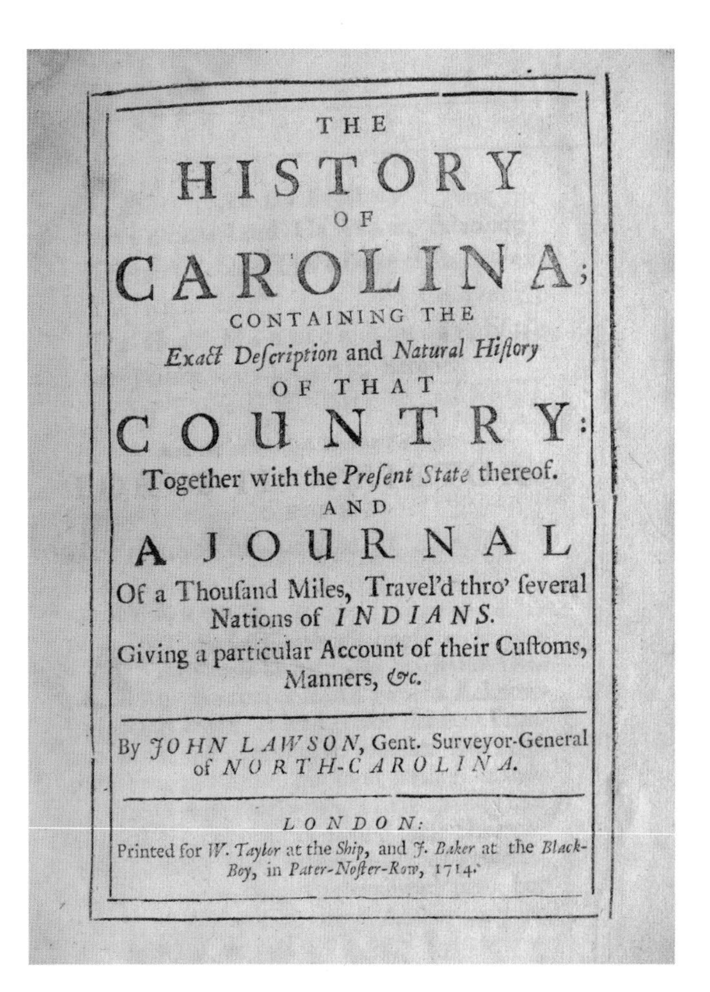

THE
HISTORY
OF
CAROLINA;
CONTAINING THE
Exact Deſcription and *Natural Hiſtory*
OF THAT
COUNTRY:
Together with the *Preſent State* thereof.
AND
A JOURNAL
Of a Thouſand Miles, Travel'd thro' ſeveral
Nations of *INDIANS.*
Giving a particular Account of their Cuſtoms,
Manners, &c.

By *JOHN LAWSON*, Gent. Surveyor-General
of *NORTH-CAROLINA.*

LONDON:
Printed for *W. Taylor* at the *Ship*, and *J. Baker* at the *Black-Boy*, in *Pater-Noſter-Row*, 1714.

FIGURE 6-2. Title-page of John Lawson, 1714, *The history of Carolina*. In 1714 and 1718, Lawson's *New voyage* was re-issued with a different title but consisting of original sheets that remained unsold from the 1709 printing. Catesby specifically noted the 1714 version when citing Lawson's observations. (Courtesy of the North Carolina Collection, University of North Carolina Library at Chapel Hill.)

He remained only briefly in Charleston, where he met Robert Ellis (fl. 1700–1705) and Edmund Bohun (1672–1734), collectors of "natural curiosities" for James Petiver of London. An apothecary and Fellow of the Royal Society, Petiver's prodigious output of scientific writings relied heavily on the continuous flow of specimens and data from his global network of correspondents and collectors.[6] Lawson's contact with Ellis and Bohun might not have been coincidental, suggesting that his "Intention" may have involved more than a simple travel adventure. If Lawson's ambitions included contributions to natural history, he might have learned about Petiver while visiting these two local enthusiasts. Perhaps he already knew of Petiver's activities before leaving London and intentionally sought out Ellis and Bohun for assistance and advice.

On 28 December 1700 Lawson and ten companions set out from Charleston on a journey of approximately 550 miles (885 kilometers) into the upper Piedmont and Carolina backcountry and then east to the lower Pamlico River area of present-day North Carolina, where he arrived in late February (figure 6-3). Eventually settling in present-day Craven County, North Carolina, Lawson engaged in survey work and began compiling natural history observations.[7]

Writing to Petiver on 12 April 1701, Lawson mentioned Ellis and Bohun, at the same time indicating his willingness to "use every means I possibly can to procure" and send "Animals Vegitables . . . shells . . . butterflies, & other insects [and] fish." He further promised that, "God willing," his first shipment of such desiderata would come that October. Lawson was apparently responding to one of Petiver's many "Advertisement" documents, which described methods for collecting and preserving plants and animals to be shipped to London.[8] Lawson also offered to send Petiver the "journal of my Voyage through Carolina," a manuscript eventually included by Lawson in the first installment of his *New voyage*.[9]

Petiver evidently overlooked or misplaced Lawson's 1701 letter, or else his reply never reached Lawson. Consequently, their collaboration was delayed

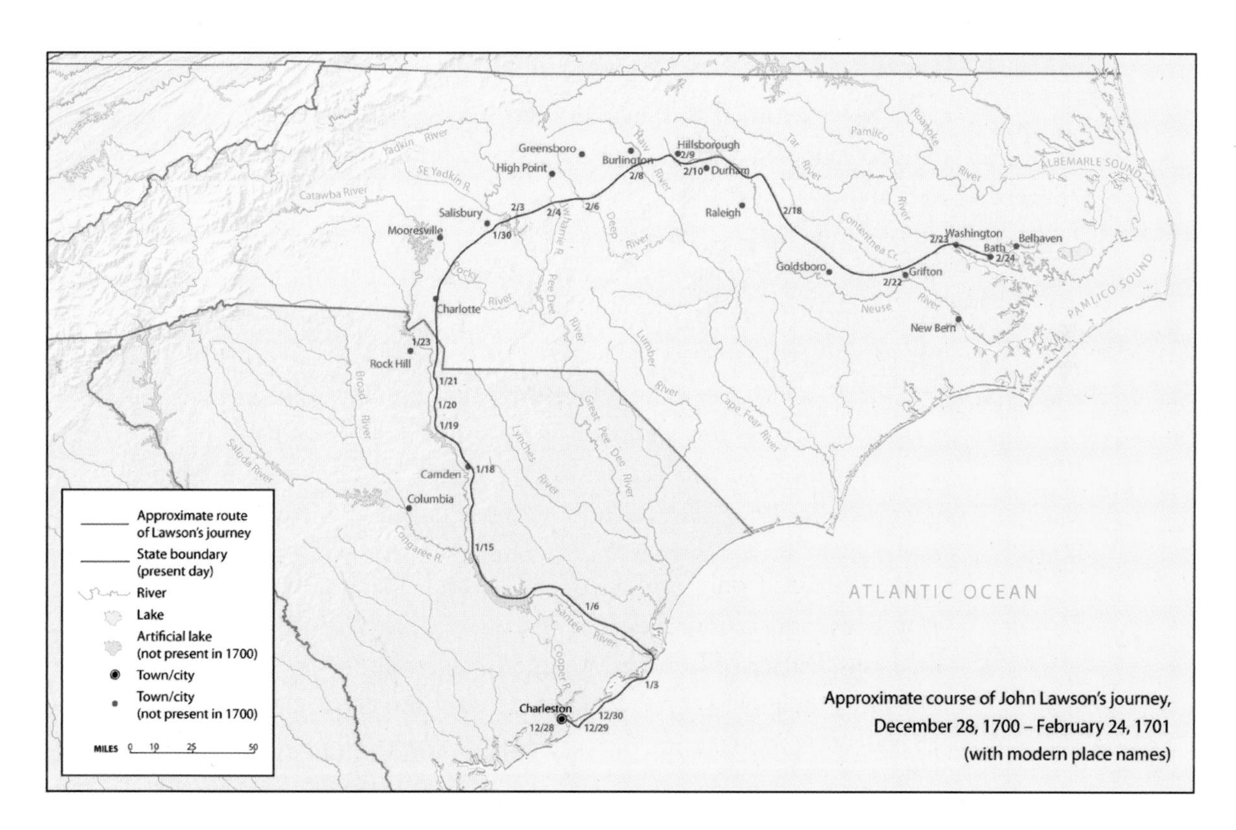

FIGURE 6-3. Probable route of John Lawson's 1700–1701 journey through the Carolina backcountry. Lawson departed Charleston on 28 December and, after a trek of about 550 miles, arrived in the Pamlico Sound area on 24 February. Lawson's account of this exploration, entitled "Journal of a Thousand Miles Travel among the Indians," formed the first fascicle of his *A new voyage*. (© David Walbert.)

until 1709, an eight-year gap of lost opportunities for both men. This lapse is particularly curious in light of Petiver's interest in Britain's Atlantic coast colonies and his own desire to produce a natural history of the Carolinas.[10] Had their relationship flourished from 1701, Lawson's *New voyage* would likely have contained information based on his specimens and notes sent to Petiver and perhaps involving communications with other members of the English community of virtuosi. Equally important would have been the possible effects on Catesby's work in America. Lawson's collecting activities might well have affected Catesby's ability to obtain financial support sufficient for his travels, plant collecting, and artwork.

Undeterred by Petiver's apparent silence, Lawson pressed on, pursuing his ambitions as a naturalist. His work as a surveyor provided not only employment but also ideal opportunities for observation and collecting throughout the colony. In *A new voyage*, Lawson succinctly revealed the modus operandi for his years of fieldwork in Carolina:

> Having spent most of my Time, during my eight Years Abode in Carolina, in travelling; I not only survey'd the Sea-Coast and those Parts which are already inhabited by the Christians, but likewise view'd a spatious Tract of Land, lying betwixt the Inhabitants and the Ledges of Mountains. . . . I have, in the following Sheets, given you a faithful Account thereof. . . .[11]

Lawson returned to England late in 1708 and remained there until mid-January 1710. During this busy year in London, he was appointed Surveyor-General for North Carolina and designated as a member of the survey commission to resolve the long-standing boundary dispute between Virginia and Carolina. Lawson also became a major participant with Christoph von Graffenried (1661–1743) in relocating hundreds of Palatine refugees in settlements in North America.[12] Despite this crowded agenda, Lawson managed to publish *A new voyage* during the spring and summer of 1709.

Lawson arrived in London at a propitious time for an author in search of a publisher. In the two decades following the "Glorious Revolution,"[13] resurgent public interest in travel, exploration, and natural history fueled an expanding market for books such as Lawson's *New voyage*. Prominent booksellers, including James Knapton (fl. 1680–1736), specialized in this literary genre, and Lawson was able to have his manuscript incorporated into a collection of travel books already being produced by Knapton and five of his associates.

Entitled *A new collection of voyages and travels*, the planned six-volume anthology was issued in monthly parts beginning in December 1708.[14] Publication in serial form encouraged readers to acquire books that they might not otherwise be willing or able to purchase while enabling printers to assess demand for a new work before committing resources to the entire book. Lawson's *A new voyage* was among the earliest natural science books produced in serial form, a strategy later modified and adopted by Catesby for publishing his *The natural history of Carolina. . . .*[15]

No evidence has been traced to indicate whether Lawson made any arrangements for publication before returning to London. His manuscript must have been in reasonable order upon his arrival, however, as the printers were impressed sufficiently to interrupt the planned sequence for *A new collection of voyages and travels* in order to produce the Lawson volume. Knapton's group opted to publish the first part of *A new voyage* in April 1709 rather than the previously scheduled book. Three more parts were issued over the ensuing five months to complete publication.[16]

The decision to insert Lawson's book into a project already well under way reflects Knapton's literary judgment and previous success with William Dampier's famously popular *A new voyage round the world* and similar works by William Hacke, William Funnel, and Lionel Wafer.[17] Lawson's experience with leaders of the publishing industry would have been invaluable in later years had he chosen to issue his proposed "Compleat History" in book format.

Major sections of *A new voyage* include Lawson's backcountry journal from 1700 and 1701 and his chapter on the Indians of North Carolina.[18] Lawson's Indian material remains a particularly valuable contribution. His accounts of customs, society, and vocabularies are among the best records of North American native culture from the early eighteenth century. Important natural history content includes the annotated list of more than one hundred species of native plants, many of which are easily identifiable,[19] and various fish, invertebrates, reptiles, amphibians, and both terrestrial and aquatic mammals.[20]

The most extensive zoological commentary is devoted to birds.[21] Lawson listed, depending on interpretation, between 125 and 130 species, the identity of which can be surmised for more than half.[22] Birds of particular interest are the now-extinct Carolina parakeet and passenger pigeon. Lawson may have been the first to describe the ivory-billed woodpecker, as the largest of his four woodpeckers, having a "white Cross on his back," a classic field mark for the species. Unfortunately, Lawson did not mention the distinctive beak, and the widely accepted earliest account of this extinct bird is credited to Catesby.[23]

Two copper-plate illustrations were included in the 1709 issue of *A new voyage*. The large folding map of Carolina was signed by John Senex (1678–1740), one of the eminent cartographers and globe-makers of the time.[24] The artist and engraver of the other plate, which included fourteen "beasts" of Carolina, remain unknown. Crafted with simple, direct lines, the drawings illustrate behaviors reported in the text. Adjacent to each figure is the page number where the species or behavior is described (figure 6-4).

In the years following its publication, the value of *A new voyage* and its author's abilities were recognized by leading naturalists, including Petiver, Catesby,

FIGURE 6-4. The "beasts" plate from *A new voyage*. Lawson's book contained only two illustrations, a map of the province and this page of animal images. Most of the figures illustrate behaviors reported in the text. Note the page numbers next to each animal that indicate where the behavior is described. (Courtesy of the North Carolina Collection, University of North Carolina Library at Chapel Hill.)

Sloane, and the zoologist and historian Elliott Coues (1842–1899), as well as a number of plagiarists,[25] who copied extensively from Lawson's work.

Petiver became aware of Lawson's activities when the parts appeared in the summer of 1709. In a letter written to George London (c. 1650–1714) in September 1709, Petiver commented that he had "lately obtained an Acquaintance with one Mr. Lawson Surveyor General of Carolina . . . a very curious person" who "hath lately printed a Natural History of Carolina." Considered "the most renowned" British gardener of the period, London's interest in the flora of America could have proven valuable to Lawson's initial efforts.[26] Petiver characterized Lawson's accounts as done with "a great deal of Judgment & accuracy." Lawson and Petiver soon developed plans for a more comprehensive "Compleat History" of Carolina, which, upon Lawson's return to America, progressed quickly to an exchange of supplies, data, and specimens between Carolina and England.

When cataloging specimens that he later bequeathed to the British Museum, Hans Sloane often identified animals using Lawson's descriptions from *A new voyage* and cited pages from the book.[27] More than 160 years after its publication, Lawson's book, having lost none of its luster, was hailed by Coues as "one of the most notable faunal lists of American birds of the last century, comparable to Bartram's on Florida Birds, Belknap on those of New Hampshire."[28]

Catesby may have discovered Lawson through Sloane, Petiver, or William Byrd II, who was a frequent host to Catesby during his early years in Virginia. Byrd was well acquainted with Lawson through their involvement with the boundary commission, and he knew the grim details of Lawson's death at the hands of the Tuscarora tribe less than a year before Catesby's arrival in North America. Whatever the circumstances, Catesby was quite familiar with *A new voyage* by the time he began production of *The natural history of Carolina, Florida and the Bahama islands* in the late 1720s, and he clearly respected the value of Lawson's work.[29]

Printing costs imposed significant constraints on the length of his text, yet Catesby often acknowledged publications relevant to his particular subject. Frequently cited examples included works by Hans Sloane, John Ray, and Francis Willughby (1635–1672).[30]

Nevertheless, Catesby's primary goal was the illustrations, not an encyclopedic compilation of sources that would have been required in a more technical work. With reference to his section on the native Americans, Catesby stated:

> Mr. *Lawson*, in his Account of *Carolina*, printed *Anno* 1714, has given a curious Sketch of the natural Dispositions, Customs, &c. of these Savages. As I had the same Opportunities of attesting that Author's Account as he had in writing it, I shall take the Liberty to select from him what is most material, which otherwise I could not have omitted from my own Observation.[31]

Nowhere within his text on North American Indians did Catesby identify observations recorded by Lawson. Furthermore, other passages in *The natural history of Carolina* . . . appear closely linked by wording or concept to accounts in *A new voyage*, also without attribution to Lawson. In some instances, however, the relationships may be to an as yet unrecognized common source used by both authors.

Catesby's descriptions of plants appear only infrequently connected with Lawson's work, a typical example being that the bald cypress was "the tallest and largest in these parts of the world." Lawson had noted that the tree was "the largest for Height and Thickness, that we have in this Part of the World."[32]

Parallel content in the zoological material often involves behavior. Of the rough greensnake, Lawson noted that "every one makes himself very familiar with them, and puts them in their Bosom, because there is no manner of Harm in them." Catesby described the species as "becoming tame and familiar, and are very harmless, so that some People will carry them in their Bosoms."[33]

Parallel texts about birds are also found in accounts such as the ruby-throated hummingbird, Carolina parakeet, northern mockingbird, and red-headed woodpecker. As to the passenger pigeon (figure 6-5), both authors mention the huge flocks that came in winter, often breaking the limbs of oaks while roosting. Lawson further commented that the pigeons eat "all before them, scarce leaving one Acorn upon the Ground, which would, doubtless, be a great Prejudice to the Planters that should seat there, because their Swine would be thereby depriv'd of their Mast." Catesby repeated this observation, with some minor rewording: "Where they light, they so effectually clear the Woods of Acorns and other Mast, that the Hogs that come after them, to the determinant of the Planters, fare very poorly."[34]

Similarities are also apparent in their reports of various mammals, including raccoon (*Procyon lotor*), mountain lion or cougar (*Felis concolor*), gray wolf (*Canis lupus*), eastern cottontail (*Sylvilagus floridanus*), bobcat (*Lynx rufus*), and American bison (*Bison bison*). Close parallels in wording are noted in the accounts of American black bears (*Ursus americanus*) and white-tail deer (*Odocoileus virginianus*). Lawson explained that female bears were thought to hide in secret places after "Conception," such that they were never encountered while pregnant. Lawson believed that this accounted for the curious fact that "no Man, either Christian or Indian, has ever kill'd a She-bear with Young." Catesby repeated the observation as "no Man, either Indian or European ever killed a Bear with young."[35]

Notable similarities occur also in their descriptions of whitetail deer. Lawson noted that "Fallow-Deer in Carolina, are taller and longer legg'd, than in Europe; but neither run so fast, nor are so well haunch'd. Their Shingles are much longer, and their Horns stand forward, as the others incline backward." Catesby reported that

> they differ from the Fallow Deer in England, in the following Particulars, viz. they are taller, longer legged, and not so well haunched as those of Europe: their

FIGURE 6-5. Passenger pigeon ("Pigeon of Passage"); plate 23, M. Catesby, 1730, *The natural history of Carolina . . .* , volume I. Now extinct, yet one of the most abundant species in North America during Lawson's time. (Digital realization of original etchings by Lucie Hey and Nigel Frith, DRPG England; courtesy of the Royal Society ©.)

Horns are but little palmated, they stand bending forward, as the others do backward, and spread but little. Their Tails are longer.[36]

Catesby seldom used Lawson's material on fish, but parallel descriptions are identifiable in the "devil fish" incident at Charleston harbor and in a few other accounts, such as red drum (*Sciaenops ocellaltus*) and Atlantic herring (*Clupea harengus*). Their details of herring migration reflect a promotional tone of natural abundance in Carolina. Lawson reported that "They spawn there in March and April, running up the fresh Rivers and small fresh Runs of Water in great Shoals, where they are taken." In *The natural history of Carolina . . .* , Catesby noted that "Herrings in March . . . run up the Rivers and shallow Streams of fresh Water in such prodigious Sholes, that People cast them on Shore with Shovels."[37]

Such borrowings demonstrate that Catesby was quite familiar with *A new voyage*, so he probably noticed Lawson's stated plans for future research. At the end of the book, Lawson succinctly outlined his intention to continue his efforts in natural history.

> I do intend (if God permit) by future Voyages (after my Arrival in Carolina) to pierce into the Body of the Continent, and what Discoveries and Observations I shall, at any time hereafter, make, will be communicated to my Correspondents in England, to be publish'd, having furnish'd myself with Instruments and other Necessities for such Voyages.[38]

Despite this announcement, Catesby was probably not aware of the broad scope of Lawson's plans or of the notes and specimens he sent to London after returning to North America. Accompanying a large group of Palatine refugees, Lawson sailed from Portsmouth, probably on the *Princess Anne*, on 12 January 1710 and arrived in Virginia about 25 March. While awaiting their smaller companion vessel, Lawson began collecting birds and plants in Norfolk County. By the end of May he had departed for New Bern to assist in settling the Palatines. He continued gathering plants along the route past Roanoke Island and on the Trent and Neuse Rivers near New Bern until at least 20 July.

In that same month Lawson sent Petiver a large shipment consisting of insects, birds, snakes, lizards, and many dried plants. Preoccupied with the Palatines and the boundary commission, Lawson apparently ceased collecting until late January 1711, when he resumed intermittent efforts at numerous locations between New Bern and the Virginia boundary. His final documented shipment to Petiver, in July 1711, contained the many plants gathered that year.[39]

In a detailed letter to Petiver, Lawson elaborated on his plans for the "Discoveries and Observations" mentioned at the conclusion of *A new voyage*. Lawson's comments reveal quite clearly that the experiences during his London year had transformed his concept of natural history research. His perspective now reflected a more deliberately empirical approach, as promulgated by the Royal Society and by the tenets of Baconian rationalism embedded in the "culture of curiosity" in England.[40]

Lawson informed Petiver that "I hope transactions hereby faithfully communicated to you & such Ingenious Gentlemen of ye Royal Society wth. their remarks on ye same will be a foundation towards a Compleat History of these parts, wch. I heartily wish I may live to tell you."[41]

Lawson never described how his "Compleat History" might be presented, but many options were certainly available to him. *A new voyage* had demonstrated his ability to create a full-length work and to establish a valuable relationship with leading publishers such as James Knapton. Lawson's "Discoveries and Observations" could also have appeared as papers and notes in scientific journals, such as the Royal Society's *Philosophical transactions*, or in Petiver's various publications, including *Memoirs for the curious* and *Gazophylacii naturae et artis*.[42]

The shift in Lawson's approach to natural history is revealed in his plans for botanical research, which show an increased focus on environmental details and life-cycles, but also retain attention to agricultural and medical ("physicall") uses:

> To make a strict collection of all plants I can meet withall in Carolina . . . giving an account of ye time & day they were gotten, when they first appearing, wt soil of ground, wn. the flower seed & disappear & wt. individuall uses the Indians or English make thereof & to have it enough of the same & to let me know how near they agree to the European plants of ye same species & how they differ . . . I would send seeds of all ye physicall plants & flowers to be planted in England. As for the trees the time the bred flower bring their ripe fruit & soil.[43]

Lawson's *New voyage* included more on birds than on any other group of animals, and he clearly intended to expand on this previous work in his nascent "Compleat History." The letter to Petiver contained a checklist of his ornithological plans couched in empirical, observation-based terminology. Lawson's attention to life-cycle details, habitat selection, and behavior, including migration, reflects concepts probably acquired during his year in England. His reference to illustrations suggests a desire to include birds:

> Birds to procure all of this place both land & water fowls from ye Eagle to the wren, to know if possible the age they arrive to, how & where they build their nests, of what material & form, the coulour of their eggs and time of their Incubation & flight, their food, beauty & colour, of wt. medicinall uses if any. If rarely designedd to the Life, this would illustrate such a history very much, their musical notes & cryes must not be omitted, wch. of them abide with us all ye year & wt. strange birds tempestuous weather winds unusual seasons & other evidence affords us.[44]

Lawson's proposals for studying insects also reveal his awareness of their feeding habits, life-cycle, and metamorphosis, topics of growing interest among entomologists in England and Europe during this period. "Insects the months they appear to us in the places of their resort, how they breed & wt. changes they undergo, their food, makes & parts. . . ."[45]

The letter elaborated on plans to report fish, fossils, shells, stones, metals, minerals, soil, agriculture, orchards, gardens, pasturage, grain, exotic plants, grapes, fruits, timber, peaches, apples, figs, prunes, cattle, and sheep. Lawson also intended to provide more information on Indians, waters, climate, and weather.

These ambitious plans obviously required extensive support from England, as colonial North Carolina lacked the necessary resources to preserve, ship, classify, and interpret such a diversity of material. Lawson's relationship with Petiver was crucial in providing this link, ensuring a two-way flow of equipment, supplies, specimens, and information between Carolina and London. Petiver sent quires of paper to preserve plants, vials and boxes for insects, jars and preservatives for birds and reptiles, and publications, the latter to assist Lawson in identifying his specimens.

Considering the extraordinary conditions under which Lawson worked upon his return to Carolina, the extent of his collecting was remarkable. During the seventeen months following his arrival, he was heavily involved with moving and settling hundreds of Palatine refugees at New Bern, in his role as Surveyor-General to resolve the boundary dispute between North Carolina and Virginia, and in suppressing the Cary rebellion, an uprising aimed at overthrowing the government of North Carolina.[46]

Nevertheless, Lawson managed to collect and send more than three hundred plant specimens and a number of birds, reptiles, and insects to Petiver. Plant specimens attributed to Lawson are in the Natural History Museum, London, having been acquired by Sloane from Petiver's estate. More than one hundred species are represented in the collection, contained in four bound volumes. Representative examples of the species include resurrection fern (*Polypodium polypodioides*), fire pink (*Silene virginianum*), eastern redbud (*Cercis canadensis*), flowering dogwood (*Cornus florida*), American devilwood (*Cartrema

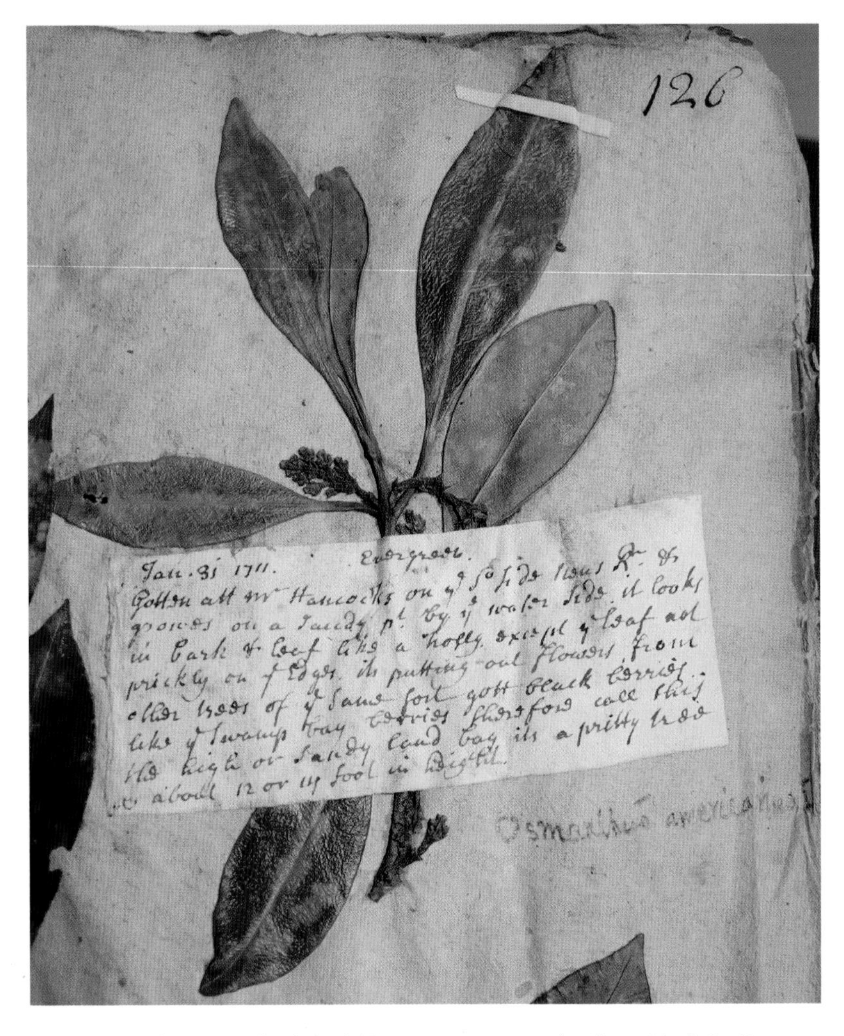

FIGURE 6-6. American devilwood (*Cartrema americanus*) collected by John Lawson (Sloane Herbarium 242, fol. 126, Natural History Museum, London). At least two large parcels of Carolina plants were sent by Lawson in 1710–1711 to James Petiver in London. Petiver's specimens were subsequently acquired by Sloane. Lawson clearly intended that these plants would be an important part of his "Compleat History."

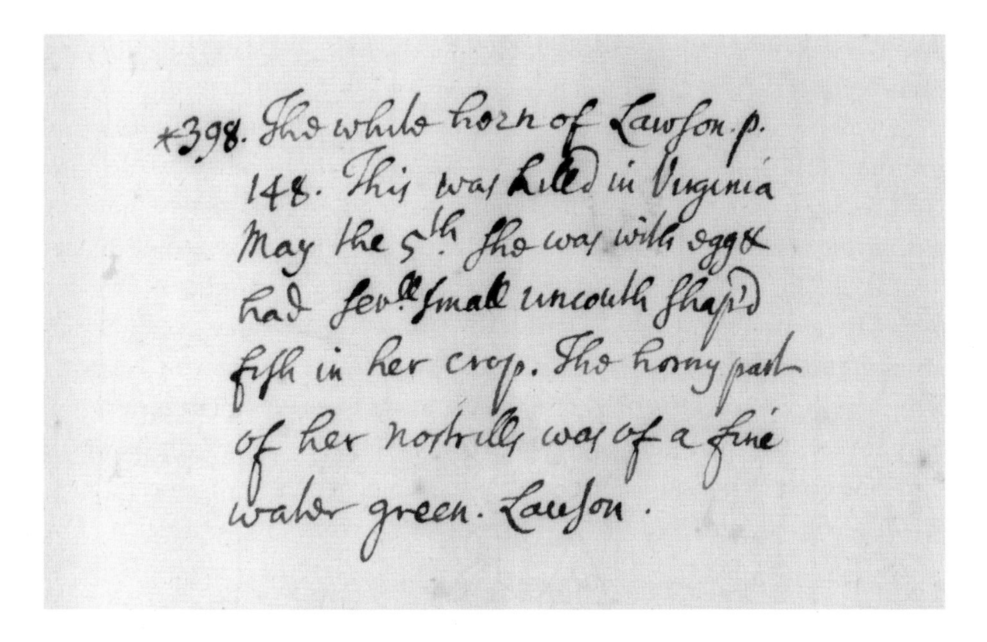

*398. The white heron of Lawson p.
148. This was killed in Virginia
May the 5th she was with egg &
had sevll small uncouth shap'd
fish in her crop. The horny part
of her nostrills was of a fine
water green. Lawson.

FIGURE 6-7. Entry for Lawson's great egret (*Ardea alba*) specimen in Sloane's fossil catalog (vol. 5, fol. 175, no. 398, Natural History Museum, London). This was one of twenty birds attributed to Lawson in the Petiver collection. Sloane's text appears to have been taken from Lawson's accompanying notes, describing the date and location, as well as the intestinal contents and the bill morphology of the bird. Unlike the plants, apparently none of the animal specimens collected by Lawson has survived.

americanus), and sweet bay (*Magnolia virginiana*). Many of the plants have been annotated in various hands, including Sloane's, with references to Ray's *Historia Plantarum* and other works. Associated notes and labels from Lawson occasionally give dates, locations, descriptions, and habitat information (figure 6-6).[47]

The thirty-four zoological specimens recorded in the catalogs have long since been lost or discarded, but Sloane's notations reveal that these included great egret (*Ardea alba*) (figure 6-7), Carolina parakeet, red-headed woodpecker, and eastern bluebird. The reptile and amphibian specimens were not described sufficiently for positive identifications, but the catalog entries indicate that snakes, lizards, a fish, and a frog were among the specimens sent by Lawson.[48]

Despite these efforts, plans for a "Compleat History" ended abruptly in the late summer of 1711. Lawson was captured and executed by the Tuscarora tribe in September during the early days of their war against the British colonists. Although Lawson retained notes and duplicates of the plant specimens he sent to England, these have not been traced and perhaps were destroyed during the war. At the time of his death, Lawson owned "several writings to him belonging," which might have included records or drafts intended for his planned "Compleat History."[49] No evidence has been discovered about the fate of these documents. The only publication known to have been written by Lawson was

A new voyage, which was re-issued without changes in 1714 and 1718 under the title *The history of Carolina*.

Lawson was uniquely positioned to achieve his stated goals. His appointment as Surveyor-General provided him not only with income but also with the advantage of government connections that would expedite his shipments to England. As the colony's surveyor, he could explore and collect while undertaking his official duties. His access to the region and extensive knowledge of the colony were probably unsurpassed at that time. Regardless of whether Lawson had any botanical training or artistic skills, he was in a strong position to provide a reliable supply of dried plants and seeds from Carolina to the leading botanists and gardeners of England. With his untimely death, however, Lawson's contributions to natural history remain defined almost entirely by his *A new voyage to Carolina*.[50]

7

Mark Catesby's world: England

JANET BROWNE

Mark Catesby's achievements connect him so closely to the early history of North America that it is hard to think of him as an Englishman, born and bred, and generally resident in England. His career, prospects, and financial livelihood were firmly based in England, where he lived through a period of great religious and political upheaval, set into motion by the abdication of the Catholic King James II, who was replaced by William and Mary; they, in turn, were followed by three other Protestant sovereigns. The Anglican Church was then established as the state religion. During this time England and Scotland were united into Great Britain, and the War of Spanish Succession was fought in the British colonies of North America. England lost much of Florida and Canada to rival European nations. There is little of this in Catesby's writings, but it nevertheless formed the backdrop to all his endeavors at home and abroad.

It is also sometimes difficult to summarize natural history in England during this period, which in Catesby's time covered a very wide spectrum of topics, including the study and collection of antiquities and much interest in mythology, languages, and anthropology, as well as the observational sciences of astronomy, meteorology, comparative anatomy, and geology. Natural history as an area of investigation was both local and international in scope as well as both individual and collective. It involved many different sorts of people with different aims, social standing, religious affiliation, and education. However, whether focused on plants, animals, the earth, or the skies, natural history ranked high among scientific enterprises. This was because, in England, naturalists felt that one of their primary tasks was to find and catalog the world's resources. Many naturalists were therefore indispensable agents of mercantile economic development. Their expertise contributed to the success of trading corporations such as the East India Company. It also played a key part in the efforts of European nations to expand their geographic reach by acquiring overseas colonies, such as those in North America.

Of course, not every English naturalist was commercially engaged. The majority had been educated as physicians or theologians, and it was not part of their daily business to link nature directly with commerce. Instead, their work was often conceived as serving the newly consolidated Protestant creed of natural theology in which the faithful described the world in terms of God's wonders and His beneficent design. For these reasons, from as early as the sixteenth century the subject of natural history was generally regarded by English men and women as one of the most valuable of the useful arts. Commercial enterprise was felt to be entirely compatible with the widespread adoption of the Anglican form of natural theology. Reliable knowledge of plants and animals was highly prized information that could lead to improvements in horticulture, animal husbandry, mineralogy, medicine, and mining while simultaneously revealing the moral truths of nature and encouraging human virtue, all of which were considered basic resources for a nation's economic success and the happiness of its inhabitants.

This promise of material gain, both individual and national, supported by lofty philosophical ambitions, prompted the English to invest enormous amounts of time and effort in obtaining precise information about nature. Botany, for example, was an essential cog in the economic engine of the growing British empire and a fundamental element in medical training. It also attracted wealthy landowners, gardeners, and horticulturists who yearned to acquire unusual plants, as well as members of the learned elite willing to sponsor collectors or support natural history publications. Natural history was the "Big Science" of England and also offered an attractive range of pursuits and interests to the public.[1]

Corresponding naturalists

How did Catesby manage to gain a foothold in this world? At this period it was not enough just to be talented or eager to travel.[2] Catesby, like so many other ambitious naturalists, needed some kind of introduction to the patronage systems of the day and offers of financial support.[3] Natural history in England during Catesby's lifetime was a tightly integrated social network comprising individuals of varied backgrounds, skills, and education. Many of these individuals were connected one way or another through the medium of correspondence.[4] Indeed, letters provided naturalists with a crucial social cement. Letters do much more than communicate news and views. They are an integral part of the organization of literate societies, one of the main means by which people establish and transmit common value systems; and letters and correspondence networks had a large part to play in consolidating the social relationships and infrastructure necessary for the rise of modern science. Ideas and business matters were shared and exchanged by letter among the European scholarly community, and they crossed the Atlantic in both directions. Correspondence was an essential component of the natural history world that Catesby aimed to join.

For example, correspondence used to be one of the main ways that natural philosophers collected, processed, and disseminated data. Charles Darwin has been a popular figure for attention in the nineteenth century, but other individuals were just as expert in generating and using correspondence networks as scientific tools. Philip Miller (1691–1771), the head gardener of the Chelsea Physic Garden during the middle years of the eighteenth century, transformed the Society of Apothecaries' private garden into one of the greatest botanical gardens in Europe through an extensive network of correspondents.[5] His correspondence generated a wide exchange of plants, many of them cultivated for the first time in Britain at Chelsea. Miller wrote a number of significant botanical works, including *The gardeners dictionary*, first published in 1731, which became the standard reference work for gardeners in Britain and North America. A later work of Miller's even provided a source of botanical illustrations to be copied onto local Chelsea tableware. He corresponded with Carl Linnaeus and invited him to visit the Physic Garden in the 1730s, although he was personally reluctant to use Linnaeus's innovative nomenclature in his publications. Miller thus positioned himself at the center of an information network that not only served him personally but also opened British botany to transformative new knowledge.

Miller was neither the first nor the only corresponding naturalist. Other botanic gardens were following suit. Those in Oxford and Edinburgh were founded in 1670, and both became important centers for the cultivation and international exchange of plants. Individuals, too, made a mark. Less well known than Miller, Henry Compton, Bishop of London, had a botanical correspondent in John Banister, who went first to the West Indies and then to Virginia. Before his untimely death, Banister sent Compton drawings, seeds, and perhaps living specimens for the bishop's garden in Fulham, and from which John Ray compiled the first published account of North American flora, issued in Ray's *Historia plantarum* in 1688. Peter Collinson, a London-born Quaker merchant who was a friend of Catesby's, engaged in an extended correspondence with John Bartram in Philadelphia and was responsible for importing many boxes of seeds from the colonies on a syndicated system intended to enrich the gardens of wealthy English owners.[6] Initially Collinson focused on establishing a regular trade between English naturalists and the American colonies. Later he transmitted botanical information to the Royal Society of London and in return relayed the Royal Society's findings back to America. The collected correspondences of Collinson, Ray, Petiver, Sloane, Richard Richardson, and others indicate the extraordinary reach of circles of letters moving across Britain and the colonies. Ray believed that Petiver had "the greatest correspondence both in East and West Indies of any man in Europe."[7] As an early editor of the Royal Society's *Philosophical transactions*, Sloane received and preserved thousands of letters.

By writing letters, Miller, Collinson, Sloane, and others were actively participating in a traditional form of scientific communication dating from the sixteenth century onwards. The hundreds of letters that survive in the archives of Conrad Gessner, the Swiss naturalist, for example, testify to the existence of an early republic of letters connected by the common language of Latin. Only a little later, the French savant Marin Mersenne used correspondence to put himself at the center of an extensive information network. Similarly, the English natural philosophers Henry Oldenberg and, afterward, James Jurin at the Royal Society of London intentionally made themselves nodal points in the traffic of ideas, calling themselves "intelligencers" and writing to figures they designated as corresponding members of the Society. In Sweden Linnaeus's global natural history enterprise required many letters, nearly five thousand of which are extant. Catesby was one of Linnaeus's many correspondents.

Before modern scientific journals had come into being, naturalists and philosophers intended their correspondence to be shared, copied, and read aloud at informal meetings. The practice served as a form of publication. Letters often provided the first step in generating a scientific fact, and communication networks brought those potential scientific facts to the attention of others. At the same time, correspondents turned increasingly to the use of the vernacular in their letters: English and French being the most popular forms of expression. Many of these early networks took shape through the exchange of information and specimens at coffee-houses (figure 7-1), for example, the one at the Temple in London. Sir Hans Sloane and James Petiver were closely associated with this group of naturalists, although much about the so-called Temple

FIGURE 7-1. Interior of a coffee-house (Anonymous, c. 1700). (© Trustees of The British Museum.)

Coffee-house club, even its name and location, remains mysterious.[8] Meetings in convivial metropolitan spots, such as inns or coffee-houses, certainly marked the beginning of botanical field networks and societies.[9] As time went by, these activities consolidated into the much more structured system that took place in the meetings of scientific societies for reading and evaluating information supplied by strangers. Writing and handling letters, it seems, were precursors to the basic procedures of modern scientific publishing, these being peer-review, authentication, and dissemination.

Networks are normally made up of people with unequal and irregular ties to each other, and these patterns of communication provide benefits and duties depending on where each person is situated in the network.[10] An unknown naturalist in the late seventeenth century, for example, could only join the network through an appropriate introduction, and much of the validity of what might be said in a letter depended on who was prepared to vouch for its accuracy. Overlapping circles of correspondence were therefore significant, allowing personal movement through the circles by introduction.[11] If an individual had some special knowledge, this could be very helpful in gaining access to the network. So was the act of gift-giving. Botanists, in particular, were active in developing a sophisticated process of gift-giving up and down the social scale. Two packets of seeds from a garden in Philadelphia might be exchanged for duplicate herbarium sheets from Chelsea Physic Garden, or they could smooth the process of introduction to a higher-ranking botanist.[12]

So although little of Catesby's own correspondence has survived, we can nevertheless track by inference his movement through the network of English naturalists. These well-organized and self-contained infrastructures, held together by correspondence and exchange, need to be factored into current understanding of Catesby's career path.

Getting into the network

Catesby is thought to have first made himself known to naturalists in England through his uncle Nicholas Jekyll. Somehow, probably through Jekyll, Catesby became acquainted with Samuel Dale (1659–1728) (figure 7-2), an excellent naturalist renowned for his knowledge of local geology, plants, and bird life. Dale was an acquaintance of the Reverend John Ray.[13] Through Dale's extensive network of contacts, Catesby was put in touch with two significant figures of the period. First was William Sherard (1659–1728) (figure 7-3), a well-connected botanist and patron of other botanists who returned to England in 1717 after serving as British Consul at Smyrna and later a generous benefactor to Oxford University and its botanic garden. Sherard gave Ray extensive data about European plants after his travels on the Continent; and his brother James (1666–1738), a London apothecary, grew a magnificent collection of plants on his estate near Eltham that was recognized as one of the finest private gardens in England. These two brothers were liberal patrons of the German systematist Johann Jacob

FIGURE 7-2.
Samuel Dale.
(Courtesy of the Wellcome Library, London.)

Dillenius (1687–1747) and instrumental in recruiting him to Oxford University. Catesby acknowledged his gratitude to William Sherard in the preface to *The natural history of Carolina*. . . . The other figure influential in Catesby's life was the nurseryman and horticulturist Thomas Fairchild (c. 1667–1729) (figure 7-4), based in Hoxton, now a suburb of London. Fairchild was one of the leading gardeners of the time and a fine experimental botanist, being the first recorded person to create an artificial hybrid by deliberately cross-pollinating a sweet william (*Dianthus barbatus*) and a carnation (*Dianthus caryophyllus*) to produce a new and highly marketable garden plant dubbed Fairchild's mule.[14] He is thought to have been a member of the Society of Gardeners and instrumental in publishing their *Catalogue of trees and shrubs both exotic and domestic which are propagated for sale in the gardens near London* in 1730. Fairchild's work *The city gardener* (1722) was a list of plants that grew well in London. In his will, he left money to fund an annual sermon at the church of St. Leonard's, Shoreditch.

Catesby's first visit to the English colonies between 1712 and 1719 was carried out primarily to escort his sister Elizabeth to Williamsburg, Virginia. She was joining her husband, William Cocke, who had recently begun medical practice in the area. This first journey enabled Catesby to meet several of the influential settlers of the region. He explored and collected in the countryside, made a number of watercolor illustrations of birds, and traveled to the English colonial islands of the West Indies. He seems to have been a congenial and cultivated man, easy to get to know, and able to empathize with the intellectual changes taking place in the Atlantic world.[15]

On this first expedition, Catesby sent specimens to Dale for forwarding to others. A ticket on a herbarium sheet from Charles Dubois's collection in the University of Oxford herbarium includes the remark "Rais'd from Bulbs

FIGURE 7-3.
William Sherard.
(Courtesy of the Department of Plant Sciences, University of Oxford.)

FIGURE 7-4.
Thomas Fairchild. (Courtesy of the Department of Plant Sciences, University of Oxford.)

sent from Virginia by Mr Catesby, anno 1715." And after Catesby's return from this journey, the judicious submission of more seeds, and probably some living plants, to Dale and Fairchild opened the doors of sponsorship for a second, more systematically organized expedition to the North American colonies.[16] Through William Sherard, Catesby acquired the financial support of twelve notables who were listed in the preface to *The natural history of Carolina* . . . , including that of Sir Hans Sloane who would prove to be an outstanding patron to Catesby. In 1727, the year after Catesby returned from North America, Sloane became President of the Royal Society and actively promoted Catesby's intellectual and publishing concerns. Five of Catesby's supporters for his second expedition were Fellows of the Royal Society and evidently personal acquaintances of both Sherard and Sloane. It is probable that Sherard firmed up the promise of a stipend for Catesby, amounting to £20 per annum, that had been offered by the incoming Governor of Carolina, Francis Nicholson. In this regard, Catesby also acknowledged the personal assistance of William Byrd II, who was a subscriber to the eventual volume. In a letter to Richard Richardson dated 7 December 1721, Sherard wrote: "I believe Mr. Catesby will be going to Carolina in a month. I have procured him subscriptions for near the sum he proposed."[17] These sponsors no doubt expected living specimens of New World plants or possibly paintings for their own collections. It is not clear that they also anticipated a lavishly illustrated book to be afterwards produced.

Also, at some point, either before or during his travels, Catesby decided to produce a volume of natural history illustrations. Such a book as he envisaged required energetic application to the network of naturalists and potential subscribers to raise funds and generate interest in the publication. In this, Catesby depended on increasingly larger circles of introduction and the opportunity to engage patrons through the Royal Society of London. Eventually Catesby's scientific and social connections, probably through Sloane, put him on a path that allowed him to approach the royal family for permission to dedicate his first volume to Queen Charlotte and, later on, the second volume to the Princess of Wales.[18] A sign of his increasing reputation was that he was ushered into the royal presence by Lord Carteret, who was one of the Lords Proprietors of Carolina.

Mark Catesby was elected a Fellow of the Royal Society on 26 April 1733. About this time, again probably through Sloane, Catesby met Georg Dionysus Ehret (1708–1770) (figure 7-5), who was to become the greatest botanical artist of the era and who enjoyed Sloane's patronage.[19] Ehret, in turn, was a source of introductions for Catesby. Ehret drew several of the plates in *The natural history of Carolina . . .* , including the one of *Magnolia grandiflora*. The watercolor drawings of it were made by Ehret from a tree that produced a single blossom in Sir Charles Wager's garden at Parsons Green, then a village west of London, in August 1737. Several versions of this image are known today. Ehret said he walked there every day from his home in Chelsea to study each stage of its unfolding and "drew every part of it in order to publish a perfect botanical study of it."[20]

Commercial Considerations

The commercial aspects of Catesby's work in this period should not be separated from his professional botanical aspirations and connections. Even though there is very little information available about his financial state at this point, it is fair to assume that he probably retained enough of the property inherited from his father to be independent financially. Yet after 1726 he began to work with several London nurserymen to encourage an interest in North American trees and shrubs. Horticulturists and landowners were prominent members of English botanical circles. It seems that Thomas Fairchild had earlier dispatched garden plants to Catesby while in Virginia to distribute to American clients.[21] He was ready to assist in developing a market for American trees and shrubs in England, which was a valuable form of support. It appears that Catesby worked with Fairchild for some years, for in a little-known prospectus for Catesby's *The natural history of Carolina . . .* , he advised potential subscribers that they could visit him at Fairchild's (later Stephen Bacon's) in Hoxton to examine the preparatory drawings for the plates.[22] Later, Catesby joined another horticultural specialist, Christopher Gray, in Fulham, west of London, who grew magnolias and other plants that Catesby had introduced to England from America.[23] Some of these plants were described and illustrated in Catesby's *Hortus Britanno-Americanus*, published posthumously in 1763 and re-issued with a slightly different title in 1767. *Magnolia grandiflora* was the first species discussed in the latter work. By this time it was probably Catesby's "signature" plant. Catesby evidently intended in this work to disseminate information about the proper storage and propagation of the choice American plants he introduced to England. An indication of the extent of his participation in

FIGURE 7-5.
Georg Dionysus Ehret.
(By permission of the Linnean Society, London.)

horticulture at this period can be found in his remark that he lost several hundred seedling magnolias in a severe frost in 1740. The preface to the 1767 edition includes this statement:

> By a long acquaintance with the trees and shrubs of America, and a constant attention since for several years to their cultivation here, I have been enabled to make such observations on their constitution, growth and culture, as may render the management of them easy to those who shall be desirous to inrich their country, and give pleasure to themselves, by planting and increasing these beautiful exotics; and I shall think myself very happy, if this little work may excite any to what in my opinion is evidently a public good.[24]

Christopher Gray's broadsheet *Catalogue of American trees and shrubs* (figure 7-6), published about 1740, used an engraving by Mark Catesby of Ehret's *Magnolia grandiflora* image as the centerpiece. This catalog, written in both English and French, made it clear that the market for American trees and shrubs extended to the rest of Europe, as well as Britain.

Indeed, Catesby's illustrations in *The natural history of Carolina . . .* and other texts were, in a way, substitutes for living specimens for those people who did not have access to real plants, in the sense that the illustrations were conveniently collected together in a series and distributed in book form almost as if they were the equivalent of a visit to a botanic garden placed between two covers. Such illustrations could serve as a tempting showpiece or visual catalog of striking new plants that were available to gardeners through English nurserymen. Taken all together, Catesby's publications reveal a fascinating integration

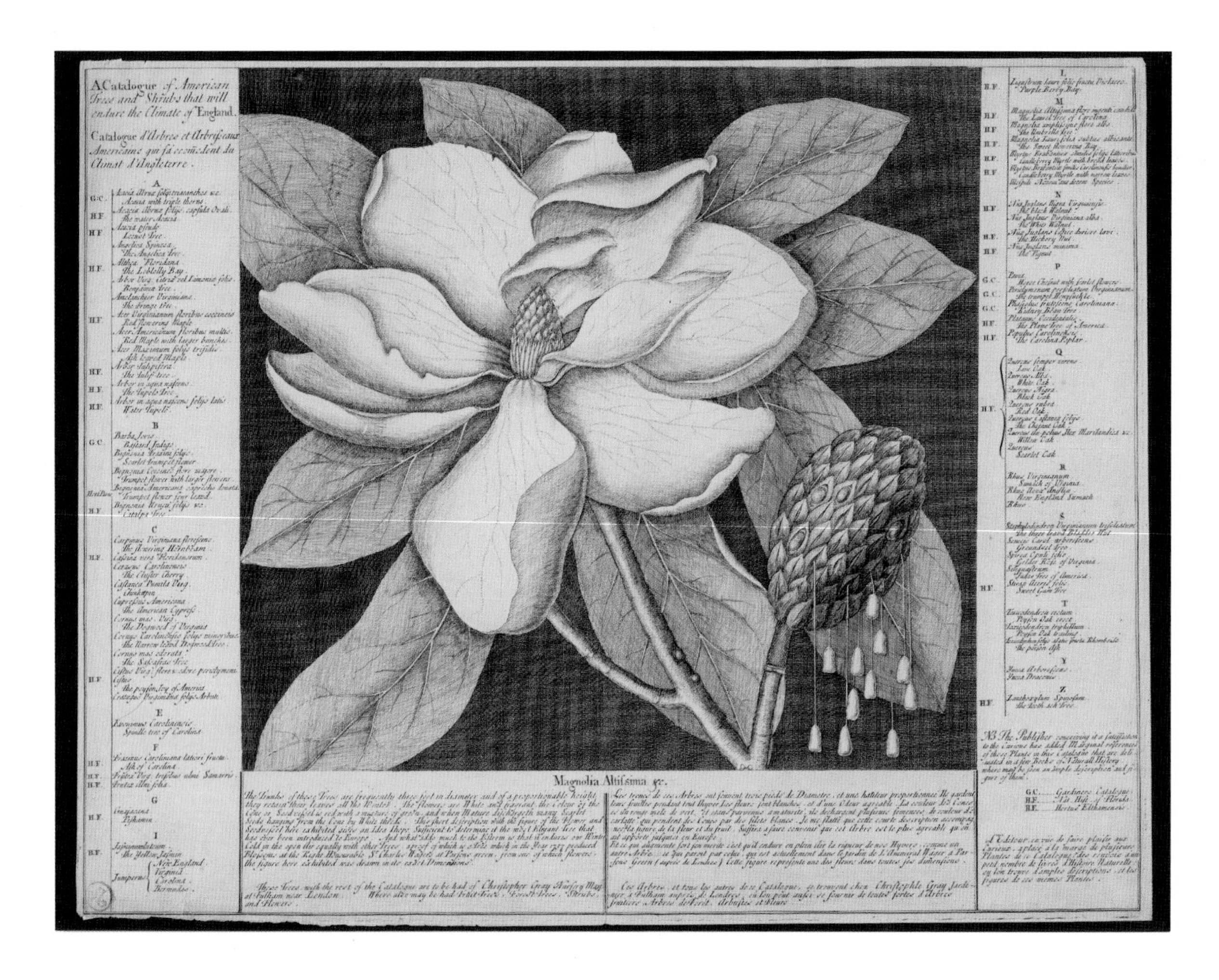

FIGURE 7-6. "*Magnolia Altissima*"; Christopher Gray, c. 1740, *A catalogue of American trees and shrubs that will endure the climate of England.* (Courtesy of John Johnson Collection of Printed Ephemera, Seed Catalogues, Bodleian Library, Oxford.)

between the expansion of academic learning, the marketing of new plants, and shrewd self-promotional material.

So Catesby's movement among the learned botanists and gardeners of the nation, his activities with horticulturists and wealthy individuals who would eventually purchase living American plants, and his efforts to obtain the necessary support for his artistic endeavors were all facilitated by the existence of a circle of botanical correspondents who welcomed him and took him up as a figure well worth their attention.

Such social links tend to be taken for granted, yet these connections were powerful enough that Catesby, as a man with ambition, a liking for natural history, and an entrepreneurial frame of mind, could finance a four-year visit to England's North American colonies and on his return successfully enter the scientific world as an artist and naturalist. His success in that enterprise directly emerged from his extraordinary talent and ingenuity. But with hindsight, it can be seen that each of those steps depended on the existence of a network of individuals living in England and the North American colonies who were connected by correspondence, gardening expertise, commercial interests, and the exchange and marketing of specimens.

8

Mark Catesby's world: Virginia

SARAH HAND MEACHAM

"*Virginia* was the Place," Mark Catesby later explained,[1] "(I having Relations there) [which] suited most with my Convenience to go to. . . ." When Catesby's sister Elizabeth and two of her children needed to travel to Virginia to join her husband in 1712, Catesby seized the opportunity to join them. As he described, "my Curiosity was such, I soon imbibed a passionate Desire of viewing as well the Animal as Vegetable Productions in their Native Countries."[2] His route in Virginia remains a little hazy. It is clear, however, that Catesby began his explorations in eastern Virginia, starting first in Williamsburg and then at William Byrd's plantation in May of that year. Byrd invited Catesby to join him and Lieutenant Governor Alexander Spotswood (figure 8-1) at Pamunkey Indian town for a day. In 1714 Catesby traveled from the James River to the Appalachian Mountains. Later that year he traveled to the West Indies, primarily Jamaica.

Mark Catesby encountered a strange new world in Virginia when he arrived there in 1712. The landscape, including the housing and system of farming, was foreign to him. He had never before encountered plants and wild animals like those of eastern North America. For instance, he had never seen a hummingbird except possibly as a stuffed specimen. The food was strange as well, particularly the breads and porridges made of maize. Even in Williamsburg (figure 8-2), the capital of English America, the difficulty and cost of importing goods meant that even well-off residents did without the household conveniences and the food and drink of their counterparts in England.

Little direct evidence remains regarding Catesby's time in Virginia. If he kept a journal, drew in a sketchbook, or wrote letters during his stay, they are not known. Still, the herbarium labels on the plants he sent from Virginia to Samuel Dale do survive, as do the entries that Virginia planter William Byrd II recorded in his diary when Catesby visited him. While in Virginia, Mark Catesby lived in Williamsburg with his sister and brother-in-law, Dr.

FIGURE 8-1. Alexander Spotswood. Catesby traveled to Williamsburg because his brother-in-law, Dr. William Cocke, had moved there to serve as Alexander Spotswood's physician. (Courtesy of Colonial Williamsburg Foundation.)

FIGURE 8-2. Sketch of Duke of Gloucester Street, Williamsburg, in 1836 by Thomas Millington. When Catesby was in Williamsburg, the town was far more rural than even this sketch of 1836 indicates. (Courtesy of Swem Library, College of William & Mary.)

William Cocke, about whom much is known. Cocke was an important person in Williamsburg. He had earned his medical degree at Cambridge in 1693 and come to Virginia with Spotswood in 1710 to serve as his doctor. Cocke soon turned to politics, being sworn in as Secretary of the colony soon after his arrival. Within the year, he was made a councilor.

Previous biographers of Catesby have claimed that life in Williamsburg in 1712 would "not have disappointed" him. They have pointed to a statement by promotional author Hugh Jones in 1724 that

> Williamsburgh is now incorporated and made a Market Town. . . . Here dwell several very good Families, and more reside here in their own Houses at publick Times [quarterly court sessions]. They live in the same neat Manner, dress after the same Modes, and behave themselves exactly as the Gentry in London.

Jones, however, was not an impartial observer. He was a proud Englishman who wrote with patriotic zeal. Like most of the authors of promotional literature who preceded him, he exaggerated, knowing that English people still linked Virginia with the starving time of the winter of 1609–1610 and the Indian massacre of 1622, when Chief Opechanconough, in response to an English settler murdering his advisor Nemattanew, led a group of Powhatan Indians in a one-day campaign of surprise attacks on at least thirty-one English plantations and settlements. One-third of the white population, about four hundred colonists, died in the attacks. Jones's readers, if they believed him at all, would have read his statement to mean only that Virginia colonists had not adopted Indian ways. Even Jones admitted that Virginia was "capable of great Improvements still, and requires several Alterations."[3]

One of those required alterations was the creation of towns. English people, including the Englishmen in Virginia who served as council members, desired towns. They feared that English people living without towns would degenerate and turn into the very Indians whom they hoped to transform into English servants. The Virginia General Assembly passed five acts throughout the seventeenth century instituting towns, though to no effect. Until the 1690s Jamestown remained Virginia's only town.

Planters and farmers believed that they could make profits in Virginia only on tobacco (figure 8-3). Most farmers held at least two hundred acres. As a result, even farmers who owned the smallest landholdings still viable as tobacco farms were spread at a minimum one-third of a mile apart. Few colonists lived within sight of another household. While elites worried that English people in America would degenerate without towns, few colonists were willing to abandon the profits of tobacco farming for town life.

FIGURE 8-3. Great Hopes Plantation. Producing tobacco was so labor-intensive that colonists had little time to build permanent houses or fence their lands as they would have done in England. (Courtesy of Colonial Williamsburg Foundation.)

On 7 June 1699, only thirteen years before Catesby arrived, the General Assembly tried again and agreed to make John Page's Middle Plantation the new capital, to be named Williamsburg; to build a new statehouse there; and to lay out streets. Francis Nicholson, who helped to sponsor Catesby's appointment to Carolina in 1722, had recently been appointed Governor of Virginia. He urged that the capital be moved to Williamsburg, hoping this would help him gain some power against a council hostile to English governors. Nicholson had long been interested in architecture, landscape design, and town design. He had previously guided the plan for the city of Annapolis, Maryland. In all likelihood, Nicholson did not design the layout of Williamsburg, but he chose its name and as its patron pressed for a classical design that gave clear authority to the governor.[4] As an additional benefit, Williamsburg was healthier than the swamp-like Jamestown it replaced, and it offered creek access to both the James and York Rivers. Planters depended on these rivers. Captains sailed their ships to planters' landings, bought their tobacco, and brought them goods ordered from merchants in England.

Catesby found himself in Williamsburg, a recently formed town with fewer than twenty buildings. When he arrived in 1712, there was a small Anglican church, which would be expanded and reopened in 1715, perhaps four houses, the Capitol, the College of William and Mary, the recently finished Governor's Palace, a tavern, several stores, two mills, a smith's shop, and a small grammar school. While the Palace, Capitol, and College made the town more dramatic than an ordinary English village, the town was only sparsely settled and had no market days or fairs (figure 8-4). Even the new government buildings were built out of brick rather than stone, as they would have been in England, which reduced their stature in English eyes. Any Englishman would have looked on the town, which was supposed to be the capital of British America, with some excitement, as well as some concern.[5]

No evidence remains of Cocke's house or furnishings. However, four other houses went up while Catesby was in Williamsburg, and the limited architectural evidence from these offers clues about what sort of house the Cockes likely had. In addition, the 1732 will of Colonel Thomas Jones of Williamsburg suggests the sort of furnishings typical for a family like the Cockes.

Early eighteenth-century houses were rougher than the genteel late eighteenth-century houses (figure 8-5) that stand in Colonial Williamsburg today. There is a good chance that Cocke's house was built of wood and was not plastered on the inside. The plain back rooms likely did not have baseboards. Still, the household may have had enough beds for the adults, requiring only a few of the children to share beds, and it probably had enough cutlery and crockery that no one needed to share eating utensils. At least three white adults and six children lived in the household at the time. Thomas Jones, who was of similar wealth and status to Cocke, died owning six beds, thirty sets of knives and forks, thirty-six glasses, twelve teacups, and an unknown number of

FIGURE 8-4. Near Williamsburg, by Lefevre James Cranstone (c. 1860). This painting shows that the simple "Virginia House" remained the typical residence for most farmers and that Williamsburg, the capital of English America until 1780, remained rural: the regional economy relied on tobacco. (Courtesy of Colonial Williamsburg Foundation.)

FIGURE 8-5. The late eighteenth-century "Virginia House" pictured here is more elaborate than most farmers' houses of the early eighteenth century. The houses Catesby saw would not have had window glass or brick chimneys; most would have contained only one room of sixteen by twenty feet and featured dirt floors and wooden chimneys. (Courtesy of Colonial Williamsburg Foundation.)

plates and bowls. Wills can give a mistaken view of what people had during the majority of their lifetimes, since people tend to amass more as they age. It is necessary to be cautious about concluding that the Cockes had the same number of items in 1712 as the Joneses did in 1732, although the families' means were similar. Still, people tended to buy bedding, crockery, and eating utensils before they purchased other goods such as mirrors, books, fine tables, and extra chairs, and the conclusion that Catesby probably had his own bed, mug, fork, and crockery at Cocke's house is fair.[6]

Once Catesby left Cocke's house and Williamsburg's one street, he saw how meagerly most Virginians lived. They did without because they grew tobacco for their livelihoods, and growing tobacco both demanded constant labor and spread colonists too far apart to be able to purchase goods at local markets or trade for them. In England, the majority of land designated as farmland had already been cleared and plowed. In contrast, in Virginia, yeomen raising tobacco had to clear the land of trees, burn the underbrush, and mix ashes into the clay soil for fertility before they could prepare and plant the seedbeds. These seedbeds had to be covered with leaves and boughs to protect them from frost, this covering being removed during the daytime for optimal sunlight. Planters had to water the plants twice a day if the weather was dry, which required them to haul water in buckets, and given that Virginia was much warmer than England, the planters may have needed to water more frequently. They had to weed and thin the plants; transplant the growing seedlings to small hills made with hoes; replace plants that had died; crush the hornworms that infested the plants; top and sucker the plants; and then cut, cure, pack, and transport the tobacco leaf for sale. Unlike in England, there were no markets or specialists such as millers, so all farmers had to shell and grind maize, haul water, slaughter hogs and cattle, salt and smoke the meat, replace damaged fruit trees, and chop firewood. In England, women could specialize in trades such as brewing, wet-nursing, laundering, and dairying, relieving other women from performing those tasks if they were willing to pay for them instead. In Virginia, again because tobacco farming spread households much farther apart, women had to be Jacks- or Janes-of-all-trades. They had to sew the clothes; nurse their babies; tend their children; make the food; preserve the fruits, vegetables, and meats; dip the candles; press the butter and cheese; clean the home; plant, weed, thin, and hoe the gardens; wash the laundry; and make cider, generally without assistance.

The demands of making a tobacco crop, especially for the vast majority of colonists who could afford one slave at most, necessitated reducing labor elsewhere and the barest subsistence farming. The results looked slovenly to English visitors like Catesby who were accustomed to lands that had long been farmed. In England, land was relatively scarce, so farmers cultivated fields and enclosed farms with hedgerows and stone walls. In order to maximize land for farming, laborers there pulled out tree stumps, added manure to the land,

and penned livestock. In contrast, in Virginia, there was plenty of land and not enough people to work it, so Virginians chose labor-saving methods of farming. They let tree stumps remain in the ground and planted around them (figure 8-6). They put up the fastest, simplest fences, called "worm fences," of rough split rails set in a zigzag pattern. And they often fenced their vegetable gardens rather than their livestock to reduce the amount of labor spent in fencing. The free-range animals then fed themselves in nearby forests. Virginians did not have plows, and they cleared land and planted tobacco only with hoes. Plows were expensive and required teams of oxen to pull them, adding to their cost. Planter Landon Carter relied on hoes for his fifty thousand acres of land in 1770:

> Oxen are not the thing to plow with, they are slow at best, and tire every hot day. And the plowmen unmercifull [*sic*] to them by constant beating them; . . . and if I get horses, then they are rode out in the nights by the negroes. So that on all accounts hoes are surest and best way of tending.

FIGURE 8-6. *An overseer doing his duty, near Fredericksburg*, by Benjamin Henry Latrobe, 1796. European visitors were dismayed by the Virginia practices of leaving tree stumps in the ground and building worm fences. They feared that colonists were turning into the very "savages" they were meant to be "civilizing." This sketch is one of four known to have been created in Virginia during the eighteenth century. (Courtesy of the Maryland Historical Society.)

The farmers did not manure the land; instead, they planted tobacco for several years, changed for a few years to maize, and then let a field lie uncultivated for several years to restore nutrients. As a result, much of the area Catesby saw was slowly returning to forest, which many visitors, who misunderstood the demands of growing tobacco, took as a sign of laziness.[7]

The farmers' homes that Catesby saw were much rougher in Virginia than they were in England. In England, houses often were constructed with wooden frames and mortise-and-tenon joints, thatch and tile roofs, and brick chimneys. In Virginia, farmers saved time and labor by framing houses around posts put directly in the ground and covering the framing with rough clapboards. They slapped tar on the roofs to keep out the rain and clay or mud between the clapboards to reduce drafts. While well-off yeomen built houses with brick chimneys, most of the other houses had wooden chimneys. Traveler Edward Kimber described an Eastern Shore community in 1746 in which "the church and all the houses are built of wood, but [only] some of them have brick stacks of chimneys: some have their foundations in the ground, others are built on puncheons or logs, a foot or two from earth." These "Virginia houses" were often only sixteen by twenty feet, with one or two rooms. Floors were often of dirt. Servants and slaves ate and slept with the family. This apparent slovenliness appalled most visitors.[8]

The diet in Virginia differed from what Catesby was accustomed to. In England, porridge and bread were made from wheat, barley, or rye, and beer was brewed with hops. Among other foods, English people in England ate pottage made from dried peas, and consumed sheep's and cow's milk and cheese, and sometimes mutton. In Virginia, however, it was too warm to grow early modern peas or wheat, and wolves preyed on sheep. Instead, colonists turned to maize, which was easier to plant and required less tending than wheat, and pigs, which could be branded and allowed to roam and feed themselves, which was more practical than husbanding sheep or cows. Making cider was less labor intensive than making beer, which required growing hops and barley or oats in the inhospitable Virginia climate, and so cider replaced beer as the staple drink.

Wealthy English colonists in Virginia tried to import British foods and drink, but doing so was unreliable and expensive. "Your convict ship arrived safe with the goods, if one may call that safe where everything is damaged and broke to pieces," fumed William Byrd. "We unpacked the beer that came from England," noted Byrd on another occasion, "and a great deal was run out."[9] Catesby himself noted the high cost of imports from abroad. His observations on the cost of beer in Carolina in 1722 hold true for his time in Virginia as well. "This place is within a trifle as dear as the west Indies few European Goods are Sold for less than 300 p Cent and oftner for 400—or 500— . . . Beer from London I have not known cheaper than 10 [pence per] Bottle which is equivalent to 2 [pence] Sterling."[10] While beer and other imports cost somewhat more in South Carolina in 1722 than they did in Virginia because of exchange rates, the expense and irregularity of imports forced colonists in both colonies to rely on what food and drink they could grow and make themselves. Virginia's climate and the difficulties inherent in shipping meant that, unlike Englishmen in England, most Virginians survived on maize, pork, and cider.

The labor system that Catesby saw was also startlingly unlike what he was accustomed to. In fact, the demographic makeup that Catesby encountered in the Chesapeake was quite different from what he would have found twenty or thirty years earlier, although it is unlikely that he knew that. There were fewer white indentured servants and more black slaves than previously. The region witnessed population growth from 1670 to 1700, even as the number of American Indians declined dramatically due to disease and migration. By the mid-eighteenth century, enslaved and free blacks made up more than a third of the population, while in Williamsburg, they made up half.[11]

Dr. William Cocke had been at Felsted Grammar School in Essex at the same time as William Byrd II (figure 8-7), so they were old classmates and friends. Byrd had been born in Virginia and sent to Felstead when he was seven years old. He developed a lifelong interest in harnessing the natural world. In 1696 he became a Fellow of the Royal Society of London for Improving Natural Knowledge. In later years, he led surveying expeditions along Virginia's borders and promoted a Swiss settlement in Virginia's southwest as well as iron-mining ventures in Germanna and Fredericksburg. He was vital to both Virginia and

FIGURE 8-7. William Byrd II. Mark Catesby spent at least a month with William Byrd, and Byrd noted some of Catesby's activities in his *Secret diary*. (Courtesy of Colonial Williamsburg Foundation.)

FIGURE 8-8. William Byrd's Westover plantation house was built of wood rather than brick when Catesby visited. Although the Westover of 1712 was large for Virginia at the time, it likely had only four rooms. (Courtesy of Colonial Williamsburg Foundation.)

England, serving as a member of the King's Council for thirty-seven years, three long stints as Virginia's official agent in London, as a representative in the House of Burgesses, and on Virginia's Council of State. Sometime after he returned to Virginia to manage the enormous properties he inherited when his father died in 1705, he planned the future cities of Richmond and Petersburg on lands he owned. At his seat, Westover, he built the largest library in colonial America, with more than three thousand volumes, and he engaged in agricultural experimentation. Years later, back home in Virginia, Byrd kept a journal, which has survived. Entries reveal that Cocke and Catesby went to visit Byrd at his estate, Westover, only a week after Catesby arrived in Virginia. This was a deliberate visit. Westover (figure 8-8) was thirty-two miles from Williamsburg, a trip that took between four and nine hours depending on whether the horse was trotting or walking.

Although the Westover of 1712 was not as elaborate as the Westover that visitors can tour today, it was much fancier than the house that Catesby stayed in at Williamsburg. The original Westover was likely a frame house, not a brick one, and had leaded casement windows. A detached library was constructed between 1709 and 1712.[12]

Byrd's journal indicates clearly that he enjoyed showing Catesby the local flora and fauna. Byrd had a lifelong interest in Virginia's natural history and how native plants and animals could be made productive. He had been writing to Hans Sloane since at least 1706 about Virginian plants. When Catesby, Cocke, Cocke's wife, Elizabeth, and their daughter arrived at Westover on 24 May 1712, Byrd welcomed them with glasses of Canary wine and "some cakes to stay their stomachs." After dinner (at three o'clock in the afternoon), "the daughter, Mr. Catesby, and I went in to the swamp to see the nest of a humming bird and the Doctor followed along. However we found a nest with one young and one egg in it." Perhaps it was this walk that sparked Catesby's interest in hummingbirds and led to his sketch of the ruby-throated hummingbird with the trumpetcreeper (figure 8-9).[13] Over the course of Catesby's one-month visit, Byrd took him to church, invited him to dine with three ship masters, and routinely drank a bottle of wine with him after "the women went upstairs by themselves." They regularly "walked in the garden." Dr. Cocke had to leave Westover after four days, but Catesby and Mrs. Cocke were able to remain until mid-June.[14]

FIGURE 8-9. (*Opposite*) Ruby-throated hummingbird and trumpetcreeper; plate 65, M. Catesby, 1731, *The natural history of Carolina . . .* , volume I. When Catesby arrived at Wes tover, Byrd first gave him cakes and wine and then took him to see a hummingbird nest. (Digital realization of original etchings by Lucie Hey and Nigel Frith, DRPG England; courtesy of the Royal Society ©.)

When Catesby walked the lands at Westover, he would have seen that Byrd followed the French style of gardening, with topiary. He advised Byrd on his gardens. Byrd recorded that "Mr. Catesby directed how I should mend my garden and put it into a better fashion than it is at present." Catesby may have recommended that Byrd plant a new garden adjacent to the riverbank, as Byrd would do in 1720. And perhaps Catesby suggested enclosing the gardens with hedges or brick walls, placing garden seats among the plants, and growing small orange trees in a little greenhouse, as Byrd did by 1736.[15]

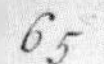

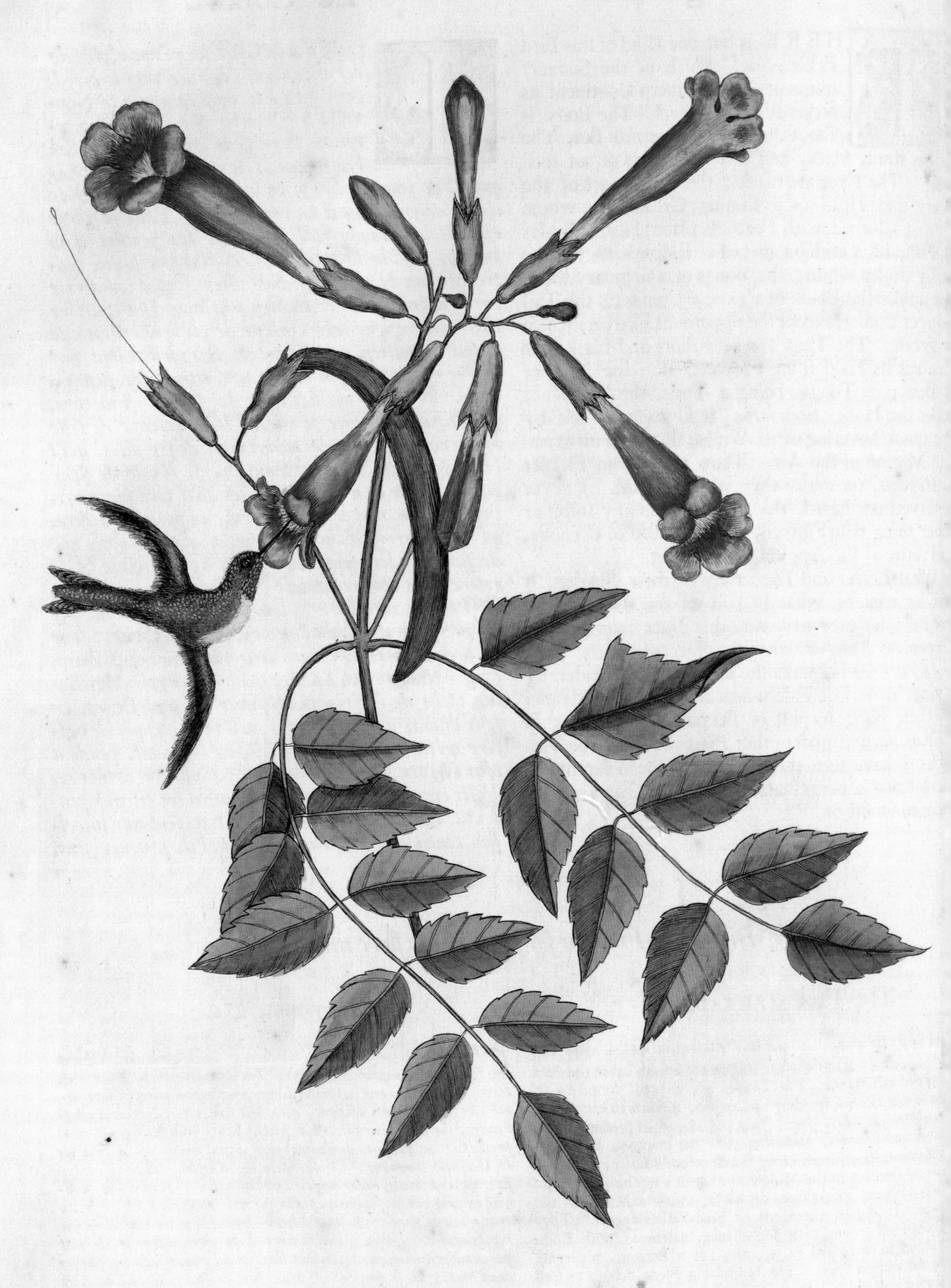
Mellivora &c.
Bignonia &c.

On 9 June, Byrd and Catesby traveled to Williamsburg and dined with Spotswood and Cocke. After dinner, Cocke was sworn in as Secretary of the colony. Byrd and Catesby must have enjoyed their time together, because Catesby returned with Byrd to Westover when he could have returned more easily to his sister's house in Williamsburg. The next night the two men drank so much "and were so merry that Mr. Catesby sang." It appears that Catesby left Westover on the evening of 15 June, although Byrd's diary is vague on this detail.[16]

Catesby returned to Westover three months later. When he was there, Byrd received a letter stating that Governor Spotswood expected Byrd to attend him at "Pamunkey Town." It appears that Spotswood went to the Indian town because a faction of Chickahominies claimed part of a three-thousand-acre reservation that had been sold to the Pamunkey. Byrd recorded that at the Indian town the rain was so heavy they "were forced to stay in one of the Indian cabins all day." One Indian man made an impression on Byrd, if not also on Catesby. Byrd recorded meeting "an Indian . . . who has now his 20 wives." These were not the only Indians Catesby saw during his time in Virginia, though it may have been the only time he saw an Indian village there. During Williamsburg's early years, twenty Indian boys attended the Indian school at the College of William and Mary and lived in various homes in Williamsburg. The colonists hoped to teach the boys the English language, religion, and way of life. They were not successful for the most part. Indians only sent the boys to school because doing so meant that they were released from paying tribute money to the governor. As soon as they could, the boys returned to their native villages.[17]

Catesby once again returned with Byrd to Westover after his visit to Pamunkey town. He continued to explore the flora and fauna at Westover, where he "killed two snakes in the pasture." A few days later, he killed a larger creature: Byrd was in his library at night when "Mr. Catesby came and told me he had seen a bear. I . . . went with a gun and Mr. Catesby shot him. It was only a cub and he sat on a tree to eat grapes. I was better with this diversion and we were merry in the evening."[18]

FIGURE 8-10. (*Opposite*) Southern flying squirrel; plate 77, M. Catesby, 1739, *The natural history of Carolina . . .* , volume II. William Byrd and Mark Catesby must have discussed Virginia's flying squirrel, a creature that fascinated Byrd and many other visitors to the region. (Digital realization of original etchings by Lucie Hey and Nigel Frith, DRPG England; courtesy of the Royal Society ©.)

Although Byrd did not record showing Catesby a flying squirrel (figure 8-10), in all likelihood the two men discussed the creature.[19] Flying squirrels fascinated Byrd. He later wrote in his *Natural history of Virginia*, published in 1737, that the flying squirrel "has no wings to fly with, but rather a tender, hairy little skin. . . . It [may] become very tame," though he did not recommend taming such animals.[20] Many visitors to Virginia were intrigued by the flying squirrels. Traveler William Hugh Grove recorded in his journal in 1732 that "Flying Squirrells are of the Colour of a Common Squirrel and differ in two Wings and the Tayl, All broad and thin films covered with hair Like other parts of their Body. [They] fly from tree to tree, but not farr."[21] Visitor and botanist Pehr Kalm described the flying squirrels in 1750, writing that "By the additional skin with which Providence has provided them on both

T. 77.
Sciurus
Viscum

sides, they can fly from one tree to another. They expand their skins like wings, and contract them again as soon as they can get hold of the opposite tree." Kalm further noted that the flying squirrels "were easily tamed. The boys carry them to school, or wherever they go, without their ever attempting to escape; if even they put their squirrel aside, it leaps upon them again immediately, creeps either into their bosom, or their sleeve, or any fold of the clothes, and lies down to sleep."[22]

Later in his life, Catesby criticized himself for not accomplishing more during his visit. "In the Seven Years I resided in that Country, (I am ashamed to own it)," he reflected, "I chiefly gratified my Inclination in observing and admiring the various Productions of those Countries."[23] He did make paintings of birds, which Samuel Dale used to argue on Catesby's behalf for support for Catesby's work. It is likely that Catesby was also kept busy helping his sister and his brother-in-law. William Cocke left Williamsburg and traveled to England on business for the colony of Virginia in 1716. Cocke stayed in England until 1718. It was typical to ask for a male relative's or a friend's assistance in such situations. Cocke may well have requested his brother-in-law's help with maintaining his affairs in Williamsburg. William and Elizabeth Cocke had seven children by 1707 and owned eight lots in Williamsburg, forty acres of woodland, and other holdings. In addition, Cocke was a member of the colony's Indian trading company. It is reasonable to conclude that Cocke would have wanted someone with some personal interest in his family assisting them in his absence.

Catesby and Byrd remained friends for the rest of their lives. In 1736, Byrd asked Captain Thomas Posford, a mutual acquaintance, to check on Catesby. "I wish you would be so kind as to call upon my freind [*sic*] Mr. Catsby [*sic*] now and then," Byrd requested, "to know if he have any letter or commands for me. He is such a philosopher, that he needs a monitor to put him in mind of his freinds [*sic*]." The following year, Byrd penned a lengthy letter to Catesby, discussing Virginia grapes, ginseng, and snakes. Byrd also wrote to Peter Collinson expressing the hope that Catesby would draw the images for a book Byrd intended to publish. "I intend this next winter to cover this dry [skele]ton [journal], and make it appear more to advantage," Byrd wrote. "And as I shall occa[sional]ly mention several plants and animals, I should be oblig'd to my [frie]nd Mr. Catesby if he'll be so good as to add the figures of [them]." There is no record, however, that Catesby did so.[24]

Catesby's *The natural history of Carolina* . . . had more of an impact in Virginia than he knew. More than thirty years after Catesby's death, Thomas Jefferson complimented Catesby's work as the only "complete, reliable, illustrated natural history of America." Jefferson used Catesby's studies to compile his own table of common North American birds in *Notes on the State of Virginia*. Finally, Meriwether Lewis and William Clark consulted Catesby's *The natural history of Carolina* . . . before heading west on their own explorations.[25]

9

Mark Catesby's Carolina adventure

SUZANNE LINDER HURLEY

Mark Catesby was primarily an artist and a naturalist, but he was also an adventurer, a man who was willing to leave his comfortable and secure home to face courageously the hardships of a transatlantic voyage and the dangers one might encounter on the frontier of colonial South Carolina. The ocean journey took about three months, and for the most part, passengers had to provide their own food and drink. With no refrigeration or knowledge of bacteria, sanitary conditions were deplorable. An unexpected hurricane could cause shipwreck, or pirates could capture the vessel and execute the passengers and crew.[1]

The hazardous nature of the transatlantic trip is illustrated by the Anglican minister of Christ Church Parish in South Carolina, who sent his reports to the Society for the Propagation of the Gospel in six copies, each to travel on a different ship in the hopes that one report would arrive safely. With only sails for power, a ship could easily be blown off course. The ship on which Catesby sailed was approaching Charleston when a strong wind forced it back into the Gulf Stream and extended his journey by about two weeks.[2]

As Catesby sailed into Charleston harbor (figure 9-1), he was probably anticipating the interesting discoveries he would make and the beautiful subjects he would paint, but he was also aware of diseases such as smallpox, malaria, and yellow fever, which exacted a high death toll. He had some experience of the health hazards not long after he arrived, having a serious cheek infection that kept him incapacitated for three months.[3]

The motivating factors must have been strong indeed to influence an English gentleman of somewhat privileged circumstances to forgo the comforts of home and travel to the frontier of the British Empire, where the colonists were still fighting Native Americans and the danger of slave revolt was ever present. However, he had the benefit of having the new governor, Francis Nicholson, provide invaluable introductions to some of the most influential families in the Province of South Carolina.

FIGURE 9-1.
Charleston harbor.
(Courtesy of Carolina Antique Maps & Prints, Charleston, South Carolina.)

Interest in the flora and fauna of Carolina was well established in the early eighteenth century, especially among members of the Royal Society of London, who encouraged the exchange of information and specimens of plants and animals. For example, James Petiver received samples from South Carolina sent by Edmund Bohun, Robert Ellis, and George Francklin.[4]

Another correspondent was Mrs. Hannah English Williams (d. 1722), who owned land at Stony Point on the Ashley River between Charleston and Dorchester.[5] She was the first documented woman collector of the fauna and flora of the British colonies in North America. Petiver named several butterflies for Williams, one of which was "Williams Selvedge eyed *Carolina* Butterfly," now known as the Creole pearly-eye (figure 9-2). Petiver wrote to Williams, "It was with noe small pleasure [that] I received ye Collection of Butterflies . . . some amongst [them] wch I never saw before. . . ."[6]

Petiver received an especially enthusiastic response from the Reverend Joseph Lord (1672–1748), the Harvard-educated pastor of the dissenting church at Dorchester, a village about twenty miles from Charleston. Because he hoped to receive medicines or medicinal herbs to share with his congregation, Lord sent hundreds of plants to Petiver, accompanied by information as to habitat, characteristics, and dates of collection. The links forged by Petiver held together communications between naturalists in Carolina and in London.[7]

William Sherard, whom Catesby called "one of the most celebrated Botanists of the Age," was another enthusiast, keen to get botanical specimens, so he

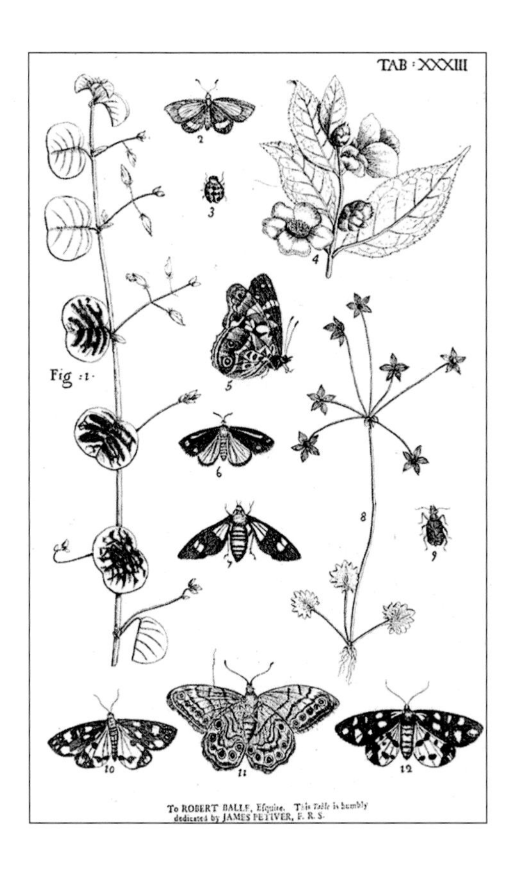

FIGURE 9-2. "Williams Selvedge eyed *Carolina* Butterfly," today called the Creole pearly-eye; plate XXXIII, figure 11 (bottom row, center), J. Petiver, 1702, *Gazophylacium*, volume 1, decade 4.

discussed with his colleagues the possibility of sending a naturalist to Carolina.[8] After seeing some paintings of birds that Catesby had done in Virginia during his first visit to North America, Sherard and his associates became Catesby's sponsors. They received encouragement from Francis Nicholson (figure 9-3), an experienced civil servant who had served as Governor of Virginia, Maryland, and Nova Scotia before becoming Governor of South Carolina in 1721. He offered a stipend of £20 per year to Catesby while he worked in the province.[9]

However, Nicholson came into a potentially difficult situation. In 1663, King Charles II had given the Province of Carolina to eight gentlemen, each of whom had some claim on his favor, and these Lords Proprietors originally exercised complete control over the province. They envisioned a system of landholding similar to feudalism, dominated by nobles known as Landgraves and Caciques. The system soon became obsolete because there was so much land available, and settlers wanted to own their land.[10]

The colonists complained that the Lords Proprietors could not provide security from attacks by the Spanish or Native Americans. The Yamassee War of 1715–1716 was ample evidence that the colonists were right. They believed the Spanish in St. Augustine had instigated the war, joining with the Indian nations against South Carolina in 1715.[11] The Yamassee and their allies massacred several hundred colonists, and the area between the Edisto and Savannah Rivers was left practically bereft of settlers. Half the cultivated land of the colony lay dormant while skirmishes with Native Americans continued until

FIGURE 9-3. Governor Francis Nicholson. (Photograph of portrait by Michael Dahl MSA SC 1621-1-590; courtesy of Maryland State Archives.)

1727. The defeat of the Yamassee was not the end of the conflict, because, except for the Cherokee and the Chickasaw, all of the Indian nations of the Southeast were united behind the leadership of the Lower Creek. The confederation of Indian nations was the greatest Indian alliance in colonial history. It had the potential to wipe out the colonies of South and North Carolina and Virginia. The fact that the Cherokee Nation stood by the settlers probably turned the tide in their favor.[12]

When Mark Catesby arrived in Charleston he found a city of tremendous ethnic diversity. In addition to English, Scots, Irish, French Huguenots, and other Europeans, there were Native Americans and Africans, who outnumbered all other groups. The population of the province in 1720 was about seven thousand whites and twelve thousand blacks.[13] The Africans came from a wide range of ethnic groups, which resulted in a multiplicity of languages and customs. Africans brought with them vegetables such as okra and yams, as well as the knowledge and skill of growing rice, which became a leading export from South Carolina.[14]

Many Native Americans, mostly women and young people, were also enslaved. Colonists had found by experience that it was not advisable to try to enslave mature males. The Commissioners of the Indian Trade issued an order to the factor at the Savannah town that he should not buy any male slave above the age of fourteen.[15] Colonists were acutely aware of the danger of slave insurrection. In May 1720, the Primus Plot, which centered in the vicinity of the upper Ashley River, had narrowly been averted due to an informant. The Board of Trade reported to King George II that the rebels had nearly succeeded, and if that had happened, the result would have been "the utter extirpation of all Your Majesty's subjects in this province."[16]

Catesby's impressions of Charleston may have been much like those of Margaret Kennett, a young woman who in 1725 wrote to her mother in England to tell her about the town. She said the houses in Charleston were of brick, "lime and hair" (meaning stucco), and timber. Most had sashed and glazed windows, and at least fifty houses were bigger than that of a prominent neighbor of her parents in England. Margaret had a shop on the bay, where she sold rum, sugar, and other products on the ground floor and lived above. She explained that she was not afraid of Native Americans coming to town, because "they have no occasion for there is Men whose Business it is to goe up in the Indian Countery 2 or 300 Miles, and Trade with them." Margaret had to rise early in the morning to greet the country people arriving in their periaguas.[17]

A periagua (from the Carib word, adopted by the Spanish, for dugout canoe) was an extended cypress canoe (figure 9-4). Periaguas were the preferred transportation of the upper class, for there were few roads outside of Charleston. South Carolina Lowcountry rivers were tidal, and if boatmen could "catch the tide" and go with the flow, their progress was both faster and easier. An affluent family might own a luxurious periagua rowed by four servants in livery and equipped with a small cabin, awnings for protection from sun and rain, and comfortable seats. Settlers had learned the technique of hollowing out a cypress log using fire and gouging from Native Americans. A hollowed log by itself was suitable for a canoe, but Europeans devised the technique of splitting the log and adding planks to widen the bottom of the boat, and sometimes they also built up the sides. Periaguas usually carried a removable mast for sailing in bays or for brief forays into open ocean. They had a shallow draft but could

FIGURE 9-4. Sketch of periagua by Philip Georg Fredrich von Reck, 1736. (Courtesy of the National Archive, United Kingdom.)

carry thirty or forty barrels of rice. Because individuals working independently constructed boats using easily available materials, the types of boats classified broadly as periaguas differed considerably in size and accouterments.[18]

The provincial legislature authorized the use of periaguas as scout boats to patrol the inland passage between Charleston and Saint Augustine to keep this route safe from Spaniards or hostile Native Americans. Catesby had an opportunity to travel by periagua and to accompany Indian traders into the backcountry, but to start with, he visited gardeners in Charleston and on plantations in the surrounding area. While in Charleston, he often stayed with a "Doctor of Phisik"—probably Dr. Thomas Cooper, a physician who had studied at Oxford and had come to the attention of Hans Sloane because of his experiments in treating snakebite. It was likely in Cooper's garden that Catesby planted the bulbs and roots that William Sherard had sent from England. At one point, Catesby and Cooper planned a trip to Mexico and sought a passport or letter of protection from Spain, but the trip never materialized.[19]

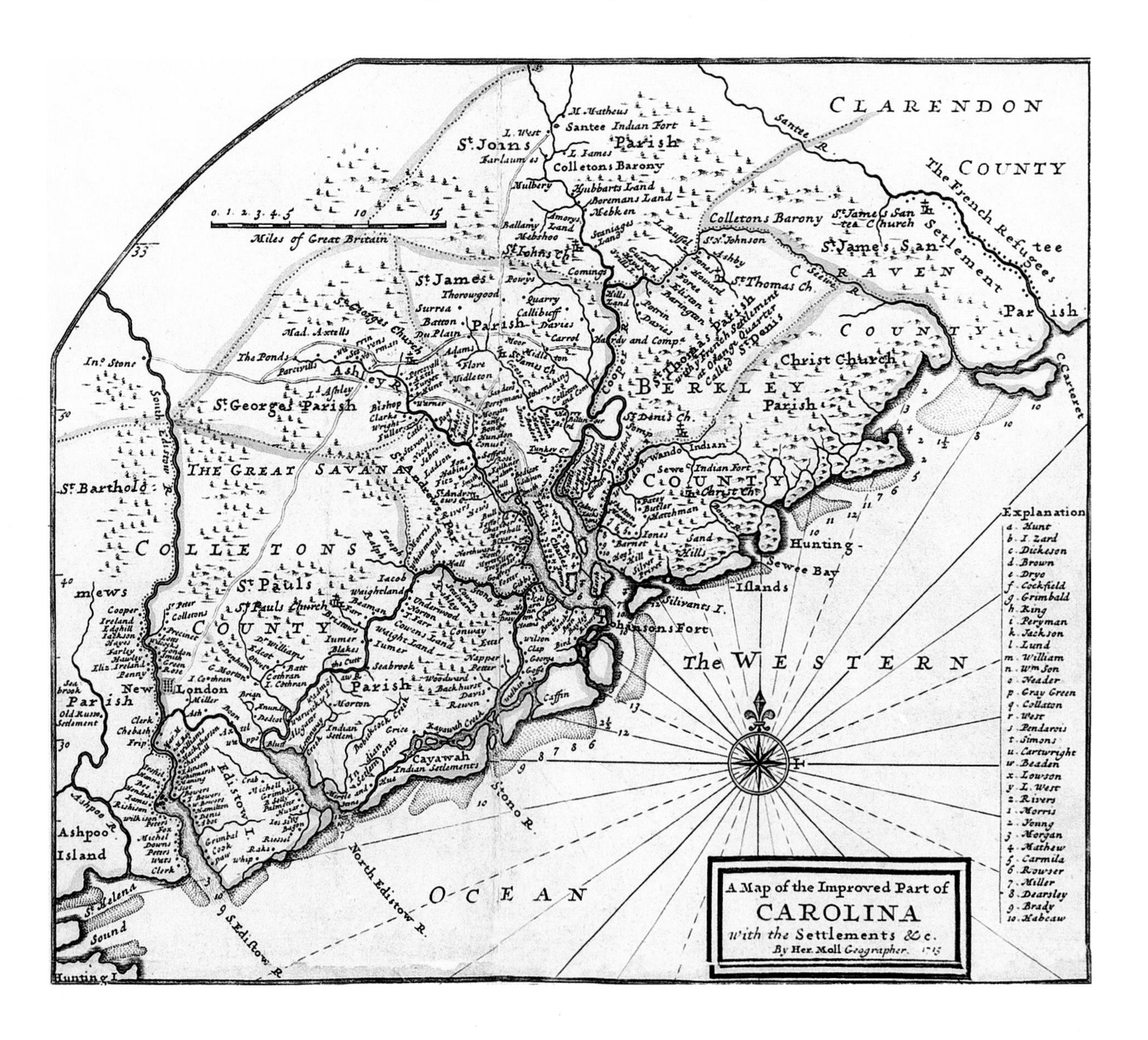

FIGURE 9-5.
Map of the improved part of Carolina with the settlements &c., 1715, by Herman Moll. (© British Library Board. [Maps 148.3.1]. DRPG.)

According to Catesby, traveling beyond the settled portion of the province involved two main dangers: the danger of getting lost in the wilderness, and the danger of attack by Native Americans.[20] The Yamassee War was ample evidence of the latter, but the war involved much more than loss of life. Many great fortunes in South Carolina were based on trading with Native Americans for skins and Indian slaves. Almost immediately after the assaults of 1715–1716, the colonists sought to re-establish the deerskin trade. The legislature set up a commission to administer a government monopoly of the Indian trade in an effort to avoid the abuse of the Indians by traders, which was a contributing cause of the war. The Commons House of Assembly closely controlled the actions of the Commission. At first, the Commissioners tried to limit trade to three trading posts, but both Indian nations and traders opposed this rule, as well as price fixing by the Commission. The Lords Proprietors repealed the act creating a government monopoly in 1718. After the overthrow of Proprietary government, Governor Francis Nicholson took over the role of controlling the traders and met delegations of Cherokee, Creek, and Chickasaw chiefs in Charleston. By October 1723, Nicholson acknowledged that it was very difficult to manage the trade. The legislature then set up a program with a sole commissioner in charge.[21]

They chose Colonel James Moore Jr., Speaker of the Commons House, for the position. Moore had commanded the army of the colonists, which defeated the Tuscarora in North Carolina and later the Yamassee. He was a very experienced Indian trader and Indian fighter. In 1716, the colonists built several forts for defense, one of which was located on the Carolina side of the Savannah River opposite the site of the present city of Augusta, Georgia. Historians believe that the fort was named in honor of James Moore. He had been an outstanding person among the "Goose Creek Men," leaders in the anti-Proprietary movement, many of whom came from Barbados. They paved the way for a royal governor, and Moore served as provisional governor until Nicholson arrived.[22]

Catesby visited Boochawee Plantation, the Moore home in St. James's Parish, Goose Creek, which adjoined St. George's Parish, Dorchester.[23] James Moore Sr., father of the provisional governor, had been heavily engaged in the Indian trade in the late seventeenth century. A principal Native American trading path passed his front door, and the elder Moore developed a lucrative business, first by trading directly with the Indians and later by financing traders who traveled deep into the interior. Carolina traders traveled at least as far west as the Mississippi River, and in 1690 James Moore Sr. explored six hundred miles into the interior. Moore purchased laden pack trains and chartered sea captains to carry skins to Europe, where they were converted into high-priced purses, vests, jackets, gloves, and other accessories. James Moore Sr. died in 1706 of yellow fever, so his son, Colonel James Moore Jr., would have met Mark Catesby. Boochawee Plantation boasted a substantial two-story brick house with numerous outbuildings, including those accommodating both African and Native American slaves. The grounds included orchards and ornamental gardens with ponds, terraces, walkways, and a variety of plants.[24] Catesby said that Colonel Moore told him that he had seen an Indian daub himself with the juice of the "Purple Bind-weed" (probably a species of *Ipomoea*) and then handle a rattlesnake without harm.[25]

Governor Nicholson's sponsorship provided Catesby with recommendations to other outstanding colonists such as Thomas Waring, William Bull, Joseph Blake, and Alexander Skene. Thomas Waring had served on the Royal Council of the Province when Colonel Moore was serving as governor. In addition, Waring's daughter Sarah was married to James Moore, son of the governor.[26]

Judging by the people and plantations Catesby mentioned, his explorations in the Lowcountry centered around Dorchester on the upper Ashley River and nearby Goose Creek. One reason for this might have been that a number of leaders of the opposition to the Lords Proprietors lived in this vicinity. They would have been among the first to welcome Francis Nicholson as governor and to entertain his protégé. Catesby mentioned that the only place he had found the "Acacia" (water locust) (figure 9-6) tree was on the plantation called Pine Hill of Thomas Waring on Ashley River near Dorchester.[27] Catesby also noted that he found "Dahoon holly" only on Colonel William Bull's plantation, Ashley Hall.[28] With Waring, Bull, and other plantation owners, Catesby was able to share European plants and also plants he had found in North America in return for gathering specimens on their plantations.[29]

In January 1723, Catesby spent time with Alexander Skene, who lived just across the river from the village of Dorchester and who later became a subscriber to Catesby's *The natural history of Carolina. . . .* Skene was a member of the Royal Council and a conscientious Anglican who offered religious instruction to his approximately one hundred slaves. He came to South Carolina from Barbados, and his unpleasant experience with proprietary government there and in Carolina made Skene a strong supporter of royal government. When Catesby was visiting Skene, they probably attended St. George's Parish Church in Dorchester, for it was required by law to attend the church of one's choice or pay a fine of five shillings. Skene, Thomas Waring, and others had been on the commission to build the impressive brick building.[30]

FIGURE 9-6. (*Opposite*) American goldfinch and water locust; plate 43, M. Catesby, 1730, *The natural history of Carolina . . .* , volume I. (Digital realization of original etchings by Lucie Hey and Nigel Frith, DRPG England; courtesy of the Royal Society ©.)

Due to its location at the head of the Ashley River, Dorchester was flourishing in the 1720s; its population comprised about five hundred whites and thirteen hundred slaves. In 1723, the legislature approved an official fair and market for the town. As a general rule, markets opened twice a week, and fairs operated for four days in May and September. Itinerant merchants might bring luxury goods not generally available, and sporting events such as shooting contests, bear baiting, horse racing, chasing a soaped pig, and pulling the head off a goose at full gallop provided an opportunity for diversion.[31]

T. 43
Carduelis Americanus.
The American Goldfinch.
Acacia abruæ folijs &c.

FIGURE 9-7. Timber rattlesnake; plate 41, M. Catesby, 1730, *The natural history of Carolina . . .* , volume I. (Digital realization of original etchings by Lucie Hey and Nigel Frith, DRPG England; courtesy of the Royal Society ©.)

Catesby had an opportunity to visit Newington, home of Lady Elizabeth Axtell Blake, widow of Governor Joseph Blake, who had been a Landgrave and also one of the Lords Proprietors. Their son, Joseph Blake, inherited the title of Landgrave and a Proprietary interest when his father died, and he received Newington at his mother's death in 1726. He would go on to make Newington a showplace of South Carolina—a huge brick mansion called "the house of a hundred windows" approached by a double avenue of live oaks (*Quercus virginiana*)—and when the Crown bought out the Lords Proprietors in 1729, he became one of the wealthiest men in South Carolina.[32]

Despite a few elegant plantations, the upper Ashley region was still a frontier. Catesby became acutely aware of this one morning when he was having tea. In the next room, the housemaid who was making his bed found a rattlesnake (figure 9-7) between the sheets where Catesby had slept. He said it was "vigorous and full of ire, biting at every thing that approach't him. Probably it crept in for warmth in the Night, but how long I had the company of [the] charming Bedfellow I am not able to say."[33] Catesby did not seem to be overly concerned, for rattlesnakes were simply native to South Carolina's subtropical coastal vegetation.

After spending some time in the vicinity of Charleston, Catesby wanted to visit the backcountry to examine a different landscape. He wrote to Sherard:

> My method is never to be twice at . . . one place in the same season. For if in the sp[ring] I am in the low Country, in the Summer I am at the heads of Rivers, and the Next Summer in the low Countrys, so alternately that in 2 years [I vis]it the two different parts of the Country.[34]

James Moore would have been the ideal person to arrange for Mark Catesby to visit Fort Moore in the backcountry. Moore and other wealthy Indian traders organized packhorse trains, utilizing from six to twenty horses (figure 9-8). Fort Moore (figure 9-9) was a trading post with storehouses and supplies. Native Americans or licensed traders could bring their goods to be shipped to Charleston. In return they received such things as guns, gunpowder and shot, iron pots and tools, hatchets, vermilion (red) paint, rum diluted with water, and calico. The journey to Fort Moore through the region hardest hit by the Yamassee War could have been extremely dangerous, and the packhorse men were known for their rough habits. A 140-mile journey on foot or horseback through largely uninhabited territory would have required both courage and endurance, and Catesby would have needed an experienced guide.[35]

James Moore's connections with the Indian trade could have enabled Catesby to negotiate with traders to guide him into the backcountry or to bring him specimens from the mountains. Catesby's relations with Native Americans were generally pleasant. While at Fort Moore, he hired an Indian to carry his box, and he wrote, "To the Hospitality and Assistance of these Friendly Indians, I am much indebted, for I not only subsisted on what they shot, but

FIGURE 9-8. Trader and packhorses by Jo Rissanen. (© Suzanne Linder Hurley.)

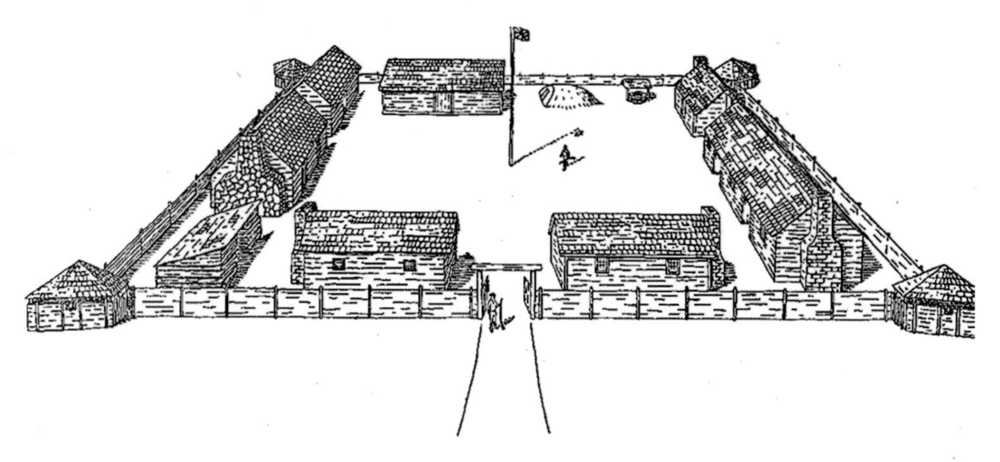

FIGURE 9-9. Plan of Fort Moore, 1724, by Larry E. Ivers. (Reproduced by permission of Larry Ivers.)

their First Care was to erect a Bark Hut, at the Approach of Rain to keep me and my Cargo from Wet."[36]

Fort Moore was on the South Carolina side of the Savannah River opposite the site of the modern city of Augusta, Georgia. Its purposes were to guard the nearby Savannah Town, an Indian village and trading center, and to protect the Savannah Path, the principal route from the Creek Nations to Charleston. Buildings inside the fort consisted of barracks, officers' houses, a guardhouse, a magazine, storehouses, a corncrib, and possibly shelter for horses. As many as forty South Carolina provincial soldiers garrisoned the fort.[37]

Despite the crude nature of the buildings, the surrounding countryside was beautiful. Catesby wrote: "It is one of the Sweetest Countrys I ever saw. The Banks of the River, perticularly where the fort stands, is 200 foot perpendicular in most places, from whence are seen large Prospects over the tops of the trees on the other side of the river."[38] Catesby and some of his companions ventured around twenty miles beyond Fort Moore on the Savannah River where the cataracts begin. They encountered a run of sturgeon and in three days killed sixteen, some of which they ate on the spot. They brought two back to the garrison, but the others, Catesby said, "to my Regret, were left rotting on the shore."[39]

Some authors have questioned whether Catesby ever reached the far western part of the province. Analysis of the plant specimens associated with Catesby—preserved in the Sloane Herbarium in the Natural History Museum, London, and linked with the species' habitat preferences and present-day distribution in South Carolina—provides indications of his wider travels. For example, a specimen of Fraser's yellow loosestrife (*Lysimachia fraseri*), a perennial herb that is only found today in the vicinity of Clemson and Keowee, about 110 miles northwest of Fort Moore, suggests that Catesby traveled farther than previously estimated and reached the Southern Blue Ridge Escarpment.[40]

From Charleston, it was possible to go by periagua through the inland passage to the mouth of the Savannah River, then up the river to Fort Moore and the adjacent Savannah Town, which was the focus of all the trails to the west and the entrepôt of the whole inland trade with Indian nations (figure 9-10).[41] Since Catesby wanted to collect specimens on his trip, it would be logical for him to travel overland on his way to Fort Moore. Burdened by his box of art supplies and a large volume of samples, he would probably have welcomed a chance to return to Charleston by water using a periagua to descend the Savannah River to the coast. Catesby's likely return route by river and the inland passage, and his visit to the North Edisto River, where he observed a "grampus" carcass being overwhelmed by sand,[42] probably took him to Willtown. Later abandoned, the town was both a trading post and a base for defense against the Spanish from St. Augustine and hostile American Indians. Some of the water-borne defenders were Scottish prisoners of war held by the British after the Jacobite rising

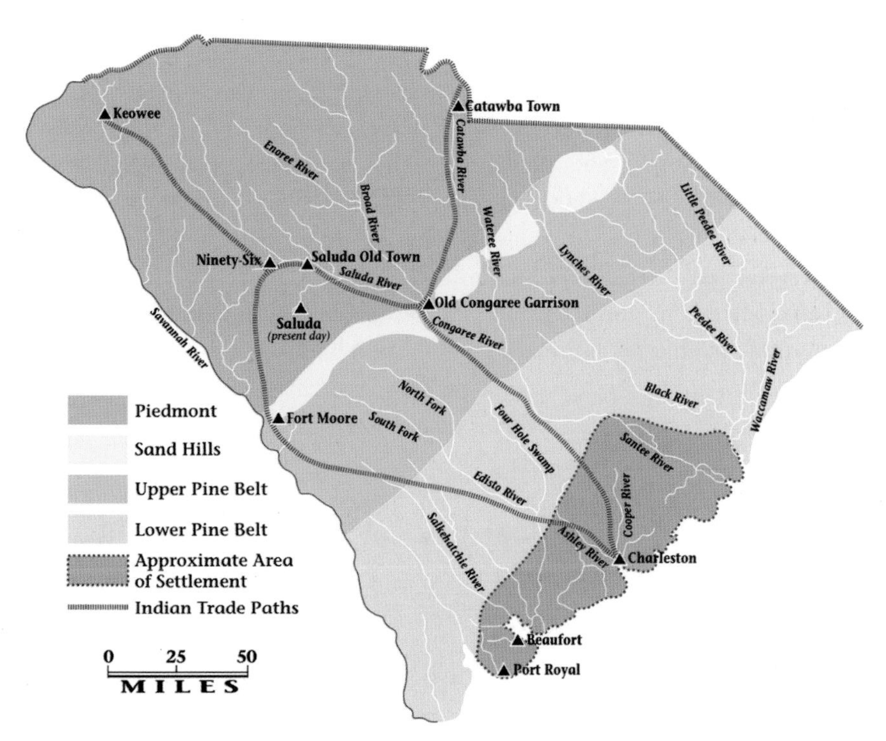

FIGURE 9-10. Map of South Carolina geographical regions and Indian trails. (Courtesy of the South Carolina Archives.)

of 1715. These prisoners were purchased by South Carolina for use as convict labor for seven years, but those distinguished by their valor and obedience were released after four years.[43]

Transporting his specimens safely was a big problem. Catesby had trouble getting brown paper and boxes to protect plants, but gourds were useful for seeds. He placed snakes and other small animals in jars filled with rum, but sometimes thirsty sailors drank the rum before the ship reached Britain. He also used rum for preserving birds, but more often, he dried them in an oven and sprinkled them with tobacco dust. It was difficult to transport enough samples or paintings to satisfy his sponsors. Catesby wrote to Sloane, "My Sending Collections of plants and especially Drawings to every [one] of my Subscribers is what I did not think would be expected from me." He explained that he needed to keep all of his drawings in order to have them engraved. Catesby's specimens have tremendous scientific value as a historic record of plants growing in the region in the 1720s.[44]

It appears that Catesby returned to the area around Fort Moore in the spring and summer of 1724.[45] Francis Varnod, the Anglican minister in Dorchester, was absent from his post for two Sundays in July 1724, which raises the question if the two men might have crossed paths. The Society for the Propagation of the Gospel had instructed Varnod to try to convert Native Americans to Christianity. He preached at the Savannah Garrison (Fort Moore), where he said that "no minister had been seen before." Varnod and Catesby were able to observe Native Americans in their natural environment.[46] This included the following adventure for Catesby:

> Some *Chigasaws*, a Nation of *Indians* inhabiting near the *Mississipi* River, being at Variance with the *French*, seated themselves under the Protection of the *English* near Fort *Moor* on *Savanna* River: With five of these *Indians* and three white Men we set out to hunt; after some Days Continuance with good Success, at our returning back, our *Indians* being loaded with Skins and Barbacued Buffello, we espied at a Distance a strange *Indian*, and at length more of them appeared following one another in the same Tract as their Manner is: Our five *Chigasaw Indians* perceiving these to be *Cherikee Indians* and their Enemies, being alarm'd, squatted, and hid themselves in the Bushes, while the rest of us rode up to the *Cherikees*, who then were increased to above twenty: After some Parley we took Leave of each other, they marching on towards their Country, and we homeward; in a short time we over took our *Chigasaws*, who had hid their Loads, and were painting their Faces, and tripping up every little Eminence, and preparing themselves against an Assault. Tho' the *Cherikees* were also our Friends, we were not altogether unapprehensive of Danger, so we separated from our *Indian* Companions, they shortening their Way by crossing Swamps and Rivers, while we with our Horses were necessitated to go further about, with much Difficulty and a long March, for Want of our *Indian* Guides. We arrived at the Fort before it was quite dark: About an Hour after, while we were recruiting our exhausted Spirits, we heard repeated Reports of Guns in the Woods, not far from us, by

> which we concluded that the *Cherikees* were come up with the *Chigasaws*, and that they were firing at each other: Nor were we undeceived, 'till the next Morning, when we were informed, that our *Indians* discharg'd their Guns for Joy that they were alive, and had escaped their enemies. But had they then known of a greater Escape, they would have had more Reason to rejoice; for the next Morning some Men of the Garison found hid in a close Cane-Swamp two large Canoas painted red: This discovered the bloody Attempt the *Cherikees* had been upon when we met them, who, with sixty Men in these Canoas, came down the River between two and three hundred Miles, to cut off the little Town of the *Chigasaws*; but from some little Incident being disheartened, and not daring to proceed, were returning back by Land when we met them.[47]

Catesby was a careful observer, and he was one of the first to describe the geographical regions of South Carolina. While the terms "outer coastal plain," "inner coastal plain," and "piedmont" (figure 9-10) did not come into common use until the twentieth century, Catesby described the area in terms approximating such a regionalization. Likewise, Catesby described different types of land that occurred in the coastal plain. By studying the abundance and geographical distribution of plants, a floristic approach, he differentiated rice land (freshwater swamp), oak and hickory land, pine-barren land, bay swamps, and shrubby oak land. This classification was a more comprehensive attempt to define the diversity of physical features on the coastal plain than anything that had been done earlier. Almost a century later, the state of South Carolina used a similar classification to appraise land for tax purposes.[48]

Mark Catesby's images of Carolina birds include several that are now extinct. The passenger pigeon (figure 6-5, p. 79), which once roosted in such numbers that they broke the limbs off trees, was extinct by 1914.[49] The Carolina parakeet (figure 9-11), extinct by 1918, was the only parrot endemic to North America.[50] The ornithologist Alexander Wilson witnessed behavior that probably contributed to their extinction. He said, "Having shot down a number, some of which were only wounded, the whole flock swept repeatedly around their prostrate companions, and again settled on a low tree. . . . [T]hough showers of them fell, yet the affection of the survivors seemed rather to increase." Although some were killed to protect crops, many more provided brightly colored feathers for the millinery trade.[51]

Perhaps the bird that most caught the attention of the general public is the ivory-billed woodpecker (figure 9-12).[52] This largest of all American woodpeckers was once a permanent resident of the forests of South Carolina. It could cover vast territory in search of wood-boring insects and grubs. According to Catesby, because its powerful bill could hew out a bushel of wood chips "in an hour or two's time," it gained the Spanish name of *carpenteros*. He also noted that

> The Bills of these Birds are much valued by the *Canadian Indians*, who make Coronets of 'em for their Princes and great warriers, by fixing them round a

FIGURE 9-11. Carolina parakeet and bald cypress; plate 11, M. Catesby, 1730, *The natural history of Carolina . . .* , volume I. (Digital realization of original etchings by Lucie Hey and Nigel Frith, DRPG England; courtesy of the Royal Society ©.)

Wreath, with their points outward. The *Northern Indians* having none of these Birds in their cold country, purchase them of the Southern People at the price of two, and sometimes three Buck-skins a Bill.[53]

Another bird described by Catesby was the "Rice Bird" (bobolink).[54] It appeared in South Carolina in May and September and severely damaged the rice crop. Catesby noted an instance in which the birds devoured so much of forty acres of rice that he doubted that harvesting what remained was worth the cost. However, the planters obtained some revenge, because they considered the birds a great delicacy. Moreover, their significance to Catesby and to science was only revealed when he moved on to the Bahamas.[55]

Mark Catesby's sojourn in Carolina was important because it gave him the opportunity to gain information both verbal and visual that would compose a large part of his published works. In South Carolina, he was able to collect samples for preservation and to paint live plants and animals. He said that he always painted plants while fresh and just gathered. He endeavored to paint animals, especially birds, from life so that he could observe their gestures. He also tried to paint birds with plants on which they fed or to which they had some relation. He described the geographical regions of South Carolina and

FIGURE 9-12. Ivory-billed woodpecker and willow oak; plate 16, M. Catesby, 1729, *The natural history of Carolina . . .* , volume I. (Digital realization of original etchings by Lucie Hey and Nigel Frith, DRPG England; courtesy of the Royal Society ©.)

classified the types of land in the coastal plain. His paintings of birds, now extinct, provide a beautiful record of ornithological detail. Mark Catesby's adventure contributed to a remarkable achievement with the publication of his *The natural history of Carolina, Florida and the Bahama islands*, a monumental work that would have been impossible without his visit to South Carolina.[56]

10

Mark Catesby's Bahamian natural history (observed in 1725–1726)

ROBERT ROBERTSON

Mark Catesby was in the Bahama Islands, east of Florida, for about the last year of his second stay in America.[1] He was the first European naturalist known to visit there, and almost one-third of the organisms he illustrated in his book are Bahamian. Some of these plates are among his best, both artistically and scientifically. However, Catesby scholars not familiar with these islands emphasize his mainland contributions.

The Bahamas are small, low, limestone rocks and islands, with the maximum height above sea level being 206 feet (62.8 meters) (Mount Alvernia, Cat Island). The number of them out of water varies with the height of the tide. The islands extend about 530 miles (850 kilometers) from northwest to southeast. Most of them are elongate, because they are solidified sand dunes. The limestone initially is white or cream-colored. It is soft and can be cut with a saw, but with weathering its surfaces harden and become gray and jagged (honeycomb rock). Only rarely are there fossil coral reefs. Particularly in the north, the islands are situated at the margins of large, shallow banks that were emergent during each of the four Pleistocene glaciations ("ice ages"), when world sea levels fell as much as 400 feet (about 120 meters), the last time only fifteen thousand years ago. Each time, the land area was vastly increased as plateaus. The bank margins are precipitous and descend into the deep blue sea (figure 10-1). There is a deep-sea embayment now called Tongue of the Ocean extending south between Andros and New Providence that Catesby sailed across. He could not have appreciated that there had been such geologically recent major changes in sea level or that there is this remarkable submarine topography, but he did mention "the unfathomable Abyss of the Ocean."[2]

Catesby described the emergent karst topography (figures 10-2, 10-3, and 10-4), with caves and some sinkholes connected underground to the sea and with their surfaces therefore rising and falling with the tides:

FIGURE 10-1.
The Bahamas from space. (Courtesy of the U.S. National Aeronautics and Space Administration.)

> Many of the Islands . . . abound with deep Caverns, containing salt Water at their Bottoms; these Pits being perpendicular from the Surface, their Mouths are frequently so choaked up, and obscured by the Fall of Trees and Rubbish, that great Caution is required to avoid falling into [them] . . . and it is thought, that many Men, which never returned from Hunting have perished in them. . . .[3]

The Bahamas are mainly tropical, the Tropic of Cancer passing through the center of the archipelago. They

> are blessed with a most serene Air, and are more healthy than most other Countries in the same Latitude. . . . [The islands] are void of noxious Exhalations, that . . . more luxuriant Soils are liable to. . . . many of the sickly Inhabitants of *Carolina* . . . retire to the [Bahamas] for the Recovery of their Health.[4]

(Catesby also alluded to "our unhealthy Sugar Islands.") In winter, especially the western islands nearest Florida are cooled by winds coming from the northwest: "At the Island of [New] Providence in December 1725, it was two Days so cold, that we were necessitated to make a Fire in the Governors Kitchen to warm us, yet no Frost or Snow ever appears there." During the rest of the year, the winds are easterly or southeasterly: "*August* and *September* are blowing Months, and are attended with Hurricanes, at which time the Winds are very

FIGURE 10-2.
Cat Island, the Bahamas. (Courtesy of the U.S. National Aeronautics and Space Administration.)

FIGURE 10-3.
Example of the Bahamas' karst topography. (© Robert Robertson.)

FIGURE 10-4.
Bahamas' beach karst topography. (© Robert Robertson.)

FIGURE 10-5.
Example of a tree growing out of fissures in rocks. (© Robert Robertson.)

changeable, shifting suddenly to all Points of the Compass."[5] (The average path of all western Atlantic hurricanes runs the length of the Bahamas. Catesby experienced one in South Carolina but not in the Bahamas.) It rains not often but violently. The northern Bahamian islands are wetter than the southern. Andros is low, wet, and full of bogs (mangroves in nearly fresh to brackish water). There are no freshwater rivers.

Catesby emphasized the paucity of soil and even shrub and tree trunks up to 4 feet (1.2 meters) in diameter growing out of rock fissures (figure 10-5): "Tho' the productive Soyl on these rocky Islands is small, the plantable Land, as it is here call'd, consists of three kinds, distinguish'd by their different Colours, as, the Black, the Red, and the White."[6] He went on to explain which plants are best suited to each of these. Crooked Island and Cat Island were reputed to have the best soils.

Catesby was in the northern Bahamas beginning in 1725, but there is little record of when or where he went; indeed, more is known about Christopher Columbus's travels in the Bahamas 233 years earlier.[7] Catesby was at sea "Midway between the *Azores* and the *Bahama*-Islands" on 20 April 1725 when the boat "was hoisted out, and a Loggerhead Turtle struck as it was sleeping on the Surface" of the ocean.[8] The earliest datable record of him in the Bahamas is in September 1725, when he was on a Native sailboat (sloop) at the island of Andros (figure 10-6). Then, as mentioned earlier, he was at New Providence in December 1725. Additionally, "For nine Months [he] observ'd a continual Succession of Flowers and Fruit" of the sevenyear-apple (*Casasia clusiifolia*) (figure 10-7), so presumably he was somewhere in the Bahamas for at least

FIGURE 10-6. Native Bahaman sloop, typical of that on which Catesby heard "Rice Birds" flying overhead at Andros Island, leading to his theory of bird migration. (© Robert Robertson.)

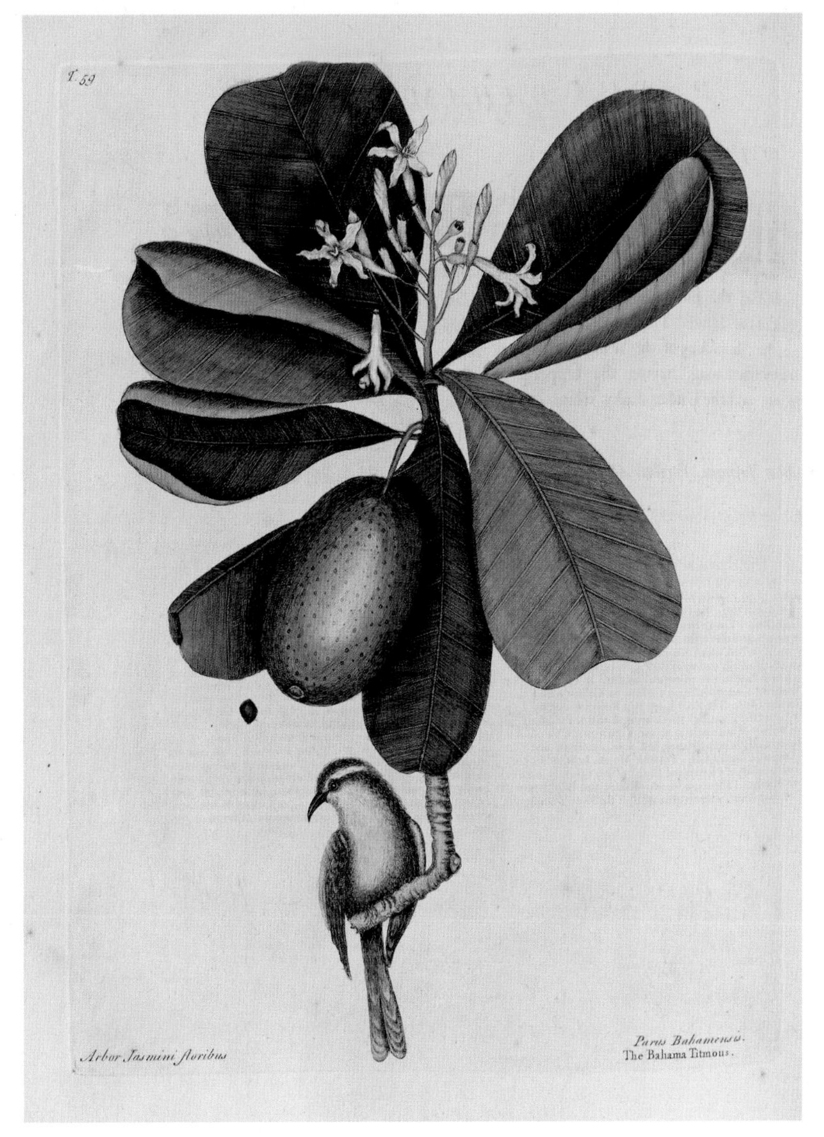

FIGURE 10-7. Sevenyear-apple; plate 59, M. Catesby, 1730, *The natural history of Carolina . . .*, volume I. (Digital realization of original etchings by Lucie Hey and Nigel Frith, DRPG England; courtesy of the Royal Society ©.)

that long.[9] Despite its name, the fruit ripens in seven or eight months. Catesby also mentioned "February" and "May" in the Bahamas without specifying the years. He collected algae at New Providence sometime in 1726, the same year he returned to England.[10] Unfortunately, no Catesby diary or letters from the Bahamas survive. Fortunately, his original paintings and pressed plants did somehow make it to England.

Catesby visited Andros, New Providence, Eleuthera, Harbour Island (a small island near Eleuthera), Abaco, and nearby islands. In 1725, there were somewhat fewer than three hundred humans on New Providence, three hundred more were said to inhabit Eleuthera, and another three hundred were at Harbour Island. The residents roved from island to island in their various pursuits. Additionally, there were about two hundred African slaves. The native Lucayan (or Taino) Indians had died out by the time Catesby arrived. According to him, in 1725 there were only about two hundred houses in the town of Nassau, then and now the Bahamian capital on the island of New Providence.[11] Most of these homes were made with plaited palmetto leaves (*Sabal palmetto*), but a few were stone-built. Catesby was relatively well housed because he was the guest of George Phenney, the second Royal Governor residing in Governor's House on a low hill overlooking the town. A fort guarded the western entrance to Nassau harbor. Pirates were the greatest threat. In 1720, the Spanish had attacked New Providence, but they were repulsed.

The main occupation in Catesby's time was farming maize or Indian corn (*Zea mays*), sweet potatoes (*Ipomoea batatas*), cassava or manioc (*Manihot esculenta*), yams (*Dioscorea* spp.), and melons. There were also fishing and hunting "guanas" (iguana lizards) and turtles. Wood was cut, ships were built, and salt was harvested and exported. The best salt ponds were at the Exuma Cays (pronounced "keys") and Crooked Island. Turtles and citrus fruits were also exported. Valuable lumps of ambergris, formed by sperm whales and used in perfumery, were once washed ashore more plentifully. Goats survived better in the climate than cattle, sheep, or horses. A cotton (*Gossypium hirsutum* var. *punctatum*) grows wild and was deemed of high quality. Bread was made from maize or wheat. Wheat flour was imported from the mainland North American colonies. Some maize, rum, and wine were imported from farther south, but water was the most general and useful of all "Liquors."

Catesby's natural history intentions were somewhat like those of the more famous John James Audubon (1785–1851) about one century later on the eastern North American mainland. Catesby has been called the "Colonial Audubon,"[12] but he was as much interested in flowering plants as he was in birds or other animals. Catesby often paired a plant and an animal in the same engraving. Sometimes his pairings reflect ecological associations, as, for example, birds with trees the fruits of which they eat, and he was perhaps the first to pair birds with ecologically relevant plants. Often, though, the pairings are incongruous, as, for example, the Bahamian manchineel tree (*Hippomane mancinella*) with smooth mistletoe (*Dendropemon purpureus*) and a pair of Spanish festoon butterflies from Cádiz.[13]

Catesby's information is nearly all we have on conditions in the Bahamas early in the eighteenth century and on some of their diverse plants and marine and land animals and their ecologies. He was particularly concerned with forest trees and shrubs because of their uses in building, carpentry, agriculture, food, and medicine. He was particularly interested in recording the uses of plant and animals, or the dangers to humans, even mentioning that the flesh of the American flamingo (*Phoenicopterus ruber*) "is delicate and nearest resembles that of a Partridge in Taste." He was not much of a conservationist and had to live off the land. Flamingos could easily be killed:

> A Man, by concealing himself from their Sight, may kill great Numbers of 'em, for they will not rise at the Report of a Gun, nor is the Sight of those killed close by them sufficient to terrify the Rest, and warn 'em of the Danger; but they stand gazing, and as it were astonish'd, till they are most or all of them kill'd.[14]

Altogether, Catesby published illustrations of about 410 identified species (excluding duplications of the same species): 246 were from the North American mainland, and 124 were from the Bahamas. Catesby also illustrated seven species from Jamaica and thirteen from miscellaneous other localities. Sixty-five of his Bahamian species were flowering plants, and all of these have now been identified.[15] Besides these plants, he figured about thirty Bahamian fishes, about fifteen birds, about five crustaceans, about four gorgonians, two sea turtles, a few insects, one gastropod shell, one lizard (an iguana), and one native mammal. Catesby illustrated more than fifteen bird species that occur in the Bahamas, but the others were not specifically mentioned from the islands; some are migrants. He recorded a few organisms from both the American mainland and the Bahamas.

West Indian mahogany trees (*Swietenia mahagoni*) grew

> to a great Height. . . . No one would imagine, that Trees of this Magnitude should grow on solid Rocks, and that these Rocks should afford sufficient Nutriment to raise and increase the Trunks of them to the Thickness of four Feet or more in Diameter; but so it is, and the Manner of their Rise and Progress I have observed as follows: the Seeds being winged are dispersed on the . . . Ground, some falling into the Chinks of the Rocks, and strike Root, if the Fibres find Resistance from the Hardness of the Rock, they creep out on the Surface of it, and seek another Chink, into which they creep, and swell to such a Size and Strength, that at length the Rock breaks, and is forced to admit of the Roots deeper Penetration, and with this little Nutriment the Tree increases to a stupendious Size in a few Years, it being a quick Grower.[16]

Catesby went on to describe and illustrate botanical details, including even a "Misleto" (*Phoradendron rubrum*) growing as a partial parasite on it. Ships were built of the mahogany wood, and it also had "Domestick Uses."

Used with caution, manchineel (*Hippomane mancinella*) was another large timber tree (figure 10-8), but it has a poisonous sap, which got in Catesby's eyes:

> The Wood . . . is close grained, very heavy and durable, beautifully shaded with dark, and lighter Streaks, for which it is in great Esteem for Tables and Cabinets . . . but the virulent and dangerous Properties of these Trees, causes a general Fear, or at least Caution, in felling them; this I was not sufficiently satisfied of, 'till assisting in the cutting down a Tree . . . on Andros Island, I paid for my Incredulity, some of the milky poisonous Juice spurting in my Eyes, I was two Days totally deprived of Sight, and my Eyes, and Face, much swelled, and felt a violent pricking Pain, the first twenty-four Hours, which from that Time abated gradually with the Swelling, and went off without any Application, or Remedy, none in that uninhabited Island being to be had. . . . Rain or Dew, falling from its Leaves on the naked Flesh, causes Blisters on the Skin, and even the Effluvia of it are so noxious as to affect the Senses of those which stand any Time under its Shade. . . . Guana's [iguanas] feed . . . on the Apples . . . without Harm to those who eat these same Guana's.[17]

A few observations like these appear not yet to have been confirmed.

The gum elemi (*Bursera simaruba*) is a tree with a brown, peeling bark. Its turpentine-like sap was used to heal wounds, especially of horses. Catesby found its "berries" in the "gizzard" of a red-legged thrush (*Turdus plumbeus*).[18]

Poisonwood (*Metopium toxiferum*) also has a poisonous sap, black when it oozes from the trunk, but it is not as virulent as manchineel. Two birds feed with impunity on its fruits, the white-crowned pigeon (*Patagioenas leucocephala*)[19] and the greater Antillean bullfinch (*Loxigilla violacea*).[20] The pigeon was one of the birds most harvested for human food, even after it had eaten poisonwood fruits: ". . . particularly while young . . . they [were] taken in great quantities from off the Rocks on which they breed."[21] The white-crowned pigeon was illustrated by Catesby on the same plate as coco plum (*Chrysobalanus icaco*), which is not eaten by the pigeon. To humans, "the Fruit . . . hath a sweet luscious Taste."[22]

Another bird frequently eaten was the yellow-crowned night-heron (*Nyctanassa violacea*):

> . . . in the Bahama-Islands, they breed in Bushes growing among the Rocks in prodigious Numbers, and are of great Use to the Inhabitants there; who, while these Birds are young, and before they can fly, employ themselves in taking them, for the Delicacy of their Food. They are, in some of these rocky Islands, so numerous, that in a few Hours, two Men will load one of their Calapatches or little Boats, taking them pearching from off the Rocks and Bushes; they making no Attempt to escape, tho' almost full grown. They are called, by the Bahamians, Crab-catchers, Crabs being what they mostly subsist on; yet they are well-tasted, and free from any rank or fishy Savour.[23]

Catesby also studied "Other Productions of Nature," such as the black form of the purple land crab (*Gecarcinus ruricola*), which tends to be

FIGURE 10-8. Manchineel; plate 95, M. Catesby, 1743, *The natural history of Carolina . . .* , volume II. (Digital realization of original etchings by Lucie Hey and Nigel Frith, DRPG England; courtesy of the Royal Society ©.)

poisonous when eaten by humans.[24] He mentioned brasiletto wood (*Caesalpinia violacea*),[25] cascarilla bark (*Croton eluteria*),[26] and wild cinnamon (*Canella winterana*), all native Bahamian species with special uses.[27] Also, he made use of joewood (*Jacquinia keyensis*):

> The Bark and Leaves of this Tree being beat in a Morter produces a Lather, and is made Use of to wash Cloaths and Linnen, to which last it gives a yellowness. The Hunters who frequent the desolate Islands of Bahama . . . are frequently necessitated to use this Sort of Soap to wash their Shirts, for want of better.[28]

Many of the native Bahamian animals and plants Catesby illustrated and described can still readily be seen. For example, land hermit crabs (*Coenobita clypeatus*) in a variety of gastropod shells rustle around in dead leaves on the ground and scavenge. Catesby did not know that these "Crabs" have to return to the sea to reproduce, their larvae being marine.[29] Catesby knew that the edible green sea turtle (*Chelonia mydas*), which is not so readily seen nowadays, feeds on turtlegrass (*Thalassia testudinum*), an abundant submarine flowering plant. He illustrated them on the same plate.[30]

Nearly all of Catesby's marine biology was done in the Bahamas. His main focus there was on fishes, and he was surprised to find how variously colored and patterned they were. Fish out of water do not retain their colors well, so he painted a succession of specimens of each species to have them fresh (figure 10-9). One wonders how he caught his fishes. Most of them were carnivores and could have been taken with a hook and line or in a fish "pot" (trap), probably made of plaited palmetto leaves (figure 10-10). His herbivores would probably have had to be netted. Catesby was not the first to record tropical fish

FIGURE 10-9. Great hogfish; plate 15, M. Catesby, 1734, *The natural history of Carolina . . .* , volume II. (Digital realization of original etchings by Lucie Hey and Nigel Frith, DRPG England; courtesy of the Royal Society ©.)

FIGURE 10-10. Bahaman fish pot (trap). (© Robert Robertson.)

poisoning, an illness now called ciguatera. But he described clearly the symptoms when toxin-containing barracudas (*Sphyraena barracuda*) and other large fishes are eaten.

> Barracudas are in great plenty in all the shallow seas of the Bahama Islands. . . . The Flesh has a very rank and disagreeable Savour both to the Nose and Palate, and are frequently poisonous, causing great Sickness, Vomiting, and intolerable Pain in the Head, with loss of Hair and Nails; yet the hungry Bahamians, frequently repast on their unwholesome Carcasses.[31]

Catesby surmised correctly that ciguatera is not caused by fishes eating manchineel fruits. The toxin is now known to come originally from a dinoflagellate (*Gambierdiscus toxicus*), which is concentrated up the food chain. Top predatory fishes are therefore prone to be the most dangerous ones to eat. The princess rockfish (*Mycteroperca venenosa*) was the most poisonous fish known to Catesby.[32] Cooking does not destroy the toxin.

Catesby also illustrated gorgonians or octocorals. These fans, whips, or plumes are not plants, as he thought, but cnidarians, simple marine animals with stinging capsules that are related to jellyfish, sea anemones, and corals. Gorgonians are strictly marine, and he portrayed some of them as incongruous backgrounds for animals out of water such as flamingos. In this way, he made the best use of the empty spaces on his plates. The seawater is "so exceeding clear, that at the Depth of twenty Fathom, the rocky Bottom is plainly seen."[33] Catesby appears not to have visited coral reefs, although he mentioned their danger to mariners.

Catesby illustrated only one molluscan shell, a whelk (*Cittarium pica*) containing the land hermit crab (*Coenobita clypeatus*).[34] He apparently consulted only one book about shells, that by Martin Lister.[35] Judging by his references to Lister's book, Catesby also had the wide-mouthed rock-shell (*Plicopurpura patula*), of interest to him because it secretes Tyrian purple, which dyes cloth that color. He also had four other common gastropods, all of which are identifiable: the bleeding tooth (*Nerita peloronta*); the beaded periwinkle (*Cenchritis muricatus*); the periwinkle (*Littoraria angulifera* living on mangroves; and a land snail (*Cerion* sp.) The "infinite Variety of beautiful Shells"[36] impressed him. Surprisingly, he did not mention the now endangered large marine gastropod called queen conch (pronounced "konk") (*Strombus gigas*). Large Lucayan to sub-modern kitchen middens of its shells record its massive use for food.[37] Although oddly expressed, Catesby's overall observations on marine molluscan ecology are essentially modern. He was way ahead of his time in deploring the study of molluscs only as shells.

Catesby illustrated the red mangrove (*Rhizophora mangle*) that he saw in the Bahamas (figure 10-11). On the same plate he showed a crocodilian life-size, thus it was juvenile. He stated that this had just hatched from an egg "no bigger than that of a Turkey" that he also illustrated. There is no clear indication that his crocodilian was Bahamian, and he mainly discussed it as the alligator that he had seen in the Carolinas, where it would have been *Alligator mississippiensis*. Conceivably, though, Catesby could have seen the Cuban crocodile (*Crocodylus rhombifer*) in the Bahamas, which is now extinct there. Pleistocene fossils of it are known from Abaco, and the species persists alive, but it is highly endangered in southwestern Cuba. Crocodiles are more seafaring than alligators, so this may be relevant to their presences and absences in the islands.[38]

The abundances of a few Bahamian species have declined greatly since Catesby's stay. The Bahamian hutia ("Bahama Coney") (*Geocapromys ingrahami*) is a terrestrial rodent that is endemic to the islands.[39] Wholly Pleistocene fossils from caves show that it once flourished nearly throughout the Bahamas.[40] Catesby saw it alive, but until recently it persisted only at a tiny south Bahamian islet not known to have been visited by him: East Plana Cay. He recorded that "their Flesh is esteemed very good, it has more the Taste of a Pig than that of a Rabbit." This hutia has recently been introduced to one or more of the Exuma Cays, where at last report it was without enemies and thriving.

Catesby did not illustrate the West Indian monk seal (*Monachus tropicalis*), but he recorded that in his time they abounded, and sealers were slaughtering them and boiling and barreling their body oil at Bimini in the northwestern Bahamas. This carnivore, up to 8 feet long (2.5 meters) and 400 pounds (365 kilograms), is now extinct.[41] At least four places with "seal" in their names are scattered around the West Indies, attesting to the former widespread occurrence of this seal species (the only one in the region). Bimini is only about fifty miles (eighty kilometers) east of present-day Miami, Florida. Between these two localities runs the swiftly northward-flowing Florida Current, which farther out in the Atlantic becomes the Gulf Stream. It is now difficult to conceive ecological conditions at Bimini nearly three centuries ago. What role did the seals then play in the shallow marine ecosystem?

Nearly all Bahamian plants are too tropical to survive outdoors in England, but that did not prevent Catesby bringing seeds of a few home with him, most notably of the lily thorn (*Catesbaea spinosa*).[42] It may seem that Catesby named this for himself, but actually the Dutch botanist J. F. Gronovius (1686–1762) suggested that he include it in *The natural history of Carolina, Florida and the Bahama islands*.

It is most curious that Catesby did not mention the endemic and therefore presumably native pine variety *Pinus caribaea* var. *bahamensis*, the prevalent tree in barrens at Grand Bahama, Abaco, Andros, and New Providence (not Eleuthera).

Catesby's observations and illustrations are remarkably accurate, but nearly all his printed images are mirror-image reversals (a few that he copied from other artists are not reversed). This can be seen by comparing the prints with the original illustrations. These reversals are of consequence only for the parts of a few flowering plants that spiral in a certain direction and for asymmetric

FIGURE 10-11. (*Opposite*) A red mangrove with an American alligator; plate 63, M. Catesby, 1739, *The natural history of Carolina . . .* , volume II. (Digital realization of original etchings by Lucie Hey and Nigel Frith, DRPG England; courtesy of the Royal Society ©.)

T. 63.
Lacertus.

FIGURE 10-12.
West Indian topsnail (or whelk); plate 33, M. Catesby, 1736, *The natural history of Carolina . . .* , volume II. (Digital realization of original etchings by Lucie Hey and Nigel Frith, DRPG England; courtesy of the Royal Society ©.)

animals. These are all reversed, and gastropod shells such as the whelk (*Cittarium pica*) are wrongly shown left-handed (sinistral) (figure 10-12). Also, the wrong claws of crustaceans come out the larger ones. All in all, though, Catesby's illustrations of some Bahamian organisms are the best ever published.

Catesby's original intention was to stay in the Bahamas for about a year, and he returned to England sometime in 1726. As a youth, I lived in the islands for much longer than this and became familiar with many of the common organisms.[43]

11

Mark Catesby's preparatory drawings for *The natural history of Carolina, Florida and the Bahama islands*

HENRIETTA MCBURNEY

Mr Catesby a Gentleman of a small fortune . . . pretty well skill'd in Natural History who designs and paints in water colours to perfection . . .[1]

This was how Mark Catesby's skills as a naturalist and artist were described by William Sherard to Richard Richardson in 1720. Around 240 of Catesby's preparatory watercolors for *The natural history of Carolina, Florida and the Bahama islands*, bought by King George III in 1768, are now housed in the Royal Library, Windsor Castle.[2] These, together with a smaller number in other collections, provide evidence of the process by which Catesby captured the likenesses of his specimens, living or dead, worked them into finished folio-sized watercolors, and transformed these into the hand-colored etched plates that illustrated his book.[3]

Materials and techniques

How did Mark Catesby make the drawings for his book? What were his materials, and what were the conditions under which he drew and painted the fauna and flora of the little-explored terrain in Virginia, South Carolina, and the Bahama Islands? He gives us tantalizingly few descriptions of the process by which he captured the likenesses of the plants and animals he observed, limiting his description of his materials to a brief general statement in the preface to his book: "I employed an Indian to carry my Box, in which, besides Paper and materials for Painting, I put dry'd Specimens of Plants, Seeds, &c. as I gather'd them," to which he added the comment: "To the hospitality and assistance of these friendly Indians I am much indebted, for I not only subsisted on what they shot, but their first care was to erect a bark hut, at the approach of rain to keep me and my Cargo from wet."[4]

The apothecary James Petiver published various sets of instructions for explorers and collectors, advising them on what equipment they should carry with them on their expeditions. Amongst the tools they should take were a "Knife, Pen and Ink, or Pencil."[5] Some of Catesby's botanical specimens preserved in Oxford University Herbaria indicate that he seems indeed to have used a knife to collect his specimens because of the clean cuts through the stems.[6] No doubt he used his knife too for sharpening his pencils. Petiver also noted the need to take "three or four Sheets of waste Paper (with some Thread) to wrap the Fruit or Seed in."[7] Being an artist as well as a plant collector, Catesby needed to carry additional loads of "Paper and materials for Painting."[8]

"Materials" for watercolor painting in the eighteenth century were not ready prepared or easily portable as we know them now. Catesby's friend, the artist and author George Edwards (1694–1773), as communicative in his writings as Catesby is reticent, provided an account at the end of his *Natural history of birds*, published in 1750, the year after Catesby's death, of how watercolor pigments were prepared to the right consistency:

> In order to procure Colours that will be exceeding fine and run very smooth . . . mix a little gummed Colour in a large Shell, and work it well with your Finger, and then thin it with Water, and let it settle a little, and by pouring a little off the Top of it into another clean Shell, you will procure a fine free working Colour, which you may make as light as you please by the Addition of Water.[9]

He also noted:

> All Mineral, or Earthy colours must be mixed with a due proportion of gum-arabick to bind them together and make them stick to the paper. If there be too little gum the colours will rub off as you pass your finger over the paper when dried. If too much, the colours will shine, crack when very dry, and sometimes peel off.[10]

The conditions under which Catesby traveled and worked on his expeditions into the "Upper Country" or foothills of the Appalachian Mountains would not have allowed him to take all the necessary materials for preparing his own colors. By the early eighteenth century, however, raw colors were available ready ground for painters; a color list in a contemporary artist's manual by an anonymous author, *The art of painting in miniature*, issued in 1729, stated that "All the Colors above-specified, with every Implement and Utensil, necessary in the Practice of Painting in Miniature, are prepared and sold at most Print-Shops in London and Westminter."[11] Catesby is likely, therefore, to have had his colors ready ground to take into the field in small wooden boxes, as well as bottles of water ready prepared with gum, and a number of seashells, or perhaps ivory dishes, for mixing the colors. The author of *The art of painting in miniature* wrote that "Gum-water must be kept in a neat Bottle corked; and never must you take any out of it with a Pencil, that has Colour upon it, but with a Quill, or some such Thing."[12]

With the exacting process of mixing and preparing his pigments to the right consistency, Catesby could hardly have intended to execute finished watercolors in the field; he is more likely to have made field drawings or sketches in pen and ink to work up later. Indeed, a number of such preliminary sketches have been revealed on the backs of his finished watercolors in the Royal Library.[13] Most of these are in pen and ink, Catesby's usual medium for preliminary drawing (figure 11-1);[14] some, however, show details filled in with watercolor (figure 11-2). When conditions were not right to use watercolor in the field, Catesby could resort to annotating his monochrome drawing with color notes as an aide-mémoire for painting the correct colors later; the pen-and-ink sketch for

FIGURE 11-1. Preliminary field drawing in pen and ink by Mark Catesby (RL 26030 verso). (Royal Collection Trust. © H. M. Queen Elizabeth II 2013.)

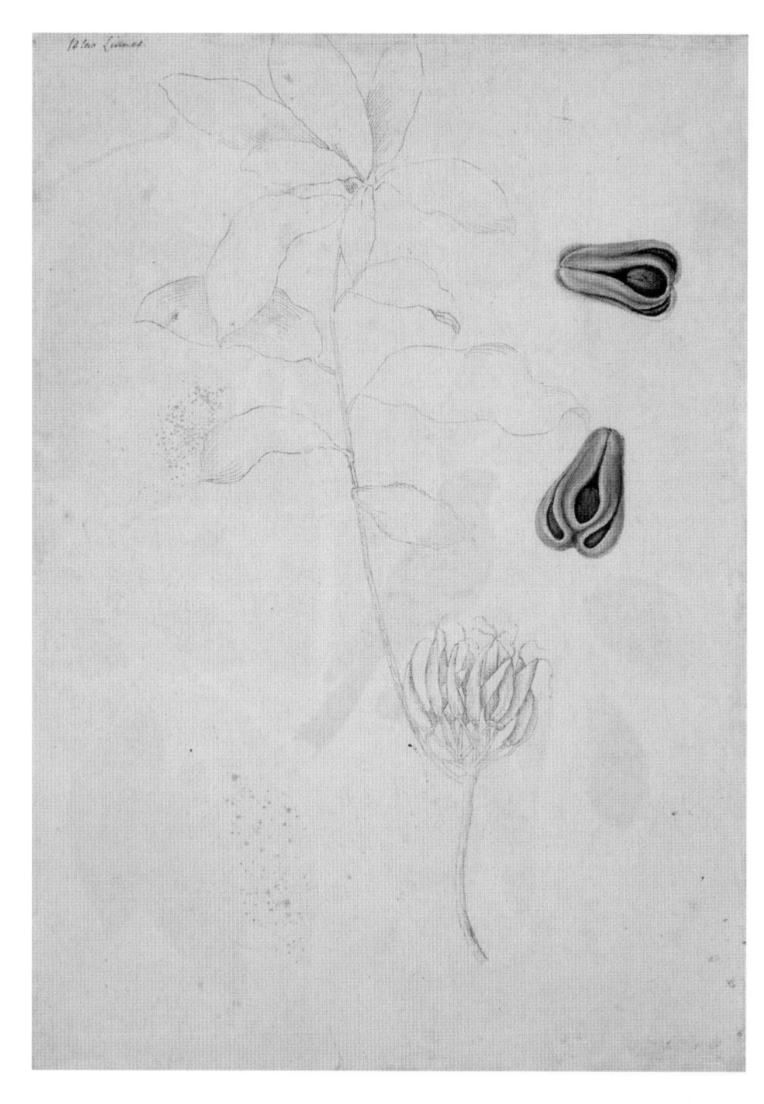

FIGURE II-2.
Preliminary field drawing, with details in watercolor by Mark Catesby (RL 25877 verso). (Royal Collection Trust. © H. M. Queen Elizabeth II 2013.)

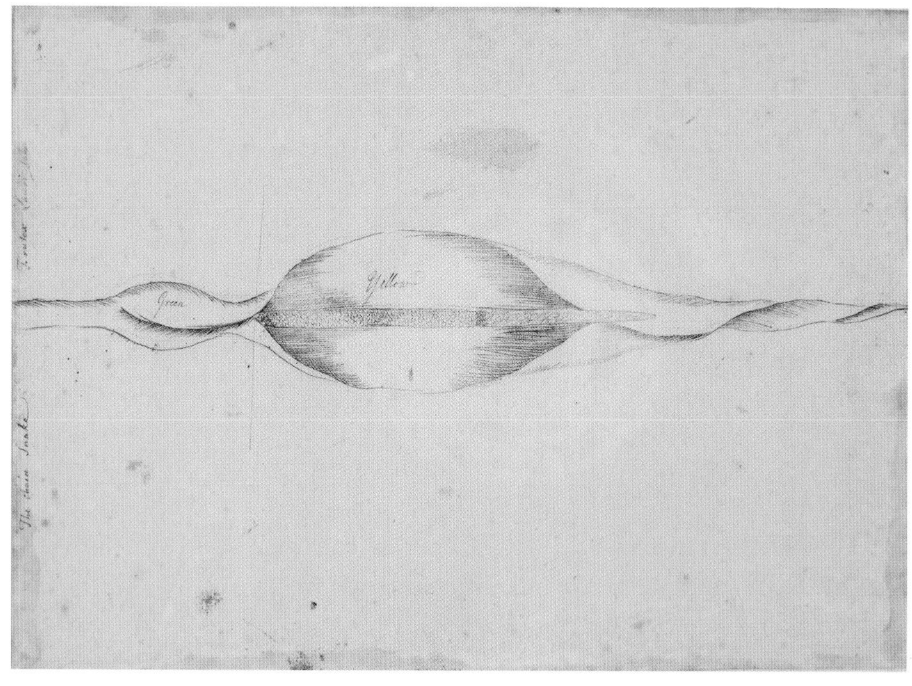

FIGURE II-3.
Preliminary field drawing with color notes by Mark Catesby (RL 26000 verso). (Royal Collection Trust. © H. M. Queen Elizabeth II 2013.)

the tannia, for example, is annotated with the words "yellow" on the inside of the spathe of the inflorescence and "green" for its outside (figure 11-3).

Nonetheless, Catesby made a point of stating that "In designing the Plants, I always did them while fresh and just gathered; and the Animals, particularly the Birds, I painted while alive (except a very few) and gave them their Gestures peculiar to every kind of Birds."[15] Clearly, it was important to him to record his subjects as accurately as possible, and training his eye to remember correctly the appearance of living specimens—which might have changed color significantly by the time he returned from his expeditions with his "cargos"—was an important skill he would have needed to hone as a naturalist and artist. Those that changed color most quickly were fishes. Most of these were painted while he was in the Bahamas in 1725, and being by the sea and nearer habitation meant he could employ local fisherman to catch and bring him new specimens at intervals so he could be sure of describing the colors accurately: "Fish, which do not retain their Colours when out of their Element, I painted at different times, having a succession of them procur'd while the former lost their Colours."[16]

It is likely, therefore, that Catesby's preliminary drawings, some colored on the spot and some not, were worked up by him after he returned from his expeditions to his lodgings in town. In Charleston, for example, we know he had a south-facing room where he dried his plant specimens; perhaps it was there that he kept his supplies of paper, pigments, and other materials, and where he might have painted up the drawings he brought back from his trips. He may also have painted during his visits to his various friends in the country, such as William Byrd II at his home on his Westover plantation in Virginia. But although Byrd recorded a number of details of Catesby's activities during his visits to Westover—including his walking around the plantation, shooting snakes in the long grass, observing a bear cub eating grapes in a tree, and swimming in the river—there is no mention of the naturalist making any drawings.[17]

For his finished watercolors, Catesby usually drew the composition in pen and ink, often over preliminary drawing in graphite, occasionally using graphite alone for the underdrawing. He then applied his layers of color, sometimes using the white of the paper to indicate the light parts but more often working from dark to light and applying white bodycolor over the other pigments for the highlights. The final details of texture and pattern were added using a fine brush and dry pigment, with almost miniaturist precision to convey details such as the anatomical parts and jewel-like markings of moths (figure 11-4). Most of his watercolors are characterized by a thick opacity obtained by the use of much bodycolor; this creates a far brighter and richer effect than the technique of "staining" with transparent washes used for coloring prints. Edwards made the distinction between the two different techniques: "There are two Ways of Painting in Water; one by mixing White with your Colours, and laying on a thick Body; the other is only washing your Paper or Vellum with a thin Water tinctured with Colour."[18]

FIGURE II-4.
Catesby's compositions combining separate drawings of different species involved cutting individual drawings from other sheets and collaging them onto the parent sheet with the plant drawing (RL 26053: moths). (Royal Collection Trust. © H. M. Queen Elizabeth II 2013.)

In addition, Catesby used gum arabic not only as a binder for his colors, as recommended by Edwards, but as a glaze for the deeper shadows and for particular details, such as the eyes of birds and other animals. Catesby wrote of how he chose his colors both for their durability and so as to achieve the greatest accuracy for the natural objects he was "illuminating": "Of the Paints, particularly the Greens, used in the illumination of the figures, I had principally a regard to those most resembling Nature, that were durable and would retain their lustre, rejecting others very specious and shining, but of an unnatural and shining quality."[19]

The care Catesby took in his choice of pigments and in their preparation was mirrored by the systematic methods he employed as a naturalist in preparing his botanical specimens for preservation and transportation. His methods were chosen to ensure that his specimens as well as his watercolors would last.

Catesby's borrowings from other natural history artists

When Catesby arrived back in England in 1726, he soon settled in London, "the Center of all Science,"[20] as he was to describe it, where he was able to prepare the materials for his book. Like other artists and naturalists, he seems to have gravitated toward Sloane's house and museum in Bloomsbury, where he would have had access to Sloane's natural history collections and to his ever-growing collection of reference books and drawings. After three years, Catesby was ready to advertise his book. In his "Proposals . . ." he referred to his "ORIGINAL PAINTINGS," which could be viewed "at Mr. *Fairchild's*, in *Hoxton*."[21] It is not known what his relationship was with the experimental nursery gardener Thomas Fairchild, whose nursery at Hoxton, then a village north of London, specialized in the raising and supply of newly imported American plants. Catesby had, however, sent Fairchild seeds and tubs of plants while he was in Virginia, and there is evidence that his new-found expertise in New World exotics was sought after by gardeners.[22] That Catesby should have chosen to exhibit his watercolors at Thomas Fairchild's nursery in Hoxton implies that he had a positive working relationship with Fairchild and that Fairchild's clientele were wealthy and "curious" enough to be potential subscribers to his book.

Sloane's library by the mid-1720s already contained more than two hundred albums listed by him as "Books of Miniature"—that is, containing "limned" or colored drawings of natural history subjects. Amongst these were rich treasures such as several volumes of original drawings by Maria Sibylla Merian (1647–1717); two volumes of Nicolas Robert's (1614–1685) exquisite bodycolors on vellum—the famous "Vélins du Roi"; a volume bought from the family of John White (1540–1593) containing early seventeenth-century copies after his drawings of plants, animals, and North American Indians;[23] and numerous albums of drawings collected or commissioned by Sloane and arranged according to their subject matter.[24] Thus the album entitled "Quadrupeds" contained images of mammals from as early as the fifteenth century, including drawings by Albrecht Dürer of an elk and a walrus, as well as contemporary drawings by artists whom Sloane commissioned to draw the exotic animals displayed in the London shows, such as St Bartholomew's Fair.[25]

Sloane made his "paper museum" of natural history images available to artists and naturalists. Catesby not only contributed drawings to Sloane's volumes but also found in them further sources of inspiration for his own work. Sometimes he was influenced by the compositions of other artists. Sometimes he found more satisfactory drawings of animals and plants than he had managed to make; he copied or even traced these drawings and used them instead of his own. He also acquired drawings by other artists—possibly in exchange for drawings of his own or perhaps as gifts. These he added to his own corpus of drawings from which he was assembling the images for *The natural history of Carolina. . . .*

Maria Sybilla Merian's work had an important influence on Catesby both scientifically and artistically.[26] On the scientific side, her pioneering illustrations of animals in the context of their habitat provided Catesby with a model for combining animals and plants in his illustrations. He said of his illustrations of birds that he painted more of them than other animals, partly because it was easier to catch them and partly because of their "having a nearer relation to the Plants which they feed on and frequent."[27] On the artistic side, Merian's bold arrangements of plants and animals are echoed in a number of Catesby's compositions, as, for instance, where he places large butterflies in the corner of sheets bearing plant studies, often so that they balance fruits at the diagonal opposites of the page (figures 11-5 and 11-6). The process of creating such compositions involved Catesby in cutting individual drawings from other sheets in which he set out his studies in the typical way of specimen sheets, and collaging them onto the parent sheet with the plant drawing (see figure 11-4).

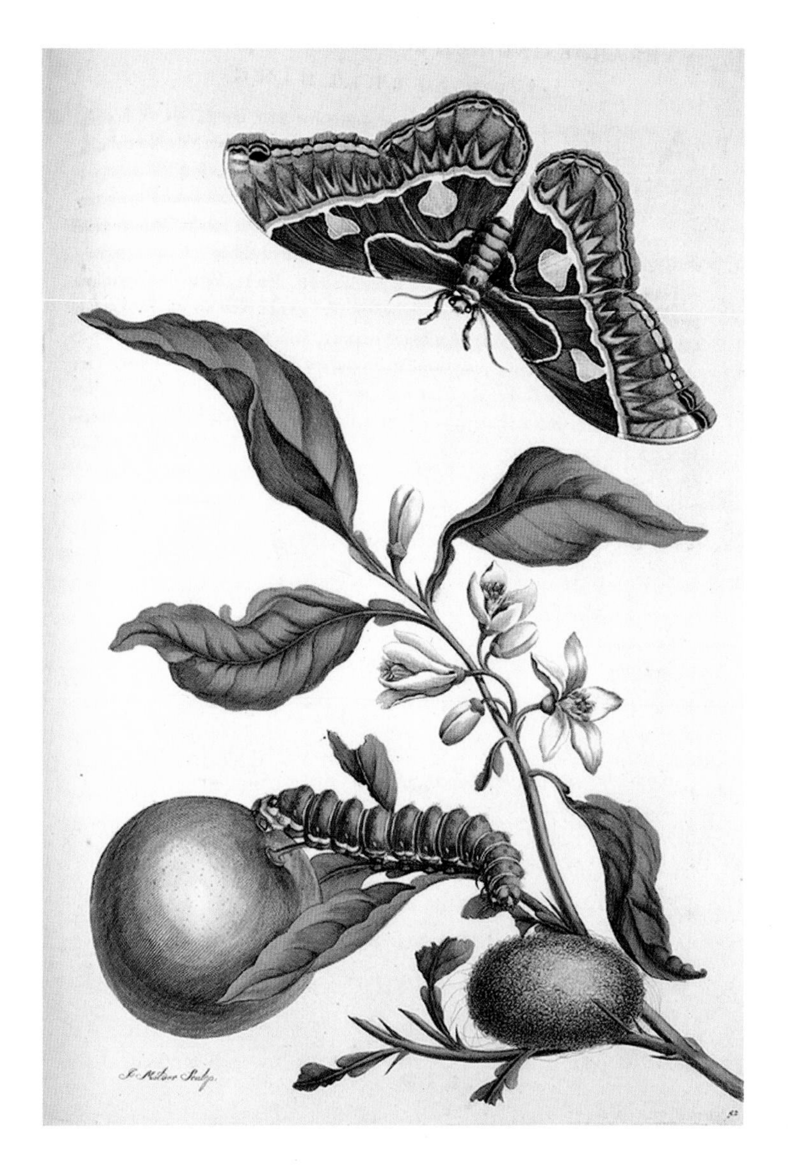

FIGURE 11-5. Maria Sybilla Merian's work had an important influence on Catesby both scientifically and artistically. Merian: orange and butterfly. (Courtesy of the Smithsonian Institution Libraries, Washington, D.C.)

FIGURE 11-6. This watercolor of the custard-apple and cecropia moth shows the result of Mark Catesby's combination of different drawings (RL 26048). (Royal Collection Trust. © H. M. Queen Elizabeth II 2013.)

Like many others, Catesby was fascinated by John White's images of New World fauna and flora, made during the expedition with Walter Raleigh to Virginia between 1585 and 1586. Thanks to the copies Sloane had had made, Catesby had access to White's images, and he made copies of seven of them—four fishes, a land crab, an iguana, and a swallow-tailed butterfly.[28] Although he is likely to have seen most of these fishes and other animals himself while he was in Virginia between 1712 and 1719, Catesby apparently made fewer drawings during that expedition (before he had resolved on making his book) than subsequently when he was in Carolina.[29] So White's images may have filled gaps in Catesby's own portfolio. In his copies of the John White prototypes, Catesby replicated the rather thin use of watercolor of the model—very different from his own technique, in which, as we have seen, he built up the watercolor in layers.

Catesby's use of White's images, however, was not simply replication of the earlier artist's work. In all cases apart from one, Catesby amalgamated the images borrowed from White with his own plant studies to create new composite images. Although the delightful and whimsical arrangement of the checkered puffer shown as if swimming between stems of red milk-pea and lancewood cannot be described as scientifically accurate, it is one that we have come to associate with Catesby's personal vision of the created world where animals are viewed as if in a natural environment, even if, in this instance, the relationships are imaginary.[30]

Sometimes Catesby's study of the drawings in Sloane's albums inspired his art in less direct ways. The prime model for Catesby's eastern chipmunk holding a hickory nut (*Carya*) is a tiny bodycolor on vellum, measuring two by three inches—literally a "miniature" (figures 11-7 and 11-8). This drawing is pasted into Sloane's volume "Drawings of Quadrupeds." We know nothing of its provenance, but it is a good example of the presence of much earlier drawings in Sloane's collection—this one being a seventeenth-century or perhaps sixteenth-century drawing, possibly of German or Dutch origin. Catesby did not copy his model exactly but adapted it to show the animal in a more life-like pose, crouching on its back legs, with its front legs raised to hold the nut.

At other times, Catesby borrowed an image without adapting it in any way. A finely worked drawing in plumbago (lead or graphite sharpened to a point) of a southern flying squirrel (*Glaucomys volans*) by the Dutch artist Everhard Kick (or Kickius—he mostly used the Latinized form of his name), clearly

FIGURE 11-7. The starting point for Mark Catesby's eastern chipmunk holding a hickory nut is this miniature pasted into Sloane's "Drawings of Quadrupeds." While its provenance is unknown, it is probably from the seventeenth century or perhaps the sixteenth century and, possibly, is German or Dutch. (© Trustees of the British Museum, register no.Sl,5261.121.)

FIGURE 11-8. Mark Catesby did not copy the preceding model exactly but adapted it to show the animal in a more lifelike pose, crouching on its back legs, with its front legs raised to hold the nut (RL 26032). This more natural pose seems to have been inspired by two further studies of a chipmunk in the same album by a different artist, Everhard Kickius. (Royal Collection Trust. © H. M. Queen Elizabeth II 2013.)

done from the life, provided the exact model for Catesby's etching of the animal (figures 11-9 and 11-10).[31] He retained the monochrome character of his model, applying just a thin gray wash over the etched lines, which, showing through the watercolor wash, describe the pattern of the fur. However, by placing the animal with its fruit on the stem of common persimmon (*Diospyros virginiana*), he conveys the impression that it has leaped into the tree to steal the fruit. Here there is a subtle fusion of a borrowed image, retained in its original state, with his own plant study.

Amongst the drawings by other artists in Catesby's own collection are four *vélins*, or bodycolors on vellum, by Claude Aubriet (c. 1665–1742), one of the most brilliant of the official French court painters. Aubriet worked with Joseph Pitton de Tournefort, the great French botanist and professor of medicine with whom Sloane studied when he was in Paris during the early 1680s. The *vélins* portray tropical American plants—Mexican vanilla (*Vanilla mexicana*), cacao (*Theobroma cacao*) which is the source of chocolate, cashew (*Anacardium occidentale*), and frangipani (*Plumeria rubra*)—specimens of which were perhaps

FIGURE 11-9. At other times, Catesby borrowed an image without adapting it in any way. A finely worked drawing of a southern flying squirrel by Everhard Kickius and clearly done from life provided the exact model for Mark Catesby's etching of the animal (RL 26034). (Royal Collection Trust. © H. M. Queen Elizabeth II 2013.)

FIGURE 11-10. Southern flying squirrel; plate 76, M. Catesby, 1739, *The natural history of Carolina* . . . , volume II. (Digital realization of original etchings by Lucie Hey and Nigel Frith, DRPG England; courtesy of the Royal Society ©.)

collected by Charles Plumier (1646–1704) and grown in the Jardin des Plantes in Paris. Although we do not know how these *vélins* came into Catesby's possession, a possible route was via Sloane and Tournefort. Unlike the majority of the images that Catesby borrowed from other artists, he did not adapt any of the four Aubriet *vélins* or combine them with images of his own; his etchings remain faithful to the original drawings, possibly because the vellums were highly prized and rare objects in their own right and he wished to reproduce them as closely as possible.

Not all the images Catesby borrowed for his work were made by earlier artists. Catesby also worked with contemporary and younger artists. In 1736, the German Georg Dionysius Ehret (1708–1770) came to London from Nuremberg while Catesby was working on the seventh part of his book. With his botanical training, intuitive sense of design, and virtuoso skills as a painter—he had been trained in Paris to paint on vellum—Ehret was to become the most outstanding botanical artist of the eighteenth century to work in England. His patrons included Sloane, the physician and collector Richard Mead, and several wealthy garden owners. Catesby, who probably met Ehret through Sloane, was clearly impressed by his work and began an association with him. During his period of working with Catesby, Ehret produced ten watercolors that were etched by Catesby for *The natural history of Carolina, Florida and the Bahama islands*, including one of smooth sumac (*Rhus glabra*), for which the preparatory study survives in the Natural History Museum, London. Ehret also executed two etchings for Catesby's book, one after a watercolor by Catesby of the sea-grape (*Coccoloba uvifera*) and the other after his own spectacular watercolor painting of the bull bay or southern magnolia (*Magnolia grandiflora*).[32]

Under the influence of Ehret, Catesby learned how to depict plants in a more rounded and three-dimensional way than his own "flat, tho' exact manner." In 1740 a specimen of mountain laurel (*Kalmia latifolia*), which Catesby acquired from Pennsylvania via Peter Collinson, finally blossomed in his garden in Fulham, and he and Ehret each drew the same stem of it. Catesby copied exactly the central group of leaves in Ehret's version but drew independently other leaves and the flower heads from an angle slightly different from that used by Ehret. Although his finished drawing lacks the enamel-like crispness of Ehret's, Catesby succeeded in conveying a far greater three-dimensional image of the plant than in his earlier plant studies.[33]

George Edwards was another artist who contributed to the drawings for Catesby's *The natural history of Carolina. . . .* Edwards was one of the principal artists employed by Sloane for his natural history albums, and Catesby met him frequently during his visits to Sloane's house. Edwards became a loyal friend of Catesby, being one of the last people to see him before he died. After Catesby's death, it seems that Edwards was responsible for helping Catesby's widow to arrange his drawings and materials in order before they were sold to Thomas Cadell, the London bookseller, later being bought by King George III.[34] It was Edwards who was responsible for seeing through the second edition of Catesby's book in 1754.[35]

Edwards stated in the inscription on the verso of his drawing that he drew the "Razor-billed blackbird of Jamaica"—now called the smooth-billed ani—from a bird in Sloane's collection brought back from Jamaica by a ship's captain named Thomas White (figures 11-11 and 11-12).[36] Edwards placed the bird in typical eighteenth-century fashion on a generic branch in a generic landscape. As for so many of the other drawings he acquired for his book, Catesby created his own composition out of the borrowed image. He removed the bird from its perch and placed it on a stem of his own elegant image of a pink lady's-slipper orchid. The contrast between Edwards's and Catesby's compositions underlines a fundamental difference between them as natural history illustrators. Catesby

FIGURE 11-11. Another source of inspiration for Mark Catesby was his friend George Edwards, who drew the "Razor-billed blackbird of Jamaica" from a bird in Sloane's collection, brought back from Jamaica by a ship's captain. Edwards placed it in typical eighteenth-century way on a generic branch in a generic landscape. As for so many of the other drawings he acquired for his book, Catesby created his own composition out of the borrowed image (RL 26034). (Royal Collection Trust. © H. M. Queen Elizabeth II 2013.)

FIGURE 11-12. Smooth-billed ani ("Razor-billed blackbird of Jamaica"); plate 3, M. Catesby, 1747, *The natural history of Carolina . . .* , volume II: Appendix. (Digital realization of original etchings by Lucie Hey and Nigel Frith, DRPG England; courtesy of the Royal Society ©.)

rejected the generic chinoiserie-style decorative background in which Edwards placed his bird and created instead a playful and inventive image out of juxtaposing the bird and the plant.[37] The scientific implausibility of the bird perching on the delicate plant stalk was of less interest to him than his perception of the underlying shared rhythms of the natural forms. As a scientist he was committed to accurate representation, as he stated in his preface. But as an artist his interest went beyond this to convey his own feeling for and imaginative experience of the natural world.

12

The publication of Mark Catesby's *The natural history of Carolina, Florida and the Bahama islands*

LESLIE K. OVERSTREET

When Mark Catesby returned to London in 1726 after his second trip to North America, he was determined to publish his observations and drawings of the plants and animals of the American wilderness. He envisioned a large folio format on a grand, almost unprecedented, scale with individual portraits of the birds, trees, and other plants and animals that he had drawn from life.[1] It would be an expensive undertaking, but the sponsorship of his travels by members of the Royal Society had come to an end. Although many encouraged him to publish, financially he was on his own.

In Catesby's day, there were no publishing houses to pay the costs of producing a book—buying the stocks of paper and ink and the copper plates for the illustrations, as well as paying the typesetters, printers, engravers, and so on.[2] An author had three choices: he could find a patron who would pay these costs, but Catesby's sponsors would not do this; he could sell the work to a printer or bookseller, who would see it through the press and own the rights to it; or he could pay the entire cost himself. This would mean that he had to provide the initial capital and take the risk of losing money if the book did not sell, but this also meant that he could retain control of it and keep the profits if there were any.

Catesby was able to choose the last option thanks to the help of his most stalwart supporter, the Quaker merchant Peter Collinson (1694–1764), a Fellow of the Royal Society and the central node in an extensive network of naturalists, travelers, collectors, and wealthy patrons. Collinson lent Catesby funds, but even with those funds, Catesby could not afford to hire expert engravers to transfer his drawings to copper plates for printing, so he sought out a professional artist, Joseph Goupy (1689–1768), to teach him to etch.[3]

Catesby's personal papers, whatever they may have consisted of, are not known to survive, and the printer of the work, whose activities or accounts might conceivably have been traced, is not identified in the book. Thus our understanding of how Catesby produced his remarkable work derives only from comments by others and from the physical evidence of the books themselves and related printed ephemera.

Proposals, for printing . . .

The first such ephemeral piece appeared in 1728 or 1729, when Catesby distributed a prospectus titled *Proposals, for printing an essay towards a natural history of Florida, Carolina and the Bahama Islands*, soliciting subscribers for the work. Almost a dozen copies of the *Proposals* are extant, and although undated, they provide textual evidence for dating the launch of Catesby's project and are an early indication of a pattern of financial caution in producing the book.[4]

Three separate typesettings of the broadsheet *Proposals* are known (figure 12-1), indicating a repeated need for more copies than Catesby first had printed. Aside from observable changes in the vertical alignment of the letters forming the text, which result from differences in the spacings between words and are virtually impossible to reproduce exactly in new typesettings, no matter how conscientiously done, they differ in several small typographical points. In what may be the earliest version, the parallel lists of supporters are bracketed with the points directed away from each other (toward the names), and Macclesfield is misprinted as "Macclesfeild." The sequence of the last four paragraphs is "As Figures . . . ," "The Encouragers . . . ," "It is intended . . . ," and "For the Satisfaction. . . ." The page ended with the information for potential subscribers that "Gentlemens Names will be enter'd . . . by the Author at Mr. Fairchild's, in Hoxton. . . ." Another version reverses the points of the brackets so that they are pointing inward toward each other and moves "The Encouragers" paragraph to the bottom. However, it retains the misprint "Macclesfeild" and "Mr. Fairchild's" as Catesby's address. What is possibly the last version keeps the brackets pointing inward and "The Encouragers" paragraph last, has Macclesfield correctly printed, and ends with the note that "These Books are to be had . . . at the Author's at Mr. Bacon's late Mr. Fairchild's, in Hoxton. . . ."

Thomas Fairchild, the commercial nurseryman who propagated many of the seeds and plants that Catesby had sent from North America, died on 10 October 1729; the nursery then passed to his nephew Stephen Bacon. Thus it may be inferred that the last version was printed after that date, while the first two were earlier. Since the plates and text of part 1 were printed and distributed by 29 May 1729 (see table 12-1) but were not advertised along with the original paintings as an inducement to subscribers, and since Catesby would have wanted to ascertain his market well in advance of launching into production on the work, it does seem reasonable to suggest that the earlier versions of the *Proposals* appeared well before May 1729 and perhaps even in the previous year.

In both versions, however, the essential information is identical: *Proposals* described the intended contents of the book, Catesby's qualifications, his

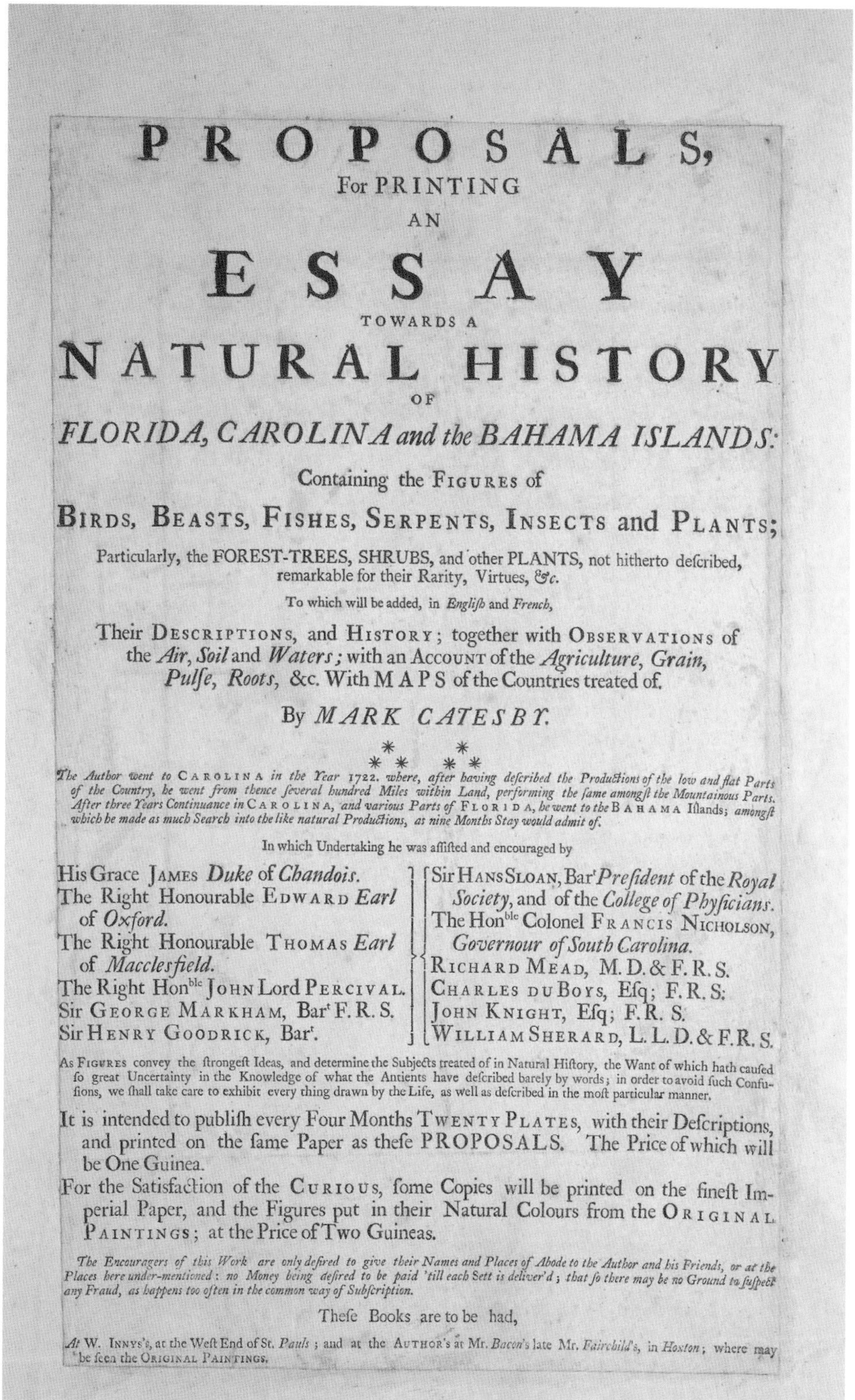

PROPOSALS,

For PRINTING

AN

ESSAY

TOWARDS A

NATURAL HISTORY

OF

FLORIDA, CAROLINA and the BAHAMA ISLANDS:

Containing the FIGURES of

BIRDS, BEASTS, FISHES, SERPENTS, INSECTS and PLANTS;

Particularly, the FOREST-TREES, SHRUBS, and other PLANTS, not hitherto deſcribed, remarkable for their Rarity, Virtues, *&c.*

To which will be added, in *Engliſh* and *French*,

Their DESCRIPTIONS, and HISTORY; together with OBSERVATIONS of the *Air, Soil* and *Waters;* with an ACCOUNT of the *Agriculture, Grain, Pulſe, Roots,* &c. With MAPS of the Countries treated of.

By *MARK CATESBY.*

* * * * * *

The Author went to CAROLINA *in the Year* 1722. *where, after having deſcribed the Productions of the low and flat Parts of the Country, he went from thence ſeveral hundred Miles within Land, performing the ſame amongſt the Mountainous Parts. After three Years Continuance in* CAROLINA, *and various Parts of* FLORIDA, *he went to the* BAHAMA Iſlands; *amongſt which he made as much Search into the like natural Productions, as nine Months Stay would admit of.*

In which Undertaking he was aſſiſted and encouraged by

His Grace JAMES *Duke* of *Chandois.*
The Right Honourable EDWARD *Earl* of *Oxford.*
The Right Honourable THOMAS *Earl* of *Macclesfield.*
The Right Honble JOHN Lord PERCIVAL.
Sir GEORGE MARKHAM, Bar^t F. R. S.
Sir HENRY GOODRICK, Bar^t.
Sir HANS SLOAN, Bar^t *Preſident* of the *Royal Society,* and of the *College of Phyſicians.*
The Honble Colonel FRANCIS NICHOLSON, *Governour of South Carolina.*
RICHARD MEAD, M. D. & F. R. S.
CHARLES DU BOYS, Eſq; F. R. S.
JOHN KNIGHT, Eſq; F. R. S.
WILLIAM SHERARD, L. L. D. & F. R. S.

As FIGURES convey the ſtrongeſt Ideas, and determine the Subjects treated of in Natural Hiſtory, the Want of which hath cauſed ſo great Uncertainty in the Knowledge of what the Antients have deſcribed barely by words; in order to avoid ſuch Confuſions, we ſhall take care to exhibit every thing drawn by the Life, as well as deſcribed in the moſt particular manner.

It is intended to publiſh every Four Months TWENTY PLATES, with their Deſcriptions, and printed on the ſame Paper as theſe PROPOSALS. The Price of which will be One Guinea.

For the Satisfaction of the CURIOUS, ſome Copies will be printed on the fineſt Imperial Paper, and the Figures put in their Natural Colours from the ORIGINAL PAINTINGS; at the Price of Two Guineas.

The Encouragers of this Work are only deſired to give their Names and Places of Abode to the Author and his Friends, or at the Places here under-mentioned: no Money being deſired to be paid 'till each Sett is deliver'd; that ſo there may be no Ground to ſuſpect any Fraud, as happens too often in the common way of Subſcription.

Theſe Books are to be had,

At W. INNYS's, at the Weſt End of St. *Pauls*; and at the AUTHOR's at Mr. *Bacon's* late Mr. *Fairchild's*, in *Hoxton*; where may be ſeen the ORIGINAL PAINTINGS.

FIGURE 12-1. One of three versions of the *Proposals*, or prospectus, dating from late 1729, that Mark Catesby issued to attract subscribers. (Courtesy of Smithsonian Institution Libraries and the Biodiversity Heritage Library.)

TABLE 12-1. Presentation of parts of *The natural history of Carolina, Florida and the Bahama islands* to the Royal Society of London

Part: contents	*Presented to the Royal Society*	*Reviewed in* Philosophical transactions *volume (number)*
Volume 1		
Part 1: text & plates 1–20	22 May 1729	36 (no. 415) for Sept./Oct. 1730[a]
Part 2: text & plates 21–40	8 January 1730	36 (no. 415) for Sept./Oct. 1730[a]
Part 3: text & plates 41–60	19 November 1730	36 (no. 415) for Sept./Oct. 1730[a]
Part 4: text & plates 61–80	4 November 1731	37 (no. 420) for Aug./Sept. 1731
Part 5: text & plates 81–100	23 November 1732	37 (no. 426) for Nov./Dec. 1732
Volume 2		
Part 6: text[b] & plates 1–20	4 April 1734	38 (no. 432) for Apr./May/June 1734
Part 7: text & plates 21–40	15 January 1736	39 (no. 438) for July/Aug./Sept. 1735
Part 8: text & plates 41–60	7 April 1737	39 (no. 441) for Apr./May/June 1736
Part 9: text & plates 61–80	7 June 1739	40 (no. 449) for Aug./Sept. 1738
Part 10: text & plates 81–10	15 December 1743	44 (no. 484) for Oct./Nov./Dec. 1747
An Account . . .	not before 1741[c]	
Appendix		
Part 11: text & plates 1–20	2 July 1747	45 (no. 486) for Feb./March 1748

[a] Mortimer presented his review to the Royal Society on 17 December 1730.

[b] Text page numbers originally were printed as 121–140; amended by hand to 1–20.

[c] Issued with, or around same time as, part 10; internal evidence indicates not before 1741.

supporters in his travels, and the emphasis on illustrations based on firsthand field observations. Most importantly, they lay out the specifics of how he intended to issue the book.

First, he planned to issue the book by subscription, which was an increasingly common way at that time of funding expensive books. To gauge his audience and ensure a buying market, Catesby asked those wishing to purchase a copy to sign up as subscribers.[5] Second, he would issue the book in parts—a relatively new mode of publication for large illustrated works. Since Catesby was writing the text, etching the copper plates, and, at least initially, hand-coloring every print, getting the book out would be a slow process. He planned to publish parts containing twenty plates and their accompanying pages of text every four months, and although he found it impossible to stick to that schedule, the subscription fees, paid upon receipt of each part, would start coming in to keep the publication going. Finally, he planned to issue the book in two versions, a regular one with the plates uncolored for one guinea per part, and a slightly larger one ("on the finest Imperial Paper") with the plates hand-colored for two guineas apiece. Virtually all known copies are fully colored, and it has been assumed that a plain version was not issued, but recent research has turned up several uncolored copies.[6]

As was common with subscription publication, Catesby included the list of subscribers in the book (figure 12-2).[7] Such lists were normally printed at the completion of the work, and this is proved in Catesby's book by, for example, the title following Martin Folkes's name: President of the Royal Society, which he became only in 1741. This list would normally be bound at the front of volume I, and the catchword at the bottom of the verso indicates that Catesby meant it to go directly in front of the Preface.

The "List of the Encouragers," as it is titled, exists in two printed states. The first includes 154 names, whereas the second has 155, adding "The Right Hon. the Earl of Iley" at the bottom of the right-hand column (below the intended last line, as indicated by the vertical rule, but fortuitously in the correct alphabetical group). Subscribers ranged from European royalty to the botanist John Bartram in Pennsylvania, as well as a few like Sir Hans Sloane who ordered multiple copies. Several names—the ones followed by long dashes—were booksellers

FIGURE 12-2.
Mark Catesby's list of subscribers was printed at the end of the book in 1743 but was intended for the front of volume I, as the catchword on the verso of the leaf indicates. (Courtesy of Smithsonian Institution Libraries and the Biodiversity Heritage Library.)

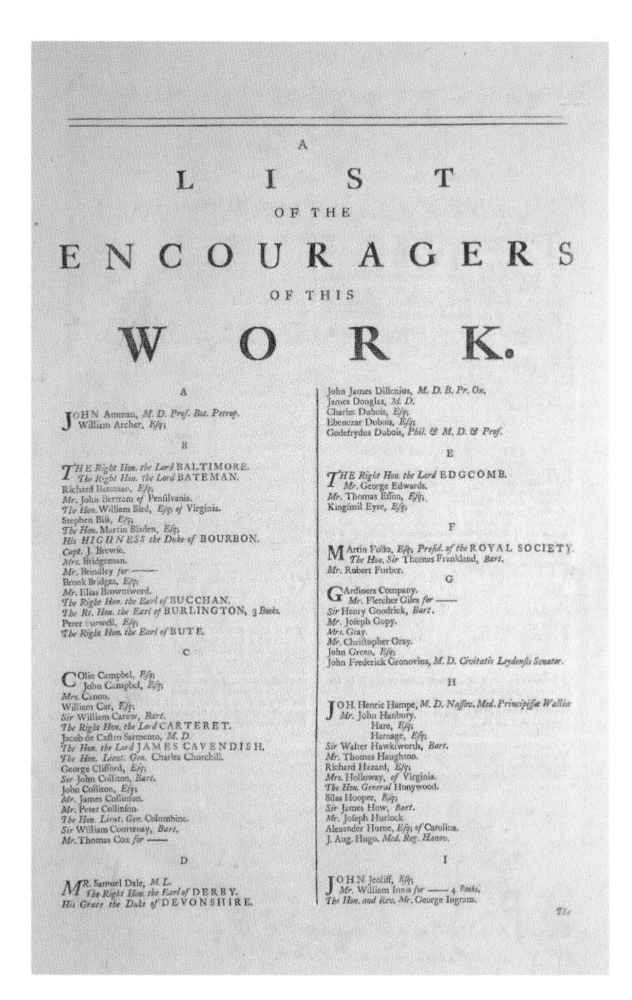

A

L I S T

OF THE

ENCOURAGERS

OF THIS

WORK.

A

JOHN Amman, M. D. Prof. Bot. Petrop.
William Archer, Esq;

B

THE Right Hon. the Lord BALTIMORE.
The Right Hon. the Lord BATEMAN.
Richard Bateman, Esq;
Mr. John Bertram of Pensilvania.
The Hon. William Bird, Esq; of Virginia.
Stephen Biss, Esq;
The Hon. Martin Bladen, Esq;
His HIGHNESS the Duke of BOURBON.
Capt. J. Brewse.
Mrs. Bridgeman.
Mr. Brindley for ——
Brook Bridges, Esq;
Mr. Elias Brownsword.
The Right Hon. the Earl of BUCCHAN.
The Rt. Hon. the Earl of BURLINGTON, 3 Books.
Peter [illegible]urwell, Esq;
The Right Hon. the Earl of BUTE.

C

COlin Campbel, Esq;
John Campbel, Esq;
Mrs. Canon.
William Car, Esq;
Sir William Carew, Bart.
The Right Hon. the Lord CARTERET.
Jacob de Castro Sarmento, M. D.
The Hon. the Lord JAMES CAVENDISH.
The Hon. Lieut. Gen. Charles Churchill.
George Clifford, Esq;
Sir John Colliton, Bart.
John Colliton, Esq;
Mr. James Collinson.
Mr. Peter Collinson.
The Hon. Lieut. Gen. Columbine.
Sir William Courtenay, Bart.
Mr. Thomas Cox for ——

D

MR. Samuel Dale, M. L.
The Right Hon. the Earl of DERBY.
His Grace the Duke of DEVONSHIRE.
John James Dillenius, M. D. B. Pr. Ox.
James Douglas, M. D.
Charles Dubois, Esq;
Ebenezar Dubois, Esq;
Godefrydus Dubois, Phil. & M. D. & Prof.

E

THE Right Hon. the Lord EDGCOMB.
Mr. George Edwards.
Mr. Thomas Esson, Esq;
Kingsmil Eyre, Esq;

F

MArtin Folks, Esq; Presid. of the ROYAL SOCIETY.
The Hon. Sir Thomas Frankland, Bart.
Mr. Robert Furber.

G

GArdiners Company.
Mr. Fletcher Giles for ——
Sir Henry Goodrick, Bart.
Mr. Joseph Gopy.
Mrs. Gray.
Mr. Christopher Gray.
John Green, Esq;
John Frederick Gronovius, M. D. Civitatis Leydensis Senator.

H

JOH. Henric Hampe, M. D. Nassov. Med. Principissæ Walliæ
Mr. John Hanbury.
Hare, Esq;
Harnage, Esq;
Sir Walter Hawksworth, Bart.
Mr. Thomas Haughton.
Richard Hazard, Esq;
Mrs. Holloway, of Virginia.
The Hon. General Honywood.
Silas Hooper, Esq;
Sir James How, Bart.
Mr. Joseph Hurlock
Alexander Home, Esq; of Carolina.
J. Aug. Hugo. Med. Reg. Hanov.

I

JOHN Jeoliff, Esq;
Mr. William Innis for —— 4 Books.
The Hon. and Rev. Mr. George Ingram.

The

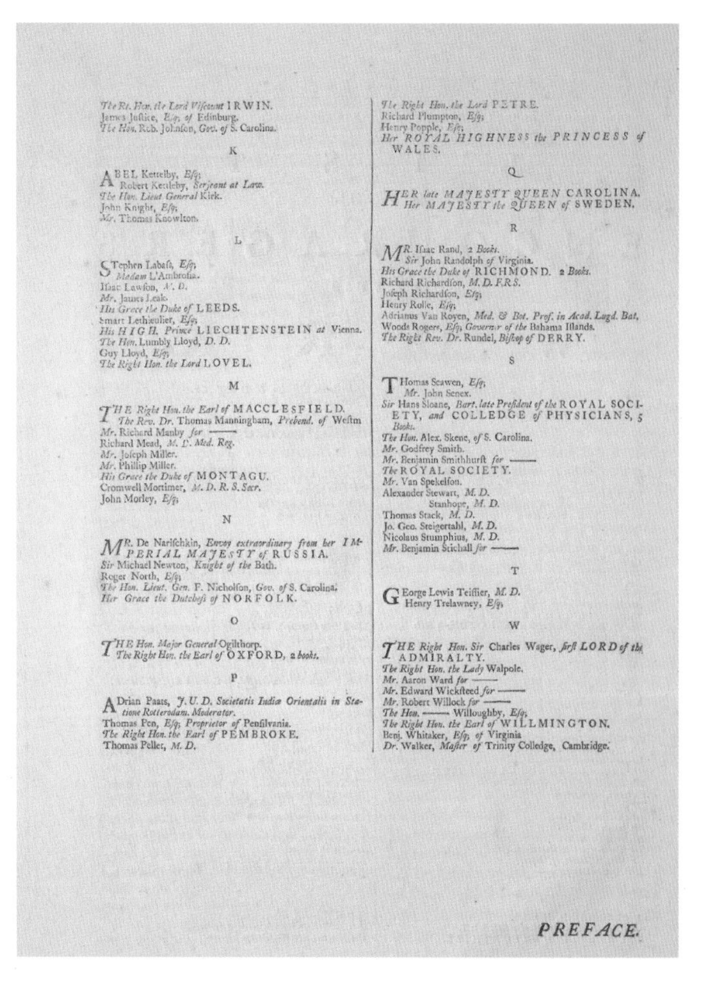

The Rt. Hon. the Lord Viscount IRWIN.
James Justice, Esq; of Edinburg.
The Hon. Rob. Johnson, Gov. of S. Carolina.

K

ABEL Kettelby, Esq;
Robert Kettleby, Serjeant at Law.
The Hon. Lieut General Kirk.
John Knight, Esq;
Mr. Thomas Knowlton.

L

STephen Labass, Esq;
Madam L'Ambrosia.
Isaac Lawson, M. D.
Mr. James Leak.
His Grace the Duke of LEEDS.
Smart Lethieulier, Esq;
His HIGH. Prince LIECHTENSTEIN at Vienna.
The Hon. Lumbly Lloyd, D. D.
Guy Lloyd, Esq;
The Right Hon. the Lord LOVEL.

M

THE Right Hon. the Earl of MACCLESFIELD.
The Rev. Dr. Thomas Manningham, Prebend. of Westm
Mr. Richard Manby for ——
Richard Mead, M. D. Med. Reg.
Mr. Joseph Miller.
Mr. Phillip Miller.
His Grace the Duke of MONTAGU.
Cromwell Mortimer, M. D. R. S. Secr.
John Morley, Esq;

N

MR. De Narischkin, Envoy extraordinary from her IMPERIAL MAJESTY of RUSSIA.
Sir Michael Newton, Knight of the Bath.
Roger North, Esq;
The Hon. Lieut. Gen. F. Nicholson, Gov. of S. Carolina.
Her Grace the Dutchess of NORFOLK.

O

THE Hon. Major General Oglthorp.
The Right Hon. the Earl of OXFORD, 2 books.

P

ADrian Paats, J. U. D. Societatis Indiæ Orientalis in Statione Rotterodam. Moderator.
Thomas Pen, Esq; Proprietor of Pensilvania.
The Right Hon. the Earl of PEMBROKE.
Thomas Peller, M. D.
The Right Hon. the Lord PETRE.
Richard Plumpton, Esq;
Henry Popple, Esq;
Her ROYAL HIGHNESS the PRINCESS of WALES.

Q

HER late MAJESTY QUEEN CAROLINA.
Her MAJESTY the QUEEN of SWEDEN.

R

MR. Isaac Rand, 2 Books.
Sir John Randolph of Virginia.
His Grace the Duke of RICHMOND. 2 Books.
Richard Richardson, M. D. F. R. S.
Joseph Richardson, Esq;
Henry Rolle, Esq;
Adrianus Van Royen, Med. & Bot. Prof. in Acad. Lugd. Bat.
Woods Rogers, Esq; Governor of the Bahama Islands.
The Right Rev. Dr. Rundel, Bishop of DERRY.

S

THomas Seawen, Esq;
Mr. John Senex.
Sir Hans Sloane, Bart. late President of the ROYAL SOCIETY, and COLLEDGE of PHYSICIANS, 5 Books.
The Hon. Alex. Skene, of S. Carolina.
Mr. Godfrey Smith.
Mr. Benjamin Smithhurst for ——
The ROYAL SOCIETY.
Mr. Van Spekelson.
Alexander Stewart, M. D.
Stanhope, M. D.
Thomas Stack, M. D.
Jo. Geo. Steigertahl, M. D.
Nicolaus Stumphius, M. D.
Mr. Benjamin Stichall for ——

T

GEorge Lewis Teissier, M. D.
Henry Trelawney, Esq;

W

THE Right Hon. Sir Charles Wager, first LORD of the ADMIRALTY.
The Right Hon. the Lady Walpole.
Mr. Aaron Ward for ——
Mr. Edward Wicksteed for ——
Mr. Robert Willock for ——
The Hon. —— Willoughby, Esq;
The Right Hon. the Earl of WILLMINGTON.
Benj. Whitaker, Esq; of Virginia
Dr. Walker, Master of Trinity Colledge, Cambridge.

PREFACE.

who handled subscriptions and post-publication sales, and one can only guess at the number that passed through their hands. But all told, it seems likely that Catesby sold as many as 180 copies.

Volume I

The first part had been issued by May 1729 (see table 1), and the other four parts followed slowly but steadily, at the latest by January 1730, November 1730, December 1731, and November 1732, respectively, according to the dates on which he presented each part to the Royal Society.[8]

It was a stunning book, with full-page illustrations and accompanying text that employed two large pictorial initials (usually depicting a bird or plants) on each page and multiple fonts of type differentiating the several textual elements. A large decorated initial and the several elements of the decorative headpiece of the Preface match those used in Johann Jakob Dillenius's *Hortus Elthamensis*, which bears a printer's colophon at the end of the second volume, and they make possible the identification of Catesby's printer as Godfrey Smith of Prince's Street, Spitalfields, London, who was one of the subscribers.[9]

Close examination of the text in different copies reveals that there were several typesettings of the first four parts (three of part 1 and two each of parts 2, 3, and 4), with changes in a variety of points, including the decorated initials, the line wraps, and in a few instances even the wording. On page 37 in volume I, for example, some copies have glued-on labels that add the citation for the plant's phrase-name and correct the English heading, while other copies have these changes made in type (figure 12-3).[10] There are many differences between the two typesettings of this page: the I initial is different, as are the line wraps (most noticeable at the end of each paragraph), the spacing between the headings and the text, the position of the signature, and so on. Clearly the page was reset and was not simply a different state or printing with stop-press corrections. Almost every page through to the end of part 4 (pages 61–80) gives similar indications of separate settings, however minute or technical.

This information suggests that Catesby, who was operating on a very limited budget, had started by keeping the number of copies printed (and therefore his costs) close to the actual number of subscribers. Thus he had to order additional copies at least once and sometimes twice to bring later subscribers up to date. The unexpected level of demand for the book and the pressures of producing it single-handedly are suggested in a letter to his niece in 1730.[11] Many people outside London, especially in North America, must have heard about the book only after it was under way. And since the different typesettings can be placed in a sequence—the earliest has various errors corrected by labels or by hand, and the last incorporates the corrections in type and, most conclusively, appears in the second edition—it seems possible, to the extent that a copy can be traced

P. 37

PASSERCULUS BICOLOR BAHAMIENSIS.

The Bahama Sparrow.

THIS is about the ſize of a Canary Bird. The Head, Neck, and Breaſt are black: All the other Parts of it of a dirty green Colour. It is the commoneſt little Bird I obſerved in the Woods of the *Bahama* Iſlands. It uſes to perch on the top of a Buſh and ſing, repeating one ſet Tune in manner of our *Chaffinch.*

Moineau de Bahama.

*IL eſt environ de la groſſeur d'un Sérin de Canaries. Sa tête, ſon coû, & ſa poitrine ſont noirs; Tout le reſte de ſon corps eſt d'un Verd ſale. C'eſt le petit Oiſeau le plus commun de ceux que j'ai obſervés dans les Bois des Iles Ba-*hama. *Il ſe perche ordinairement ſur la cime d'un Buiſſon, où il chante en repetant toûjours précíſément le meme air comme ſait notre Pinçon.*

Bignonia arbor pentaphylla flore roſeo majore ſiliquis planis. PLUM. CAT.

BIGNONIA.

THIS Shrub uſually riſes to the height of about ten Feet. From the larger Branches ſhoots forth long ſlender Stalks, at the end of every of which are five leaves fixed on foot-ſtalks an inch long. Its flower is monopetalous, of a Roſe-colour, and ſomewhat Bell-ſhaped, though the margin is deeply divided into five or ſix Sections, to which ſucceed Pods of five Inches long, hanging in Cluſters, and containing within them ſmall brown Beans.

Bignonia.

CETTE Plante s'éleve en Buiſſon, à la hauteur d'environ dix piés. Les groſſes branches pouſſent de longues tiges menuës, qui portent à leur extremité cinq feuilles attachées par des pédicules d'un pouce de long. Sa fleur eſt Monopetale. Elle a à peu-près la figure d'une Cloche; mais ſes bords ſont profondément découpés en cinq ou ſix ſections. Quand elle eſt paſſée, il lui ſuccéde des Coſſes longues de cinq pouces, attachés par bouquets: Elles contiennent de petits pois bruns.

K

P. 37

PASSERCULUS BICOLOR BAHAMENSIS.

The Bahama Sparrow.

THIS is about the ſize of a Canary Bird. The Head, Neck, and Breaſt are black: All the other Parts of it of a dirty green Colour. It is the commoneſt little Bird I obſerved in the Woods of the *Bahama* Iſlands. It uſes to perch on the top of a Buſh, and ſing, repeating one ſet Tune in manner of our *Chaffinch.*

Moineau de Bahama.

*IL eſt environ de la groſſeur d'un Sérin de Canaries. Sa tête, ſon coû, & ſa poitrine ſont noirs; tout le reſte de ſon corps eſt d'un Verd ſale. C'eſt le petit Oiſeau le plus commun de ceux que j'ai obſervés dans les Bois des Iles Ba-*hama. *Il ſe perche ordinairement ſur la cime d'un Buiſſon, où il chante en repetant toûjours précíſément le même air comme ſait notre Pinçon.*

Bignonia arbor pentaphylla flore roſeo majore ſiliquis planis. PLUM. CAT.

BIGNONIA.

THIS Shrub uſually riſes to the height of about ten Feet. From the larger Branches ſhoots forth long ſlender Stalks, at the end of every of which are five leaves fixed on foot-ſtalks an inch long. Its flower is monopetalous, of a Roſe-colour, and ſomewhat Bell-ſhaped, though the margin is deeply divided into five or ſix Sections, to which ſucceed Pods of five Inches long, hanging in Cluſters, and containing within them ſmall brown Beans.

Bignonia:

CETTE Plante s'éleve en buiſſon, à la hauteur d'environ dix piés. Les groſſes branches pouſſent de longues tiges menuës, qui portent à leur extremité cinq feuilles attachées par des pedicules d'un pouce de long. Sa fleur eſt monopetale. Elle a à peu-près la figure d'une Cloche; mais ſes bords ſont profondément découpés en cinq ou ſix ſections. Quand elle eſt paſſée, il lui ſuccede des Coſſes longues de cinq pouces, attachés par bouquets: Elles contiennent de petits pois bruns.

K

FIGURE 12-3. The numerous differences on page 37 between these two typesettings, one with printed paper slips correcting an error in the botanical headings and the other with the corrections set in type, show that Catesby had to go back and make additional copies of the early parts as the number of subscribers increased. (Courtesy of Smithsonian Institution Libraries and the Biodiversity Heritage Library.)

back to a particular owner, that it could be determined roughly at what point in the book's publication history that person subscribed and, more generally, how many subscribed at different points over the course of the book's publication. This information in turn might provide better insight not only into practical aspects of book production in the mid-1700s but also about larger issues of scientific communication networks in this period.

Only the first four parts (text pages 1–80 of volume I) were subject to numerous changes of typesetting. With part 5, Catesby seems to have become convinced that the book would sell adequately to cover the costs, and he began printing a much larger number of copies from a single typesetting, enough

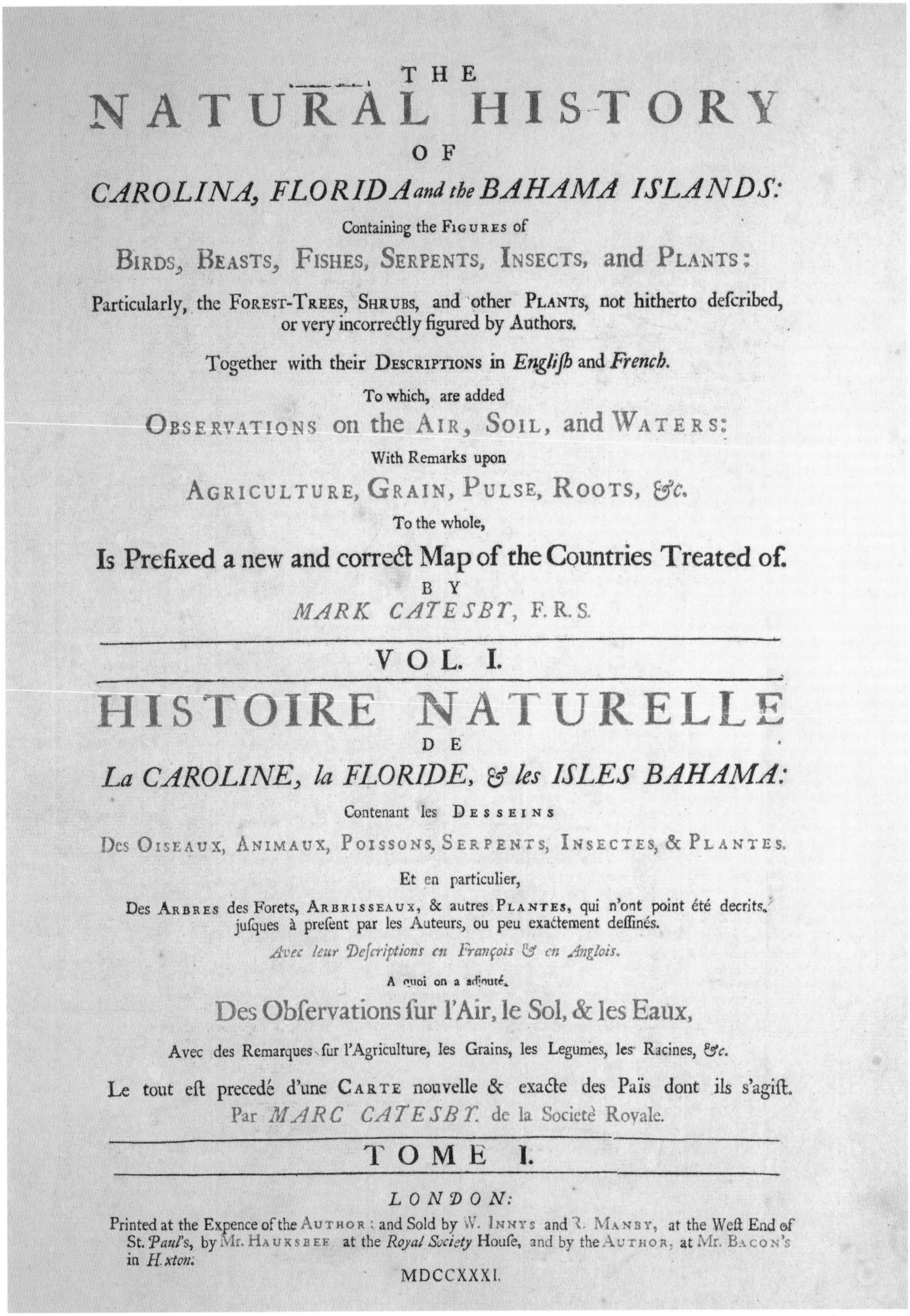

THE
NATURAL HISTORY
OF
CAROLINA, FLORIDA and the BAHAMA ISLANDS:
Containing the FIGURES of
BIRDS, BEASTS, FISHES, SERPENTS, INSECTS, and PLANTS:
Particularly, the FOREST-TREES, SHRUBS, and other PLANTS, not hitherto described, or very incorrectly figured by Authors.
Together with their DESCRIPTIONS in *English* and *French*.
To which, are added
OBSERVATIONS on the AIR, SOIL, and WATERS:
With Remarks upon
AGRICULTURE, GRAIN, PULSE, ROOTS, &c.
To the whole,
Is Prefixed a new and correct Map of the Countries Treated of.
BY
MARK CATESBY, F.R.S.

VOL. I.

HISTOIRE NATURELLE
DE
La CAROLINE, la FLORIDE, & les ISLES BAHAMA:
Contenant les DESSEINS
Des OISEAUX, ANIMAUX, POISSONS, SERPENTS, INSECTES, & PLANTES.
Et en particulier,
Des ARBRES des Forets, ARBRISSEAUX, & autres PLANTES, qui n'ont point été decrits, jusques à present par les Auteurs, ou peu exactement dessinés.
Avec leur Descriptions en François & en Anglois.
A quoi on a adjouté,
Des Observations sur l'Air, le Sol, & les Eaux,
Avec des Remarques sur l'Agriculture, les Grains, les Legumes, les Racines, &c.
Le tout est precedé d'une CARTE nouvelle & exacte des Païs dont ils s'agist.
Par *MARC CATESBY*, de la Societè Royale.

TOME I.

LONDON:
Printed at the Expence of the AUTHOR; and Sold by W. INNYS and R. MANBY, at the West End of St. *Paul's*, by Mr. HAUKSBEE at the *Royal Society* House, and by the AUTHOR, at Mr. BACON's in *H.xton*.
MDCCXXXI.

FIGURE 12-4. The title-page of volume I is dated 1731, although it was probably sent to subscribers in 1732 with part 5 (pages and plates 81–100), which completed the volume. (Courtesy of Smithsonian Institution Libraries and the Biodiversity Heritage Library.)

to accommodate all of the subscribers who signed on through the remaining fifteen years of the work's publication.

Catesby completed the five parts comprising volume I (a total of one hundred pages of text and one hundred accompanying plates) by November 1732, and with the last part he included, as was customary at the time, the title-page, printed and dated 1731 (figure 12-4), and a dedication page for the volume.[12] Once again, with the dedication there are multiple states of the text and typography; the differences seem to reflect stop-press corrections and other changes.

The "Note"

Unique to the copy at the Smithsonian Libraries,[13] attached at the bottom of an inserted leaf at the front of volume I, is a brief "Note" that was apparently sent to subscribers (figure 12-5). Although undated, it states that "a Frontice-Piece, Preface, and Maps" were yet to come, all of which would normally be intended for the first volume of a multi-volume work. Thus it seems likely that the "Note" was printed and sent out with part 5, which otherwise would have been thought to have completed the first volume. This dating is supported by the fact that the prefatory materials as projected in the "Note" differ somewhat from what was actually printed, suggesting that it reflects an early intention rather than the achieved reality, as these kinds of materials were normally printed at the end of a work's production. In order to accommodate material that would not be produced until that later time, Catesby explicitly warned subscribers not to have the volume bound yet. (Until the latter decades of the eighteenth century, virtually all books were bound only at the point of sale, with the buyer deciding what particular kind of cover he or she wanted or could afford.)[14]

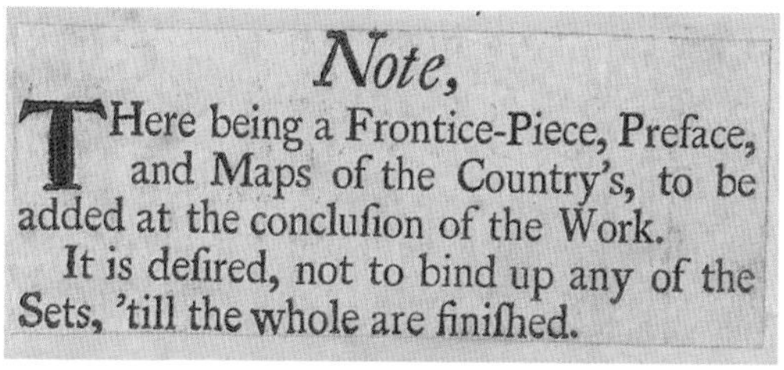

Note,

THere being a Frontice-Piece, Preface, and Maps of the Country's, to be added at the conclufion of the Work.

It is defired, not to bind up any of the Sets, 'till the whole are finifhed.

FIGURE 12-5. A unique survival in a copy at the Smithsonian, this "Note" was sent to subscribers, probably at the end of volume I, instructing them to wait for additional material before having the volume bound. (Courtesy of Smithsonian Institution Libraries and the Biodiversity Heritage Library.)

Volume II and the "Advertisement"

Catesby proceeded with parts 6 through 10, comprising the second volume, between 1734 and 1743 (table 1).[15] With the final part he would have issued the title-page and dedication for that volume and the list of subscribers and an index to the whole work, as well as the extra bits that he had promised: a preface, a map of the parts of North America he had traveled, and a forty-four-page "Account" of the country. He seems not to have produced a separate frontispiece, although some copies have the map or one of the plates bound in this position.

Again he sent subscribers an announcement saying that he planned an extra part of twenty plates to add plants and animals that remained from his own

ADVERTISEMENT.

THE Part now publiſh'd, of the *Natural Hiſtory* of *Florida* and *Carolina*, concludes 200 Plates, which are all that were at firſt deſigned; but with what remains of the Collection I brought from *America*, and an Addition of other non-deſcript Animals and Plants, received ſince from that Part of the World: I have now by me ample Materials for another Set of Twenty Plates, which, if approv'd of, I deſign to add by Way of *Appendix*. This however need not obſtruct the immediate Binding of both Volumes, for by leaving a ſmall Vacancy at the End of the Second Volume, with Guards for fifteen Sheets, they may be inſerted without any Defacement, or being perceived.

This additional Part of the Collection conſiſts of ſuch curious Subjects, that for the Reputation of the Work, I am loath to omit, and for no other Reaſon.

PARTICULARS as under.

ANIMALS.

BISON Americanus. The Buffello of *America.*
Lepus Javenſis. The *Javan* Hare.
Perdix ſilveſtris Americanus. The Partridge of *America.*
Lagopus. A Kind of Heathcock.
Penguin. A particular Species.
Alka. A Species of the Auk.
Hirundo cauda aculeata Americana.
Pittrel, or Storm Finch.
Avis Tropicorum. The Tropick Bird.
Ardea criſtata maxima Americana. The largeſt Heron.
Regulus Criſtatus Americanus. The Golden Crown'd Wren.
Anas, &c. A beautiful Duck from *New-foundland.*
Fiſh in Armour, it being covered with Bone.
A monſtrous Fiſh.
Salamandra maculata. A ſpotted Eft.

INSECTS.

Scolopendra.
Scarobæi.
Blattæ.
Veſpæ-ichneumones.
Formicæ.
Chegoes, &c.

VEGETABLES.

COCAO, or Chocolate Plant.
Volubilis ſiliquoſa Mexicana, Vanelles.
Anacardium, Acajou.
Lilium Attamaſco, *Indis dictum.*
Lilium rubrum minimum.
Lilio-Narciſſus Poliantbos flore albo.
Flos paſſionis flore elevato ſuavè rubente fructu hexagono rufo, folio bicorni abſque angulo prominente in medio.
Ficus Citrii folio fructu parvo purpureo.
Chryſanthemum Martigonis foliis, floribus ramoſis.
Iſora Althææ folio non ſerrato fructu longiore & Anguſtiore.
Calceolus flore maximo rubente purpureis venis notato foliis amplis hirſutis venoſis radice Dentis Canini.
Pſeudo-Acacia Hiſpida floribus roſeis.
Chamærhododendron Americanum.
Magnolia affinis.
Hamamelis. Gronov.

FIGURE 12-6. Announcing the conclusion of the work as it was originally intended, this "Advertisement" suggests leaving room for an eleventh part, the "Appendix," as Catesby called it, when having (only now) both volumes bound. (Courtesy of Smithsonian Institution Libraries and the Biodiversity Heritage Library.)

collections or had been sent to him after his return to England. This "Advertisement" (figure 12-6), known in only two copies, in the Smithsonian Libraries and at Trinity College, Cambridge,[16] alerted the subscriber to leave room for the additional part by having the binder insert tabs, or placeholders, for it when binding both volumes (including the first volume, finally). The copy at Trinity College, Cambridge, still has the tabs waiting for the Appendix, which ended up being bound separately.

It took another four years for the Appendix, as Catesby called it, to appear, accompanied by its own index. Thus by July 1747 the book was finally complete; eleven parts, totaling 220 plates and 220 pages of text, plus the two volumes' title-pages, dedications, list of subscribers, the Preface, map, "Account," and two indexes.

Perhaps 180 to 200 copies were produced, and about 100 survive.[17] One of the more interesting things about them is the considerable variation in the contents of the two volumes and in their sequence as bound. About sixty copies have been examined, and many are incomplete—some lack the Appendix, and quite a few lack one or another of the prefatory elements. Many copies have much of that preliminary material, usually the map and the "Account," bound in the second volume, indicating that the owners did not wait to bind the first volume, as Catesby had instructed. And there are a few copies that have the Appendix bound as a separate third volume, the owners again apparently missing Catesby's instructions for fitting it into the second volume.

FIGURE 12-7.
The armorial design with monogram on the cover of this copy at the Smithsonian Libraries shows that it belonged to Cromwell Mortimer, Secretary of the Royal Society and a subscriber. (Courtesy of Smithsonian Institution Libraries and the Biodiversity Heritage Library.)

There are only a handful of perfect copies—copies that are both complete and are bound in the correct, intended sequence. Of those few, the Smithsonian Libraries' copy[18] is the only one yet known to have all three pieces of ephemera relating to the production of the work. The unique authority of this copy is supported by the fact that the original owner, who put his armorial design and monogram (figure 12-7) on the luxurious full russia binding, was Cromwell Mortimer, Secretary of the Royal Society and a subscriber.[19] As Secretary, Mortimer was assigned the role of reviewing new publications for the Society's *Philosophical transactions* and did so for all eleven parts of Catesby's book after they were presented to the Society; he was strong in his praise, hailing it at its conclusion as "the most magnificent Work I know of, since the Art of Printing has been discover'd."[20]

The second edition, 1754

Five years after Catesby's death in 1749, his colleague George Edwards, Beadle at the Royal College of Surgeons of England, saw a second edition (1754) through the press.[21] Although the title-page says that the work was "revised by" Edwards (figure 12-8), not a word was changed, to judge from the fact that, among the dozen or so copies of this edition that have been examined, many, if not most, are made up with at least some (and in one copy, entirely with) printed sheets of text from the first edition, identifiable by their large pictorial initials.

The first-edition text sheets are used in a seemingly random mix with newly set text, which can be recognized by its different set of initials in a completely

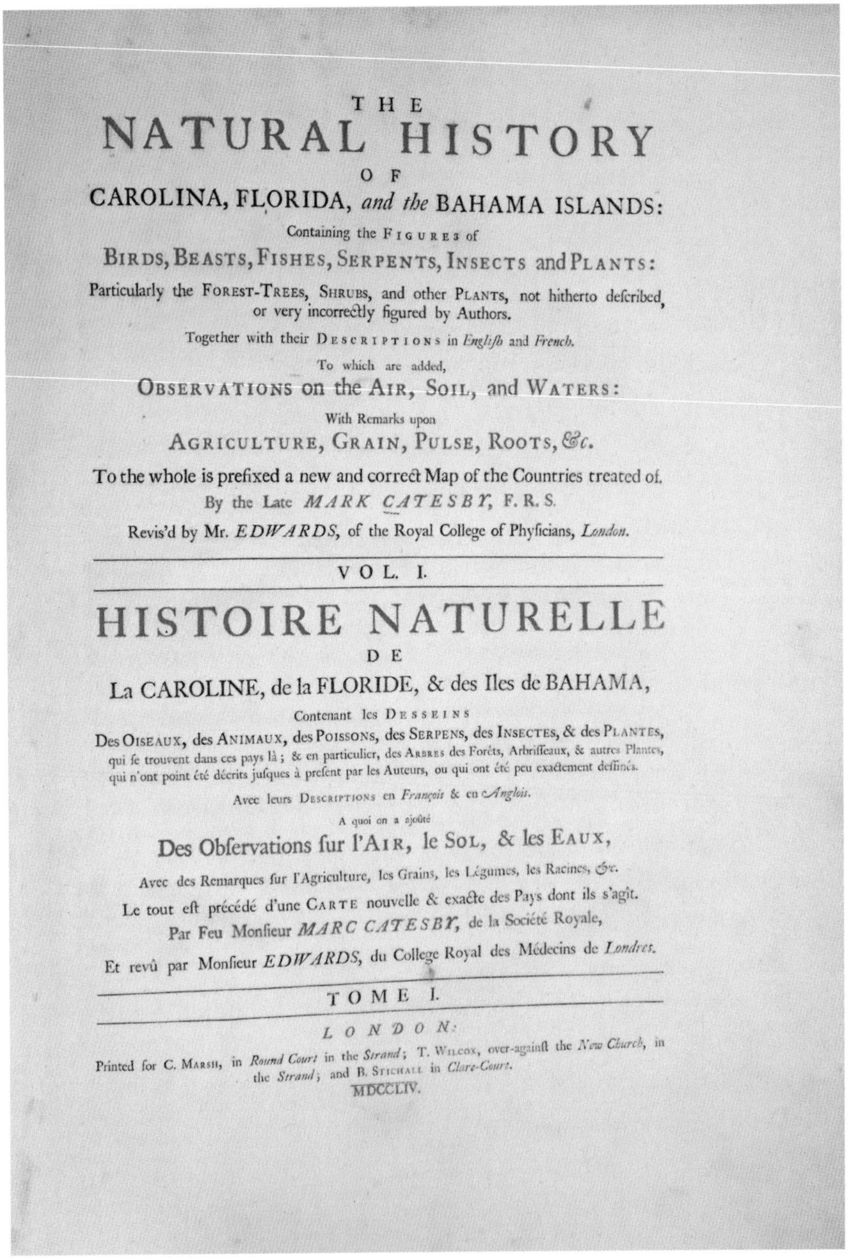

THE
NATURAL HISTORY
OF
CAROLINA, FLORIDA, *and the* BAHAMA ISLANDS:
Containing the FIGURES of
BIRDS, BEASTS, FISHES, SERPENTS, INSECTS and PLANTS:
Particularly the FOREST-TREES, SHRUBS, and other PLANTS, not hitherto defcribed, or very incorrectly figured by Authors.
Together with their DESCRIPTIONS in *Englifh* and *French*.
To which are added,
OBSERVATIONS on the AIR, SOIL, and WATERS:
With Remarks upon
AGRICULTURE, GRAIN, PULSE, ROOTS, &c.
To the whole is prefixed a new and correct Map of the Countries treated of.
By the Late *MARK CATESBY*, F. R. S.
Revis'd by Mr. *EDWARDS*, of the Royal College of Phyficians, *London*.

VOL. I.

HISTOIRE NATURELLE
DE
La CAROLINE, de la FLORIDE, & des Iles de BAHAMA,
Contenant les DESSEINS
Des OISEAUX, des ANIMAUX, des POISSONS, des SERPENS, des INSECTES, & des PLANTES, qui fe trouvent dans ces pays là ; & en particulier, des ARBRES des Forêts, Arbriffeaux, & autres Plantes, qui n'ont point été décrits jufques à prefent par les Auteurs, ou qui ont été peu exactement deffinés.
Avec leurs DESCRIPTIONS en *François* & en *Anglois*.
A quoi on a ajoûté
Des Obfervations fur l'AIR, le SOL, & les EAUX,
Avec des Remarques fur l'Agriculture, les Grains, les Légumes, les Racines, &c.
Le tout eft précédé d'une CARTE nouvelle & exacte des Pays dont ils s'agît.
Par Feu Monfieur *MARC CATESBY*, de la Société Royale,
Et revû par Monfieur *EDWARDS*, du College Royal des Médecins de *Londres*.

TOME I.

LONDON:
Printed for C. MARSH, in *Round Court* in the *Strand*; T. WILCOX, over-againft the *New Church*, in the *Strand*; and B. STICHALL in *Clare-Court*.
MDCCLIV.

FIGURE 12-8.
Although the title-page of the second edition says that it was "revised" by George Edwards, the text is unchanged, and many copies used sheets left over from the first edition. (Courtesy of Rowan Public Library, North Carolina.)

different decorative style. Another distinguishing characteristic is that the newly set text in the second edition added "Volume II" to the direction line of each page that carried a signature in that volume to distinguish the two volumes' number runs for the binder, since they were both being published at the same time and might have been mixed up.

Curiously, the newly set text re-introduced some errors that had been corrected in the first edition either by hand or in type. As a single example, in some copies of the first edition, the second bird description on page 54 has, on the eighth line, the typeset word "red" erased and "black" inserted by hand; but in the newly set pages of the second edition, the text reverts to "red" although in the plate the bird's legs are colored black, as they should be. It would seem that the typesetter was working from an uncorrected text, but the leftover first-edition text sheets that Edwards was using had it right: the word "black" is in type, along with the other corrections made in the first edition (figure 12-9). The pattern of these occasional re-introduced errors is something that needs looking into and may provide evidence—as the multiple typesettings in the first edition are evidence of Catesby's cautious approach to print runs—of how Edwards went about producing the second edition.

The red-Ey'd Fly-catcher

WEIGHS ten Penny-weight and an half. The Bill Lead-Colour : The Iris of the Eyes red. From the Bill, over the Eyes, runs a dufky white Line, border'd above with a black Line. The Crown of the Head is gray. The Reft of the Upper-Part of the Body is green. The Neck, Breaft aud Belly white; the Legs and Feet *black*. Both thefe breed in *Carolina*, and retire Southward in Winter.

The red-ey'd Fly-catcher.

WEIGHS ten penny-weight and an half. The bill is lead-colour : the iris of the eyes are red. From the bill, over the eyes, runs a dufky white line, bordered above with a black line. The crown of the head is gray. The reft of the upper-part of the body is green. The neck, breaft and belly are white; the legs and feet red. Both thefe breed in *Carolina*, and retire Southward in Winter.

FIGURE 12-9. In some copies of the first edition, the word "red" set in type was corrected by hand to "black" (*left*); in the later typesetting of this part, it was corrected to "black" in type. In those copies of the second edition that used leftover first-edition sheets, it is therefore correct, but in newly set sheets of the second edition, the original error reappears. (First edition [*bottom left and right*]: courtesy of Smithsonian Institution Libraries and the Biodiversity Heritage Library; second edition [*top left*]: Rowan Public Library, North Carolina.)

The plates in the second edition, again with some left over from the first (most likely uncolored), were printed from Catesby's original copper plates but tend to be noticeably more strongly or brightly colored.

The third edition, 1771

A third edition was published in 1771 by the bookseller Benjamin White, who had acquired Catesby's etched copper plates (figure 12-10). This edition consists of completely reset text, this time with large but plain capital letters. The text leaves bear the volume number in the direction line (at the bottom of the text)

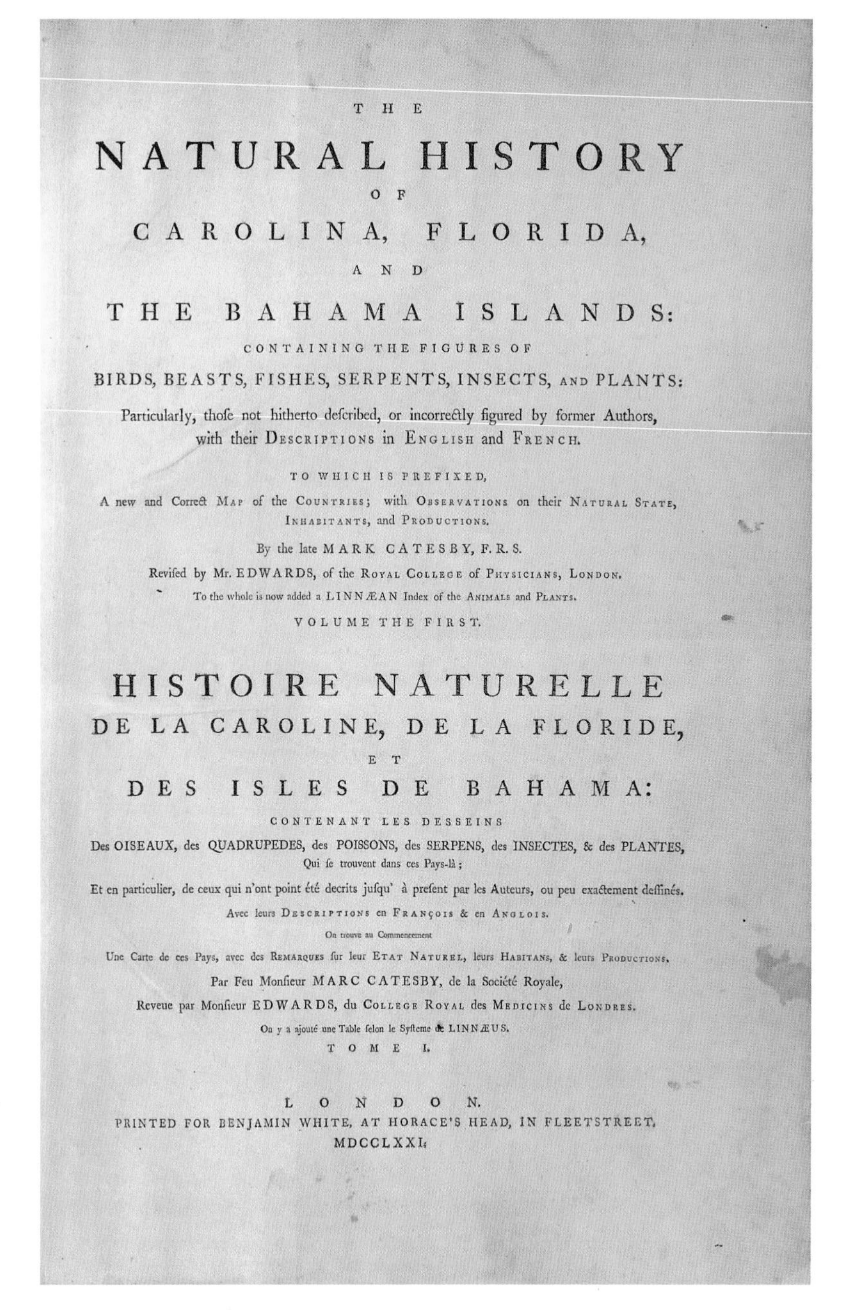
THE
NATURAL HISTORY
OF
CAROLINA, FLORIDA,
AND
THE BAHAMA ISLANDS:
CONTAINING THE FIGURES OF
BIRDS, BEASTS, FISHES, SERPENTS, INSECTS, AND PLANTS:
Particularly, thoſe not hitherto deſcribed, or incorrectly figured by former Authors,
with their DESCRIPTIONS in ENGLISH and FRENCH.
TO WHICH IS PREFIXED,
A new and Correct MAP of the COUNTRIES; with OBSERVATIONS on their NATURAL STATE,
INHABITANTS, and PRODUCTIONS.
By the late MARK CATESBY, F. R. S.
Reviſed by Mr. EDWARDS, of the ROYAL COLLEGE of PHYSICIANS, LONDON.
To the whole is now added a LINNÆAN Index of the ANIMALS and PLANTS.
VOLUME THE FIRST.

HISTOIRE NATURELLE
DE LA CAROLINE, DE LA FLORIDE,
ET
DES ISLES DE BAHAMA:
CONTENANT LES DESSEINS
Des OISEAUX, des QUADRUPEDES, des POISSONS, des SERPENS, des INSECTES, & des PLANTES,
Qui ſe trouvent dans ces Pays-là;
Et en particulier, de ceux qui n'ont point été decrits juſqu' à preſent par les Auteurs, ou peu exactement deſſinés.
Avec leurs DESCRIPTIONS en FRANÇOIS & en ANGLOIS.
On trouve au Commencement
Une Carte de ces Pays, avec des REMARQUES ſur leur ETAT NATUREL, leurs HABITANS, & leurs PRODUCTIONS.
Par Feu Monſieur MARC CATESBY, de la Société Royale,
Reveue par Monſieur EDWARDS, du COLLEGE ROYAL des MEDICINS de LONDRES.
On y a ajouté une Table ſelon le Syſteme de LINNÆUS.
TOME I.

LONDON.
PRINTED FOR BENJAMIN WHITE, AT HORACE'S HEAD, IN FLEETSTREET,
MDCCLXXI.

FIGURE 12-10. Benjamin White printed a third edition of Catesby's text in 1771; the work remained available and in print, with sets of the plates printed and colored in batches for quiring with the stock of text sheets, for forty-five years. (Courtesy of Smithsonian Institution Libraries and the Biodiversity Heritage Library.)

54

MUSCICAPA FUSCA.

The little Brown Fly-Catcher.

WEIGHS nine penny weights. The bill is very broad and flat: the upper mandible black; the lower yellow: all the upper part of the body of a dark aſh colour: the wings are brown, with ſome of the ſmaller feathers edged with white: all the under part of the body duſky white, with a tincture of yellow: the legs and feet are black.

Petit Préneur de Mouches brun.

CET Oiſeau peſe trois dragmes. Son bec eſt fort large & plat: la mandibule ſupérieure eſt noire, & l'inférieure jaune: tout le deſſus de ſon corps eſt d'une couleur de cendre foncée: ſes aîles ſont brunes, excepté que quelques unes des plus petites plumes ſont bordées de blanc: tout le deſſous de ſon corps eſt d'un blanc ſale, avec une nuance de jaune: ſes jambes & ſes piés ſont noirs.

MUSCICAPA OCULIS RUBRIS.

The Red-ey'd Fly-catcher.

WEIGHS ten penny-weights and an half. The bill is lead colour: the iris of the eyes are red: from the bill, over the eyes, runs a duſky white line, bordered above with a black line: the crown of the head is grey: the reſt of the upper part of the body is green: the neck, breaſt, and belly are white: the legs and feet red. Both theſe breed in *Carolina*, and retire Southward in Winter.

Préneur de Mouches aux yeux rouges.

CET Oiſeau peſe un peu plus de trois dragmes. La moitié de ſon bec eſt de couleur de plomb; & l'iris de ſes yeux rouge. Depuis le bec juſqu'au deſſus des yeux s'étend une raye d'un blanc ſale, bordée d'une ligne noire par en haut. Le deſſus de ſa tête eſt gris; & toute le reſte, juſqu'à la queue eſt verd: ſon cou, ſa poitrine & ſon ventre ſont blancs: ſes jambes, & ſes piés rouges. Ces deux dernieres eſpeces de préneurs de mouches ſont leurs petits à la Caroline, *& ſe retirent vers le Sud en Hiver.*

Arbor lauri folio, floribus ex foliorum, alis pentapetalis, pluribus ſtaminibus donatis.

THIS Shrub has a ſlender ſtem, and grows uſually about eight or ten feet high. Its leaves are in ſhape like thoſe of a pear, growing alternately on footſtalks of an inch long; from between which proceed ſmall whitiſh flowers, conſiſting of five petals; in the middle of which ſhoot forth many tall ſtamina, headed with yellow apices. The roots of this Plant are made uſe of in decoctions, and are eſteemed a good ſtomachic and cleanſer of the blood. The fruit I have not ſeen. This Plant grows in moiſt and ſhady woods, in the lower Parts of *Carolina*.

CET Arbriſſeau a le tronc fort menu, & s'éleve ordinairement à la hauteur de huit ou dix piés. Ses feuilles reſſemblent à celles du poirier; & ſont diſpoſées alternativement ſur des tiges d'un pouce de long. Il ſort d'entre les feuilles de petites fleurs blanchâtres, compoſées de cinq feuilles, du milieu deſquelles ſortent pluſieurs longues étamines, qui ont de petites têtes jaunes. On ſe ſert de la racine de cette Plante en décoction; & on lui attribue la vertu de purifier le ſang, & de fortifier l'eſtomac. Je n'en ai point vû le fruit. Cette Plante croît dans des bois marecageaux & couverts, dans les endroits le plus bas de la Caroline.

FIGURE 12-11. In the third edition of 1771, the text sheets—all newly set—repeat the error of "red" instead of "black" for the bird's legs, and the colorist has (incorrectly) followed suit. (Courtesy of Smithsonian Institution Libraries and the Biodiversity Heritage Library.)

on every other page in both volumes, and the Appendix is renumbered 101 to 120, continuously from the end of the main text of the second volume. This edition also had added a catalog of the Linnaean binomials for all of the plants and animals in Catesby's text.[22]

As in the second edition, some of the same textual errors reappear. On page 54, for example, "red" is back again, and in many copies, if not all, the colorist "corrected" the coloring of the bird's legs to agree with the text (figure 12-11). The plates again are quite brightly colored compared to Catesby's own in the first edition, and they also provide evidence that the third edition stayed in print for forty-five years. The text and the dated title-page are the same in all copies—White seems simply and cannily to have printed a large number of copies of the text when it was set in type in 1771 and drawn on that stock as needed. He printed and colored sets of the plates in smaller batches as the market required over the years, evidenced by the fact that some copies of the

FIGURE 12-12. In German, Dutch, and French issues of Mark Catesby's birds, the plates were re-etched and signed in the lower right by Johann Michael Seligmann: see figure 9-11, page 124, for Catesby's original etching. (Courtesy of Smithsonian Institution Libraries and the Biodiversity Heritage Library.)

edition contain plates on either laid or wove Whatman paper with watermark dates ranging from 1794 to 1816.[23] This indication of a continuing market for the work is impressive testimony of its enduring value.

European translations

As the first fully illustrated work on the flora and fauna of any part of North America, based on years of firsthand observation, Catesby's book was enormously important and influential throughout Europe for the rest of the eighteenth century.

Separately from Catesby's own book, several European works came out in the late 1700s translating his text into German, Latin, and Dutch, as well as an edition of the French text for the original bilingual edition, all with illustrations copying his images but re-etched by others. The bird illustrations from Catesby's first volume and appendix were re-etched by Johann Michael Seligmann. Published in parts, the new versions incorporated the vernacular names of the animals and plants prominently positioned within the image area and sometimes contained additional background elements. Credits to Catesby

TABLE 12-2. European editions of selected plates and text from Mark Catesby's *The natural history of Carolina . . .*

Volume 1 Birds	
1749–1770	*Sammlung verschiedener auslandischer und seltener Vogel.* Nuremberg.
1768–1776	*Recueil de divers oiseaux étrangers et peu communs. . . .* Nuremburg
1772–1781	*Verzameling van uitlandsche en zeldzaame vogelen. . . .* Amsterdam.
Volume 2 Fishes, reptiles, etc.	
1750–[?1757]	*Piscium, serpentum, insectorum, aliorumque non nullorum animalium nec non plantarum . . . Die Abbildungen verschiedener Fische, Schlangen, Insecten, einiger andern Thiere, und Pflanzen. . . .* Nuremberg.
1777	*Piscium, serpentum, insectorum, aliorumque non nullorum animalium nec non plantarum . . . Die Abbildungen verschiedener Fische, Schlangen, Insecten, einiger andern Thiere, und Pflanzen. . . .* Nuremberg.
An Account . . . [text only]	
[1755]	*Die Beschreibung von Carolina, Florida und den Bahamischen Inseln.* Nuremberg.
1770	*Histoire naturelle de la Caroline, la Floride et les isles Bahama. . . .* Nuremberg.

as the artist ("ad. viv. del.") appear in the lower left corner and to Seligmann as the engraver ("sculps. et excud.") at the right in every plate (figure 12-12). Seligmann's plates were accompanied by text in German (1749–1770), French (1768–1776), and Dutch (1772–1781). The animals of Catesby's second volume—fishes, reptiles and amphibians, mammals, and insects—were similarly re-etched by Nicolaus Friedrich Eisenberger and Georg Lichtensteger and were published with text in Latin and German (1750–[?1757]) (see table 12-2).

13

The plant collections of Mark Catesby in Oxford

STEPHEN A. HARRIS

Early eighteenth-century North America was an exotic place for British naturalists to explore. Regular links between London and the New World meant that all manner of natural history treasures, not to mention economic wealth, were brought back to London.[1] Among these riches were pressed, dried specimens of plants—herbarium specimens—which had been at the heart of botanical science since the mid-sixteenth century. *Hortus siccus* (literally, a dry garden), a phrase traditionally used to describe a collection of pressed plants, fails to emphasize that a well-labeled herbarium specimen is the link between a plant's identity and the evidence of its occurrence at a particular point in time and space. In the eighteenth century, herbarium specimens were greatly desired by European botanists as they tried to understand and classify the diversity of plants revealed by overseas exploration.[2] Wealthy botanists were willing to pay to have unusual plants represented in their personal collections. However, it was not only the curious who owned and used herbaria; in 1691, the nurseryman William Darby of Hoxton used an herbarium as a sales catalog.[3]

The motivations for plant collecting are manifold. They may be selfish or selfless, and exploration undertaken by the maverick or the institutionalized. There were those for whom plants were merely another spoil of adventure; others were interested in personal fame and fortune. Some were motivated by a desire to know or inspired by an idea. Expeditions were financed by personal fortunes, the generosity of private sponsors or institutions, the largesse of governments or monarchs or personal obsessions. Explorers traveled individually or as members of ad hoc expeditions or commissioned enterprises. Mark Catesby's New World explorations were funded by a cartel of scientists and gentlemen that included notables such as the botanist and diplomat William Sherard and James Brydges, owner of a renowned garden.[4] Scientific plant collection—getting a plant from the field to the herbarium, where it may be studied in detail—is a complex process. The skill and decisions made by a collector in the field determine a specimen's scientific value. The judgments of collection owners and curators determine a specimen's longevity and scientific relevance.

Plant exploration and species discovery are serendipitous activities dependent on the collector, the season, and the place the collector chooses to explore. The botanical explorer must be knowledgeable and make best use of often short flowering and fruiting seasons. In the field, a collector must decide which individual to preserve. The collector must also decide which parts of the selected individual to harvest; usually this will include flowers and/or fruits. Collectors also work by remembering what they have previously collected and, perhaps having a good knowledge of what is already represented in collections. Consequently, the collector filters material. Familiar plants or those known to be in collections are unlikely to be collected. Furthermore, collectors funded by sponsors tend to focus upon novelties; patrons do not want the commonplace.

The technology associated with the preparation of herbarium specimens is largely unchanged from that used in the late fifteenth century. Fresh plants, with fruits and/or flowers, are spread out between sheets of absorbent paper and dried quickly under pressure to keep the samples flat. However, considerable skill is necessary to produce a specimen that displays all of the necessary plant parts. The New England botanical activities of the eighteenth-century Englishman Thomas More (fl. 1670s–1720s) have been criticized because "very few profits to science accrued . . . , the scanty returns clearly having been the consequence of [his] carelessness."[5]

The pivotal role of collectors in the scientific enterprise and the judgments they make are sometimes unappreciated by those who use the data collected. Louis de Bougainville, in the memoir of his voyage around the world, undertaken about four decades after Catesby returned to Britain, made the point forcefully:

> I am a voyager and a seaman; that is, a liar and a stupid fellow, in the eyes of that class of indolent haughty writers, who in their closets reason in *infinitum* on the world and its inhabitants, and with an air of superiority, confine nature within the limits of their own invention. This way of proceeding appears very singular and inconceivable, on the part of persons who have observed nothing themselves, and only write and reason upon the observations which they have borrowed from those same travellers in whom they deny the faculty of seeing and thinking.[6]

The academic naturalist relies on collectors' skills, judgments, and decisions. A collector's decisions and those made by generations of curators about the specimens in their care have consequences for the scientific value of specimens in collections. Two poorly labeled volumes in the Sloane Herbarium (kept in the Natural History Museum, London) are the best-studied collection of Catesby's specimens, especially in relation to the identification of the plants illustrated in the plates to Catesby's *The natural history of Carolina. . . .*[7] Specimens Catesby gave to his friend Samuel Dale are also in the Natural History Museum.[8]

Scattered through the early collections of Oxford University Herbaria are two sets of specimens Catesby sent to his sponsors William Sherard and Charles Du Bois.[9] These specimens and Catesby's own labels will be used to explore Catesby's collecting methods and how curatorial practices may limit the permanent scientific record of his plant-collecting activities.

Catesby's botanical field practices

As Catesby was traveling in a poorly known area, he was collecting all sorts of animals and plants for sponsors with multifarious desires and little understanding of the conditions under which he was working. If Catesby was to succeed, he had to be skillful at managing his sponsors' expectations; preserving animals and plants; protecting his collections from pests, fungi, and damp; and recording what he observed. Besides the dead collections, Catesby was also making living collections, for example, in the form of seeds that he hoped might one day grow in Britain.[10]

Plant collecting is a physically challenging and dangerous activity that requires the collector to be efficient and organized. When he first arrived in North America, Catesby was a visitor reliant on the kindness of strangers to help him find interesting collecting sites. In his first letter to Sherard, after he reached South Carolina in 1722, Catesby assured his patron that "I am told up the rivers there are abundance of fossils and petrifactions and that those parts have not been searched for plants which give me hopes . . . in some measure of effecting what you perticularly desire. . . ."[11] Catesby gradually became familiar with the region, but he had to remind his sponsors of the physical difficulties he was enduring. For example, in the case of one plant, Catesby commented: "ten to one I have an opertunity of collecting seed unless I goe 100 miles or more for it."[12] In unknown areas, Catesby was faced with the dangers of getting lost or encountering wild animals and hostile humans. In early 1723, Catesby articulated these apprehensions to William Sherard: "the fear of . . . meeting with Indians as five of us hapned to doe as we were out a Buffello hunting, tho' they hapned to be those we were at peace with they were about 60."[13] In another letter, probably addressed to Sir George Markham, one of his sponsors, Catesby wrote: "Where a vein or Artery is pricked by the bite of a Rattle Snake no Antidoe will avail any thing, but Death certainly and Suddenly ensues Sometimes in 2 or 3 Minutes which I have more than once seen."[14]

Catesby left no details of the botanical equipment and practices he used in the field. Consequently, inferences must be drawn from the specimens themselves and practices of other botanists. Few botanical instruction manuals were published in the early eighteenth century. Broadsheets for natural history collecting had been published around 1700[15] by James Petiver, an apothecary and friend of Sherard. However, instructions were cursory; Petiver gave his instructions for the preservation of all animals, vegetables, and minerals in fewer than a thousand words.

> In Collecting PLANTS, Pray observe to get that part of either Tree, or Herb, as hath its Flower, Seed or Fruit on it; but if neither, then gather it as it is, and if the Leaves which grow near the Root of any Herb, differ from those above, be pleased to get both to Compleat the Specimen; these must be put into a Book, or Quire of Brown Paper stich'd (which you must take with you) as soon as gathered; You must now and then shift these into fresh Books, to prevent either rotting themselves or Paper.

Petiver was aware of the importance of gathering samples of all available species: "As amongst Forreign Plants, the most common Grass, Rush, Moss, Fern, Thistle, Thorn, or vilest Weed you can find will meet with Acceptance, as well as a scarcer Plant: So in all other things, gather whatever you meet with." But there were limits: "if very common or well known, the fewer of that Sort, will be acceptible to."[16]

In the field, Catesby realized he had to make choices: "to collect every thing [is] impossible but with many Years application."[17] Catesby's approach to collection was simple: "never to be twice at . . . one place in the same season."[18] He did return to similar areas to collect specimens of the southern catalpa (*Catalpa bignonioides*; Indian bean tree):

> The flowers hang in bunches after the manner of the hors chesnut which it resembles at a distance tho' much more beautifull 2 or 3 long pods of the Seed I Sent last year but I suspect not good there being very few to be had, this year they hang prodigiously full in clusters They grow Near Rivers tho' very remote from the Settlements.[19]

The majority of Catesby's specimens have flowers or fruits, although occasionally he collected sterile specimens. A sample of slippery elm (*Ulmus flava*) has the apologetic label: "I never could discover either Seed or blossom this tree has."[20] In the field, Catesby probably placed the collected specimens into a book rather than the "botanical box" (called a vasculum[21]) that started to denote membership of a botanical clique later in the century, since he mentions sending specimen books to his sponsors.[22] Catesby's extant specimens are well flattened, indicating that when he returned from the field, drying may have been completed in a plant press.

However, some specimens are a challenge even to the most determined plant collector. For example, large fleshy flowers can be difficult to dry quickly. Catesby successfully preserved waterlily flowers,[23] but he met his match with an American lotus (*Nelumbo lutea*) in 1722 (figure 13-1). On the reverse of a recently rediscovered pen-and-ink drawing (figures 13-2 and 13-3) that he sent to William Sherard, Catesby noted: "This seems to be Clusius his Egyptian Bean The flowr I could not preserve so have sent this Scetch. The fruit here is called Water Chinkipins which I have not seen yet. it grows in Water."[24]

FIGURE 13-1. Leaf of American lotus (*Nelumbo lutea*) collected by Mark Catesby in Carolina in 1722 and sent to William Sherard (Sher-1090-10, Oxford University Herbaria). Main label (*top right*) in William Sherard's hand, with additions in Jacob Dillenius's hand; label (*bottom left*) in an unknown hand (240 × 371 mm).

FIGURE 13-2. Mark Catesby's pen-and-ink sketch of American lotus (*Nelumbo lutea*) flower and leaf, sent to William Sherard to accompany the leaf (figure 13-1) (Sher-1090-10, Oxford University Herbaria) (311 × 382 mm).

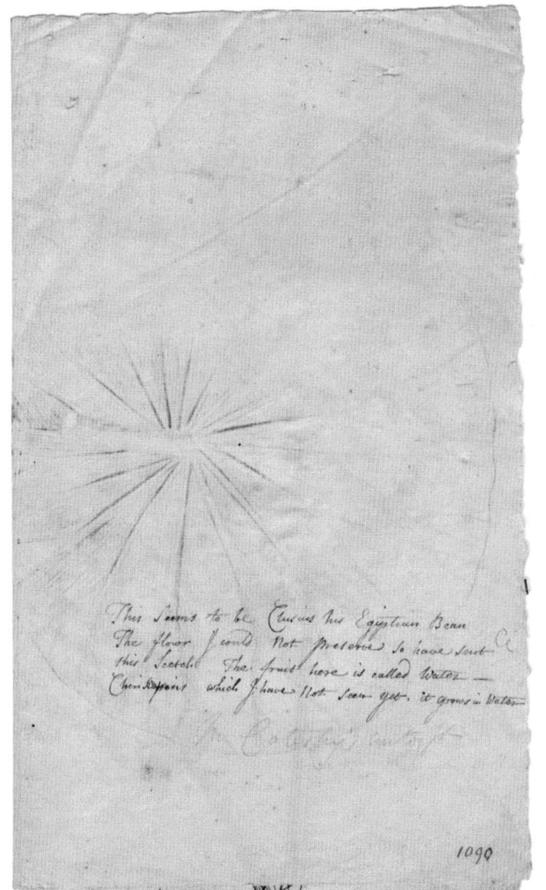

FIGURE 13-3. Mark Catesby's autograph on the verso of the American lotus (*Nelumbo lutea*) sketch (figure 13-2) (Sher-1090-10, Oxford University Herbaria) (194 × 382 mm).

Included on the drawing of the flower is a sketch of the preserved leaf he sent to Sherard at the same time. A sketch was an elegant, if not entirely satisfactory, solution to the problem of plant preservation.

Catesby had a dozen sponsors to satisfy, including Sir Hans Sloane and William Sherard; consequently, he had to collect several specimens of each different plant:

> Out of the last Sent I reserved Some which I now send So that those with the triplicates I presume may be spared, not that I know who they will be acceptable to except Mr Dale for I have sent 2 Books to Sr Hans, So pleas to doe with them what you think fitt.[25]

Furthermore, he had to keep track of the information he collected about each specimen; Sherard advised a numbering system,[26] and Catesby appears to have used this to link herbarium specimens and seed samples.[27] His original paper labels apparently had two slits in them for attaching to the cut stem of the specimen,[28] although these labels appear to have been added when specimens were sent to Britain rather than at the time of collection.

For a specimen to be most useful, information must be recorded about its appearance in life, its habitat, and where and when it was collected; a specimen without its label has limited scientific value. Petiver made the point plainly: "If to any ANIMAL, PLANT MINERAL & c. you learn its Name, Nature, Vertue or Use, it will be still the more Acceptible."[29] Most Catesby specimens have no associated information or only collection year and general collection area. However, where his specimens are labeled, information is provided on species' form and habitat. We learn that sand heath (*Ceratiola ericoides*) "grows on barren Hills usually 6 or 8 foot high Those with bare leaves was gathered in Aprile the other in September."[30] A specimen of clammy locust (*Robinia viscosa*) is a "kind of Acacia I never saw but on one Spot of high land near Savanna river This was a Shrub—occasioned by the fire The flower is a little tinctur'd with purple the Stalks not hairy . . . but very glutinous and clammy. . . ."[31] Catesby also took Petiver's advice and collected information on vernacular names and uses. Thus, May apple (*Podophyllum peltatum*) "is called here Ipecacoana" and it is used for the same purpose as ipechanha, although "3 times the quantity is given" (figure 13-4).[32] Basswood (*Tilia americana* var. *heterophylla*) "is called Wahoo or the rope-Tree from the toughness of it's Bark being much Made use of instead of Rope."[33] In the case of hairy crabgrass (*Digitaria sanguinalis*), Catesby's label is simple and direct: "Crop Grass The plague of the planters, as it is in Virginia" (figure 13-5).[34] Catesby was keen to satisfy his sponsors; six weeks after his arrival, he asked Sherard's advice as to "whether you require to know the soyl [in which] every plant and tree grows. . . ."[35]

Some specimens were clearly vouchers for seeds Catesby was sending back to Britain. For example, a specimen of a trout lily (*Trillium* sp.) is labeled "This kind of Saterion (as I think it is) is of a pale red has . . . between a Bulbose and Tuberus root. . . . The seed is N^O 8."[36] As he spent more time in North America,

FIGURE 13-4. Mayapple (*Podophyllum peltatum*), labeled by Mark Catesby (Sher-1078-3, Oxford University Herbaria) (240 × 361 mm).

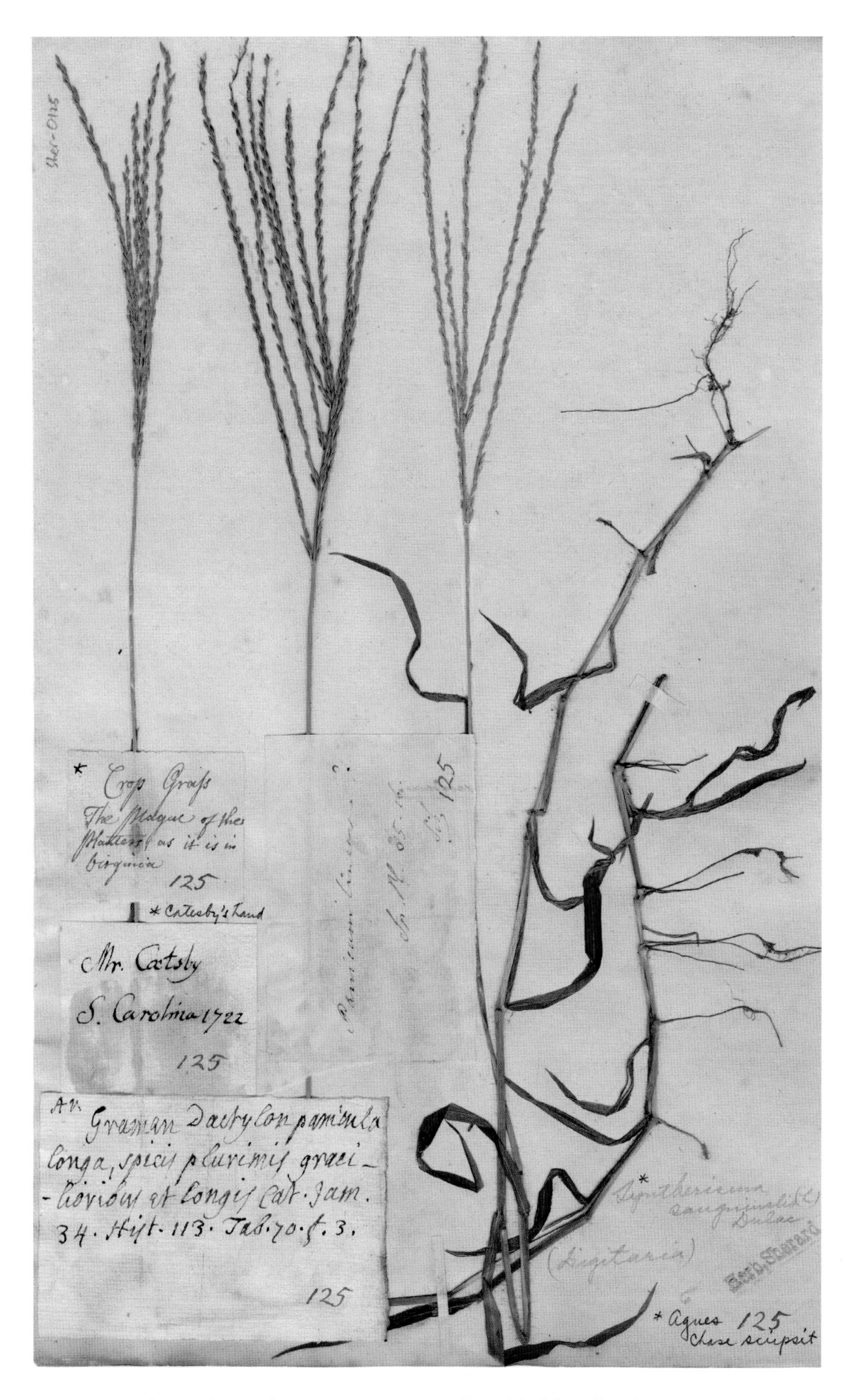

FIGURE 13-5. Hairy crabgrass (*Digitaria sanguinalis*) collected by Mark Catesby in 1722 in South Carolina, with labels by Catesby and William Sherard, together with a determination label in John Sibthorp's hand (Sher-0125-b, Oxford University Herbaria) (229 × 369 mm).

Catesby became more knowledgeable, probably refining his art to the desires of his sponsors, the realities of his working conditions, and his own interests.

Once dried, specimens had to be kept away from pests and fungi; sponsors were not interested in moldy specimens or ones eaten by "Vermin which often breed in them."[37] Some specimens were more easily preserved than others:

> In looking over the Books I now send I find 5 or 6 plants Mouldy, tho' they were perfectly cured and well dryed, when put up and are kept in a dry, upper, South Room. In turning over those for Sr Hans I find the same plants mouldy and only those, so that I find some plants tho' not Succulent are perticularly Subject to it. The Weather is indeed and has for some time been, extream Moist Cloudy and Sultry.[38]

Keeping specimens in a fit state for transport back to Britain would have been a constant concern for Catesby.

Once Catesby had collected sufficient specimens, he sent them to Sloane and to Sherard, who distributed them to Catesby's sponsors, particularly Charles Du Bois. However, returning specimens to Britain was a high-risk enterprise. Outside of the collector's care, specimens might be neglected under sail and suffer insect, fungal, and water damage. By the end of the eighteenth century, Alexander von Humboldt had summed up the collector's plight: "in seas infested with pirates a traveller can only be sure of what he takes with him."[39]

Curation practice and botanical evidence

Once Catesby successfully collected, dried, and transported his specimens to Britain, they became part of the dried plant collections of gentlemen and were perhaps integrated into the formal academic exercise of recording American plant diversity. The latter required collection owners to order and identify specimens and publish their results. Alternatively, owners might allow others to access ordered collections. The Sloane Herbarium has been continually studied, and the results from these studies published, since it was established. In contrast, investigations of the Sherard and Du Bois Herbaria have been episodic.

The Sherard Herbarium, bequeathed to the University of Oxford in 1728, is the core collection of Oxford University Herbaria. Through a network of European botanical friendships, William Sherard amassed a global herbarium to rival that of Hans Sloane.[40] Sherard studied law at Oxford, forming a strong and lasting friendship with Jacob Bobart, son of the first Keeper of the Oxford Physick Garden, through whom he appears to have developed his botanical interests.[41] Sherard rarely dated specimens in his collection, making it difficult to establish when the herbarium was started. Since an herbarium was an essential tool for a serious eighteenth-century botanist, it is likely it was started in about 1680. As a wealthy, well-connected, amiable man, Sherard had the resources to amass specimens from across the known world; his aim was to publish an account of the world's plants.[42] When he died in 1728, his herbarium, which comprised some twelve thousand sheets, was the largest and most celebrated in Europe.[43] Over the next century, it continued to grow and today comprises about twenty-one thousand specimens.

Numerous collectors, including John Bartram, John Clayton, William Clerk, and William Houstoun, contributed North American specimens to Sherard's herbarium. However, Catesby was the largest single North American contributor. In the Sherard Herbarium, 348 specimens bear Catesby's name and are localized to Virginia, Carolina, South Carolina, and Providence (Bahamas).[44] The Sherard Herbarium also contains specimens of cultivated plants, usually raised at Eltham, the home of Sherard's brother, from seeds collected by Catesby. Among these is a specimen of a boneset (*Eupatorium* sp.) labeled, in William Sherard's hand, "Mr Catsby S. Carolina. 1723.," to which Johann Jacob Dillenius, Sherard's aide and eventually Professor of Botany in Oxford, added the note: "Eltham from Carolina Seeds."[45] There is also a specimen of chickpea (*Cicer arietinum*) (figure 13-6), a European introduction, annotated as "Eltham 1725. ex sem. ad. Catesby misso."[46] At the beginning of the nineteenth century, the authority on North American plants, Friedrich Pursh, was delighted to find Catesby's plants in Oxford and thought Sherard's herbarium "the most complete collection of North American plants."[47]

During his lifetime, Sherard actively worked with the specimens in his herbarium, adding polynomial names as he constructed his *Pinax*, a comprehensive listing of plant names.[48] The herbarium was reordered by Dillenius, and the present order was introduced at the end of the nineteenth century.[49] However, the apparent order of the Sherard Herbarium belies considerable complexity, confusion even, over the labeling of individual herbarium sheets. This complexity is a consequence of its long history and the attitudes to herbarium stewardship by its curators and professors of botany over nearly three centuries. Sherard's own practices have not helped interpretation, since he appears to have habitually relabeled, in his own hand, the specimens he got from correspondents and then discarded the original labels. Many of the specimens are mounted on pre-nineteenth-century paper, but it remains unclear when they were mounted; the longer that specimens and labels are not mounted, the more opportunities there are for the two to become mixed.[50] Numbers on Sherard sheets were added by William Baxter in the late 1840s[51] over a century after the collection had been acquired by the university. Some of the confusion created by these curatorial practices can be seen in two specimens with Catesby labels.

There has clearly been confusion over labeling, making it difficult to determine which (if any) of the labels are associated with the specimen. For example, a specimen of Maryland meadowbeauty (*Rhexia mariana*) is mounted with two eighteenth-century labels (figure 13-7).[52] One label bears the inscription "Mr. Catsby South Carolina 1723," apparently in William Sherard's hand, and "Lychn?" in Dillenius's hand. The other label states: "Lysimachia siliquosa n[on] papposa, fl[ore] amplo luteo, fol[iis] Linariæ glabris?" in Sherard's hand,

FIGURE 13-6. Chickpea (*Cicer arietinum*) grown at Eltham in 1725 from seeds supplied by Mark Catesby, labeled by William Sherard (Sher-1504-10, Oxford University Herbaria) (235 × 357 mm).

to which was added "Entr."[53] in John Ray's hand. A specimen of Carolina false vervain (*Stylodon carneus*) is mounted with three labels:[54]

"Virginia" apparently in Catesby's hand.

"Verbena scoradonia fol. Spicata mariland." in Sherard's hand, to which is added "Entr." in Ray's hand.

"Ver~~onica~~bena Caroliniana spicata erecta, folijs asperis, capsula bicapsulari, facia Apariscas." and "Mr Catsby. Carolina. 1722." in Sherard's hand and "Verbena Scoradonia folio Spicata Marilandica D. Sherard Raj. Hist. III. 287. n. 21." and "not in Pluk[enet]. Pet[iver]. ~~Raji.~~" in Dillenius's hand.

FIGURE 13-7. Maryland meadowbeauty (*Rhexia mariana*) illustrating labeling ambiguities on sheets attributed to Mark Catesby, with labels in William Sherard's hand with additions by Jacob Dillenius and John Ray (Sher-0769-30, Oxford University Herbaria) (240 × 372 mm).

In both of these specimens, Catesby's labels cannot be associated with those bearing Ray's mark, since they post-date Ray's death in 1705.

However, Catesby was not above mislabeling specimens in the field. Apparently responding to a request for information from Sherard, he added the note: "This is what produces ye Square Nut you desire to know concerning, it is a Shrub"[55] to a specimen of the American snowbell (*Styrax americanus*) from South Carolina. Dillenius annotated the specimen with "this must be a mistake for ye flower does not agree. there is a Specimen of this very like if not ye same amongst ye Styrax," to which an unknown hand has added the comment "it is a mistake."

Further confusion was created at the end of the nineteenth century, when specimens were abstracted from the Sherard Herbarium to create the Dillenian Herbarium.[56] Specimens that appeared to have been grown at Eltham and used in the production of the plates for Dillenius's *Hortus Elthamensis*, which was published in 1732, were separated. However, the decision to separate the Dillenian Herbarium was flawed, since not all the specimens were cultivated, and some were collected very much earlier than the publication date of *Hortus Elthamensis*. Consequently, Catesby specimens in the Dillenian Herbarium are part of the Sherard Herbarium.

The Du Bois Herbarium was apparently given to Oxford University in the mid-eighteenth century. Between 1697 and 1724 Charles Du Bois (1658–1740), a London-based businessman, Treasurer of East India Company, and friend of the Sherard brothers, amassed an herbarium of some thirteen thousand specimens. The herbarium was originally bound in seventy-four elephant folio volumes, perhaps in a similar manner to the present-day Sloane Herbarium.[57] The Du Bois Herbarium was unbound, remounted (figure 13-8), and organized into its present form in the 1880s. This curatorial decision produced considerable confusion and was eventually regretted by the then curator.[58] Some of the American collections have been studied, but generally this herbarium has been the subject of little research. Pursh made no mention of the collection when he worked in London and Oxford in the early nineteenth century.[59]

The main collectors who contributed North American specimens to Du Bois's herbarium were David Krieg (c. 1670–1710) and William Vernon (1666/67–c. 1715), who collected in Maryland, and Mark Catesby. In Du Bois's herbarium, 458 specimens bear Catesby's name and are localized as being from Virginia, Carolina, and South Carolina.[60] Du Bois was one of Catesby's sponsors, albeit a difficult one to satisfy: "The discontent of Mr Du-Bois and the trouble he gives my Friends in receiving his Subscription is such that I had rather be without it. . . ."[61] Sherard appears to have been Catesby's intermediary with Du Bois, but many of Catesby's specimens from the Carolinas in the Du Bois Herbarium are not represented in the Sherard Herbarium; Sherard was not just sending Du Bois duplicates. Unlike other specimens in the Du Bois Herbarium, few of Catesby's collections have polynomial names. Du Bois

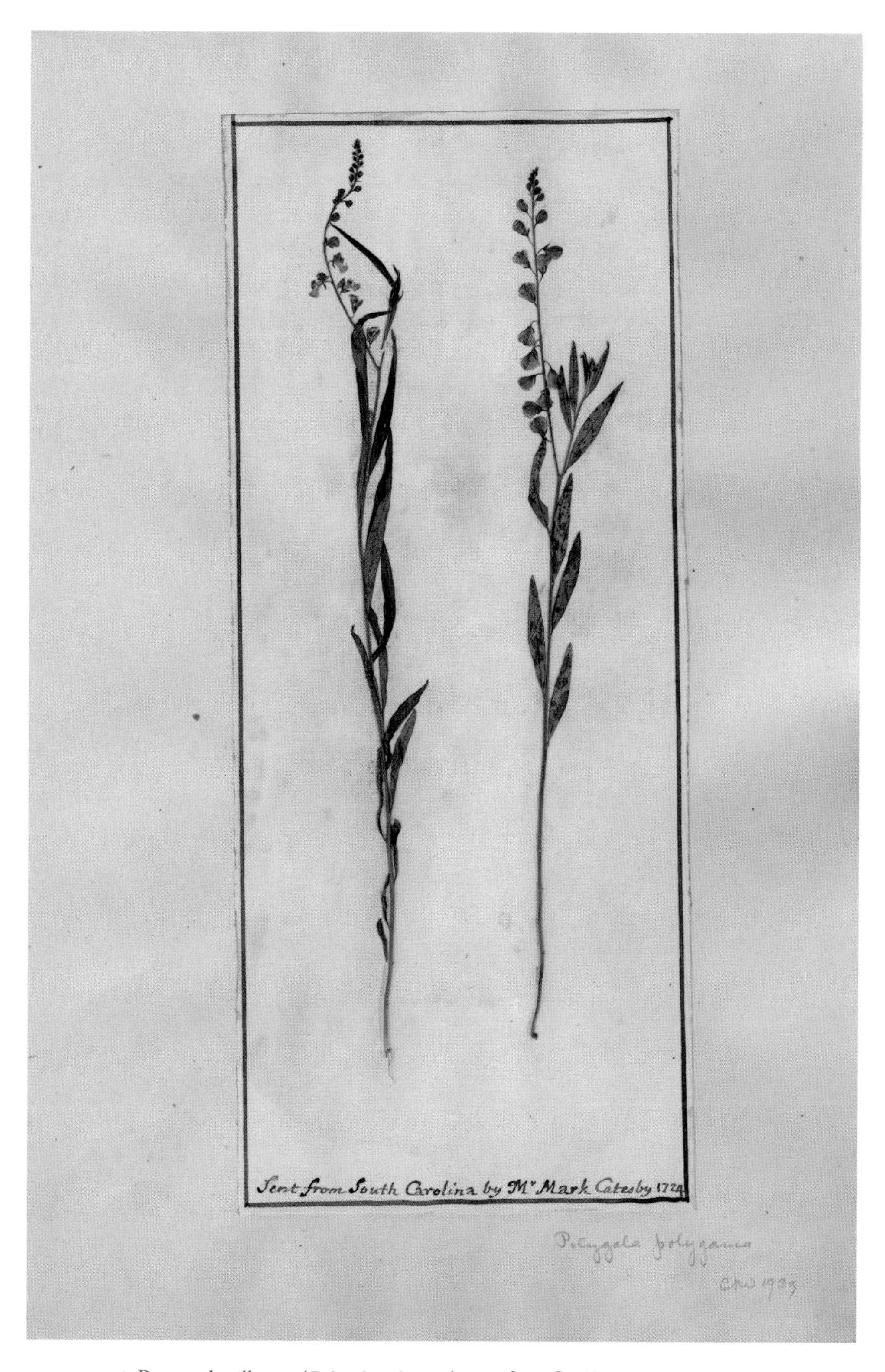

FIGURE 13-8. Racemed milkwort (*Polygala polygama*), sent from South Carolina by Mark Catesby in 1724. A typical Catesby specimen in the Du Bois Herbarium, with handwriting that appears to be that of Du Bois's amanuensis (DB-00087304M, Oxford University Herbaria) (original paper mounted on nineteenth-century paper, 254 × 381 mm).

appears to have done little work with Catesby's specimens other than having them mounted and filed.

Plant collectors who stay for short periods in any area can only collect opportunistically. Catesby had the prospect of studying plants of the Carolinas systematically at different times of the year. The evidence from Catesby's herbarium specimens in Oxford shows that he took these opportunities. However, Catesby was constrained by the demands of his patrons and the search for botanical novelties. Specimens only become part of the botanical mainstream if they are used by other researchers. Few of Catesby's Oxford specimens have been used as part of research since they were first collected; for example, his field sketch of the American lotus has been overlooked. Once Catesby had sent specimens to Sherard and Du Bois, he appears never to have seen them again. Furthermore, the curatorial history of the Oxford University Herbaria meant that the specimens Catesby sent to Sherard and Du Bois were unavailable to all but the most persistent researcher.

14

Carl Linnaeus and the influence of Mark Catesby's botanical work

C. E. JARVIS

On 2 June 1753, at Uppsala University in Sweden, a student made a public defense of a dissertation. As was the practice, the student, Jacob Bjurr (1732–1774),[1] had paid for its printing, but the content was primarily the work of his professor, Carl Linnaeus (1707–1778). In this dissertation, Linnaeus railed against botanical publications containing hand-colored copper plates because their cost was so high "that not a few sons of Botany who are reared in modest circumstances are compelled to do without such high-priced books."[2] Along with five other works, he explicitly mentioned Mark Catesby's *The natural history of Carolina, Florida and the Bahama islands* as one of the offending books which Linnaeus described as produced through patronage, wealthy men having supported Catesby's voyage to America and having paid well for his pictures. This did not mean that Linnaeus failed to make use of Catesby's book—he used it very extensively—but its high cost almost certainly put it beyond the reach of a professor in a Swedish university.

Carl Linnaeus saw it as his task to order the biological works of the Creator, and he was responsible for introducing a revolution in the way that organisms were classified, particularly in plants for which he adopted a practical system based on the number and arrangement of the floral parts.[3] However, his single most important legacy is undoubtedly the binomial system for the scientific naming of organisms, a system that is still in use today for plants and animals.[4] Introduced in 1753 for plants[5] and five years later for animals,[6] these two-part names and the publications in which they appear are the starting point for modern scientific nomenclature. Catesby's "high-priced book"—hand-colored copies had cost two guineas (£2 2s. od.) a part, that is, twenty-two guineas (£23 2s. od.) for the complete work—played a modest part in this nomenclatural revolution.

For example, the original publication (see figure 14-1) of Linnaeus's binomial name *Vinca lutea*[7] shows the generic name *Vinca* accompanied by a short

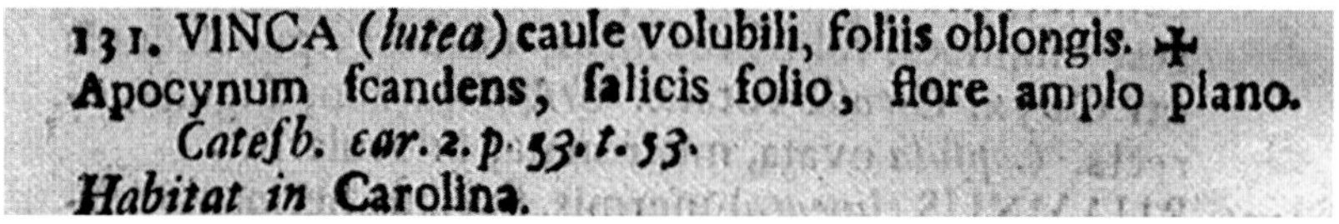

131. VINCA (*lutea*) caule volubili, foliis oblongis. ♃
Apocynum fcandens, falicis folio, flore amplo plano.
Catefb. car. 2. *p.* 53. *t.* 53.
Habitat in Carolina.

FIGURE 14-1. C. Linnaeus, 1756, *Centuria II plantarum*, 12. In Linnaeus's original publication of the Latin name *Vinca lutea*, the generic name *Vinca* was accompanied by (*lutea*) with a short descriptive phrase in Latin, as well as a reference to Catesby's *The natural history of Carolina . . .* , volume II, page and plate 53. (© Natural History Museum, London; reproduced by courtesy.)

descriptive phrase in Latin, in this case, "caule volubili, foliis oblongis," that is, with twining stems and with leaves oblong in shape. This Latin phrase served both as a formal name and as a description, allowing this species of *Vinca* to be distinguished from the two others Linnaeus knew (*Vinca major* and *Vinca minor*, which had either trailing or upright stems, and leaves of a different shape).[8] Linnaeus's innovation was to introduce a single word, the specific name or epithet, in this case, the word *lutea* (enclosed in brackets; see figure 14-1), to serve as a shorthand for the long name and in this way to separate the descriptive and naming functions. Linnaeus also cited Mark Catesby's own descriptive phrase-name "Apocynum scandens, salicis folio, flore amplo plano" and the accompanying plate (no. 53) from the second volume of Catesby's book, as well as indicating that he believed the species came from Carolina.

The type method

In deciding to which species a particular scientific name should be correctly applied, a taxonomist designates a "type" specimen to serve as a permanent reference point, usually a pressed, dried herbarium specimen for plant names. In the case of plants, types can sometimes be illustrations (for example, original drawings or published plates) and, for *Vinca lutea*, as Catesby's description and plate of this species were apparently the only ones that Linnaeus had seen, Catesby's illustration (figure 14-2) serves as the type of Linnaeus's scientific name for the plant commonly known as hammock viper's-tail.[9]

Linnaeus's knowledge of plants

From a modest rural upbringing, Linnaeus had developed a good knowledge of the flora of his native Sweden and parts of what is now Finland, and he was familiar with those species from farther afield that either would survive the harsh Swedish winters or could be nurtured in heated glasshouses or orangeries. Through his Swedish mentors, he had access to some of the standard botanical texts, although information on the "productions of nature" from the New World would not have been easy to come by in Sweden.

FIGURE 14-2. Plate 53, M. Catesby, 1737, *The natural history of Carolina* . . . , volume II. In deciding to which species a particular scientific name should be applied, taxonomists designate a "type" specimen as a permanent reference point. For *Vinca lutea* (now called *Pentalinon luteum*, hammock viper's-tail), Catesby's description and plate were apparently all that Linnaeus had seen and thereby became the type specimen for Linnaeus's scientific name for this plant. (© Natural History Museum, London; reproduced by courtesy.)

In 1735, at the age of twenty-eight, Linnaeus left Sweden for what would be three years away, with most of the time spent in the Netherlands. There he rapidly acquired a medical degree, met the notable men of botany, and marveled at the range of new plants being grown in Dutch gardens.[10] The milder climate of the Netherlands made it possible to grow many more species than was possible in Sweden, and Dutch gardens were uniquely enriched by the Dutch East India Company's activities, particularly in botanically rich South Africa and the East Indies.

Early in his Netherlands visit, Linnaeus made the acquaintance of men who possessed good botanical libraries, and it is possible that he first became aware of many New World plants through books in the possession of men such as Johannes Burman, Herman Boerhaave, Albert Seba, Adriaan van Royen, and Johan Gronovius.[11] Important early publications based on in situ observations of New World plants included those of the French monk Charles Plumier on the plants of Haiti and Martinique,[12] the remarkable Maria Sibylla Merian on the plants and insects of Surinam,[13] and Sir Hans Sloane on the natural history of Jamaica.[14] In addition, the publications of English armchair collectors like King William III's botanist Leonard Plukenet[15] and the London apothecary James Petiver[16] also yielded information on previously undescribed North American species, many from the British colonies.

Access to Catesby's *The natural history of Carolina . . .*

From September 1735, Linnaeus was employed as personal physician and botanist to the wealthy Anglo-Dutch financier George Clifford III. Clifford owned a large estate, the Hartekamp, near Haarlem, where he indulged his passion for natural history in gardens and glasshouses full of the rarest plants, as well as a menagerie. He also had an extensive library and kept a herbarium. Linnaeus was given the task of recording the plants that were cultivated in the garden, and the result was his *Hortus Cliffortianus*, written during 1736 and 1737 and published in 1738.[17] This book contained thirty-six plates, many of them engraved from drawings prepared by Georg Dionysius Ehret. The book also included a catalog of the books present in Clifford's library (figure 14-3), in which two entries (numbers 183 ["193"] and 184) confirm that Clifford possessed most of the plates of Catesby's *The natural history of Carolina . . .* that had been published.

The plates and text of Catesby's work were published in parts between May 1729 and July 1747, and, in 1737 (when Linnaeus was preparing the text for *Hortus Cliffortianus*), Linnaeus stated that Clifford's library contained 140 plates, numbers 1–100 from the first volume and plates 1–40 from the second, all of which had been published by mid-January 1736.[18] The subsequent plates 41–60, although published in London in April 1737,[19] presumably did not reach Clifford in time for Linnaeus to see them in the Netherlands. Although *Hortus Cliffortianus* features frequent references to Catesby's plates, detailed scrutiny of the book (which was completed in 1737 though not distributed until the following year) shows that twenty-six of Catesby's plates that feature plants were

193. Catesby *Marcus.*
- Hiſtoria natur. Carolinæ, Floridæ, Bahamæ. tomus 1.
- - Londini. 1731. fol. maxim. p. 100. t. 100. angl.-gallic.
184. - Hiſtoria natur. Carolinæ, &c. tom. 2dus. inceptus, ad pag. 40. t. 40.
Figuræ nitidiſſimæ, in tomo 1mo aves & plantæ, in 2do piſces & plantæ.

FIGURE 14-3. C. Linnaeus, 1738, *Hortus Cliffortianus*, confirming that Clifford possessed most of the plates of Catesby's *The natural history of Carolina . . .* that had by then been published. (© Natural History Museum, London; reproduced by courtesy.)

cited there, with two of them from the January 1736 part (*NH* II: plates 26, *Phyllanthus epiphyllanthus*, and 33, *Amyris elemifera*).[20] However, although five of the plates[21] that were published in April 1737 (*NH* II: plates 41–60) do relate to species featured in *Hortus Cliffortianus*, the accounts of these species contain no reference to Catesby, suggesting strongly that those plates were indeed unavailable to Linnaeus when he was completing the text for his book.

Comparison with citations of Catesby plates in the first part of Johan Gronovius's *Flora Virginica*,[22] published in Leiden the year after *Hortus Cliffortianus* was distributed, suggests that the third part of Catesby's second volume containing plates 41–60 (published in early April 1737) must have reached the Netherlands not long after Linnaeus left the country, because Gronovius was able to cite Catesby's illustration of "*Chrysanthemum Americanum* . . ." in the synonymy of the species Linnaeus would later name *Rudbeckia purpurea*.[23] Like George Clifford, Linnaeus's friends Adriaan van Royen and Gronovius were also subscribers to Catesby's book, their names featuring among the 155 "Encouragers."[24]

In August 1736, Linnaeus made a month-long visit to England ostensibly to acquire new plants for his employer, George Clifford, and met the pre-eminent English botanists and collectors of the day. Notable among these were Sir Hans Sloane, Professor Johann Jacob Dillenius, and the Superintendent of Chelsea Physic Garden, Philip Miller (1691–1771). He also met the English Quaker and mercer Peter Collinson, who, when his business interests (which were particularly strong in North America) allowed, was an enthusiastic gardener and entomologist. Collinson and Linnaeus evidently impressed one another, and the Englishman provided the Swede with specimens and seeds and continued to supply him with North American plants long after Linnaeus returned to Sweden in 1738. The correspondence between the two men started in 1736, and Collinson was a major conduit of information, specimens, and news to Linnaeus from North America and England until his death in 1768.[25]

While Linnaeus's own library (today housed by the Linnean Society of London) was very extensive, it lacks a copy of Catesby's book, and there is no mention of the work in the manuscript catalog of Linnaeus's library prepared by its new owner, James Edward Smith, when it reached London in 1784.[26] This is in contrast to the situation with another heavily illustrated book, Linnaeus's now absent copy of Dillenius's *Hortus Elthamensis*, present in Smith's catalog of 1784 but later sold by its new owner, and now in the University Library in Jena.[27] An observation in a letter sent by Collinson to Linnaeus seems to confirm that the Swede's library lacked a copy in May 1749.[28] Copies of publications that Linnaeus owned were frequently marked by him with the binomials that he published from 1753 onward, but, to date, no copy of Catesby's book containing such Linnaean annotations appears to have been traced.[29] It seems almost certain that Linnaeus's library never held a copy, a view also held by his bibliographer, John Heller.[30]

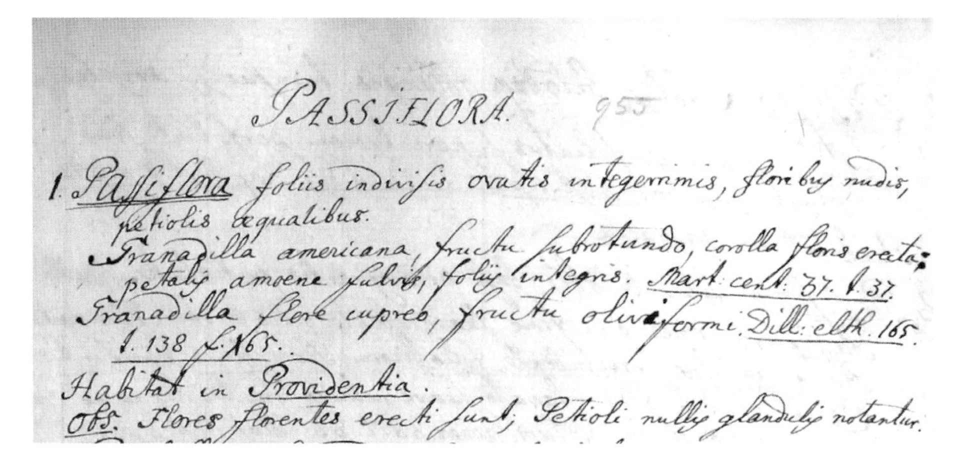
Passiflora. 955
1. Passiflora foliis indivisis ovatis integerrimis, floribus nudis, petiolis aequalibus.
Granadilla americana, fructu subrotundo, corolla floris erecta; petalis amoene fulvis, foliis integris. Mart. cent. 37. t. 37.
Granadilla flore cupreo, fructu oliviformi. Dill. elth. 165. t. 138 f. 165.
Habitat in Providentia.
Obs. Flores florentes erecti sunt; Petioli nullis glandulis notantur.

FIGURE 14-4. Linnaeus's draft manuscript of *Species plantarum* (1746–1748): entry for "Passiflora . . . Habitat in *Providentia*" (that is, the Bahamas). (© Linnean Society of London; reproduced by courtesy.)

The absence of a Linnaean copy does raise some questions about how Linnaeus was able to get access to Catesby's work after he left the Netherlands in May 1738. The presence of references to plates from the ninth part of Catesby's book in an early (1746–1748) draft manuscript of *Species plantarum* shows that Linnaeus must have seen these twenty plates in Sweden by then,[31] but possibly more significant is the lack of any direct reference to any of the plates in parts 10 and 11 (Appendix).[32] This suggests that, toward the end of 1748, Linnaeus had still not seen these last two parts, though, intriguingly, drafts for two of the species (the future *Cordia sebestena* and *Passiflora cuprea*; see figure 14-4) include "Providentia," indicating that Linnaeus believed them to occur in the Bahamas. It is difficult to see how he could have acquired this information other than via Catesby, but it is unclear why he did not cite the corresponding plates at the same time.

Although Linnaeus's comments on the high cost of the book and the absence of a copy in his library make it unlikely that he ever himself owned a copy of Catesby's book, an interesting series of letters written to Linnaeus by Collinson touch on the supply of the last three parts of Catesby's book, via Linnaeus, destined for the wealthy Swedish industrialist, and keen entomologist and natural history collector, Charles De Geer (1720–1778). De Geer, who lived at Leufsta near Uppsala, was a friend of Linnaeus, and it seems highly likely that the latter would have been able to consult Catesby's book in De Geer's library. On 14 October 1748, Collinson wrote to tell Linnaeus that he was sending "the Books that was wanting to make up Mr. Catesbys Work Compleat" via Captain Tornland on the *Assurance*, sailing for Stockholm within a few days. The "books" comprised the ninth and tenth parts of *The natural history of Carolina* . . . and the supplement (the Appendix) at two guineas for each part, which, with postage and packing, amounted to £6 8s. 0d. (see figure 14-5).[33]

Consequently, Linnaeus may have had the opportunity to see and study the final forty plates for the first time around the end of 1748, that is, after he had abandoned his early draft of *Species plantarum* (most of these last forty plates are cited in the version published in 1753).

Another set of plates that Linnaeus may have consulted is suggested by Catesby's listed "Encouragers," among whom was "Her MAJESTY the QUEEN of SWEDEN," that is, Queen Ulrika Eleonora (1688–1741).[34] Linnaeus was subsequently much occupied with cataloging the collections of her natural history–loving successor, Lovisa Ulrika (1720–1782), so he may well have had access to the Royal Library and, with it, the Queen's copy of Catesby's work.

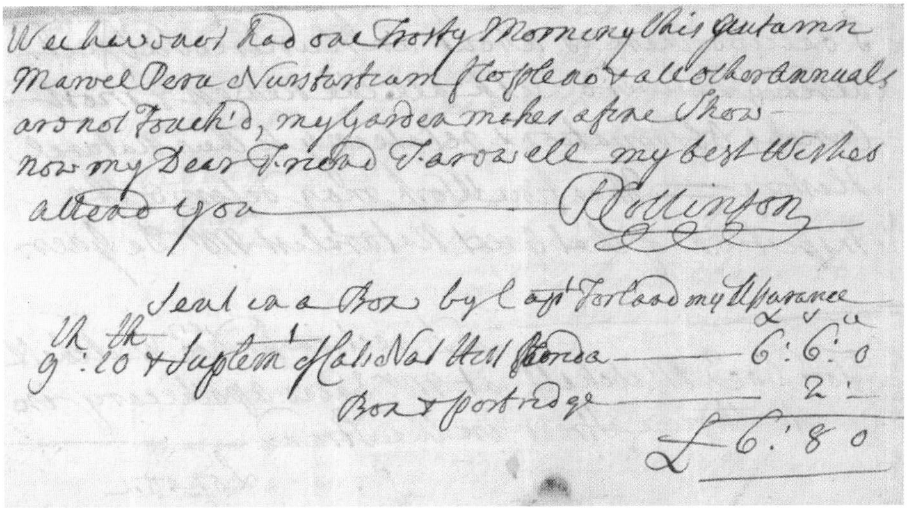
We have not had one Frosty Morning this Autumn
Marvel Peru Nasturtium ... & all other Annuals
are not touch'd, my Garden makes a fine Show—
now my Dear Friend Farewell my best Wishes
attend you —— PCollinson

Sent in a Box by Capt Forland my Assurance
9th 20 & Septem of Hist Nat Carolina —— £ s d 6:6:0
Box & portridge —— 2:—
£6:8:0

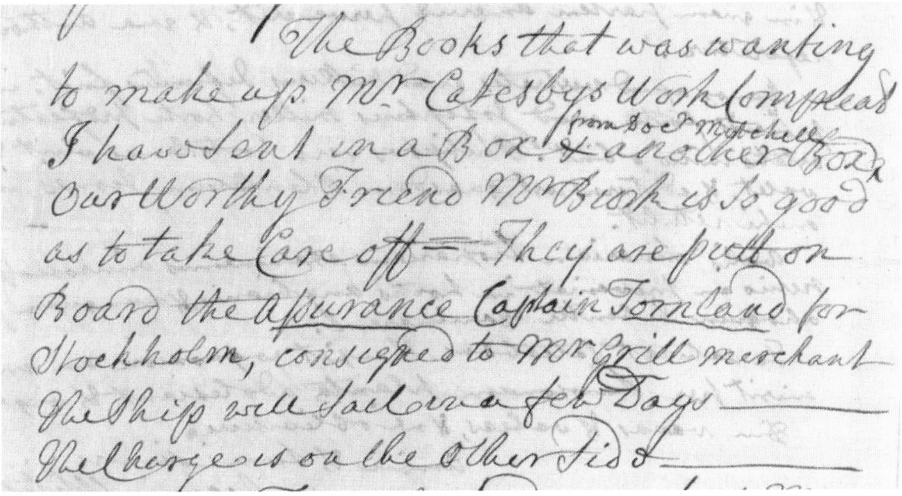
The Books that was wanting
to make up Mr Catesbys Work Compleat
I have Sent in a Box & another Box from Doct Mitchell
Our Worthy Friend Mr Brook is so good
as to take Care off— They are putt on
Board the Assurance Captain Tornland for
Stockholm, consigned to Mr Grill merchant
The Ship will Sail in a few Days ——
The Charge is on the Other Side ——

FIGURE 14-5. Extracts from a letter by Peter Collinson, dated 14 October 1748, telling Linnaeus that he was sending the last two parts of Catesby's *The natural history of Carolina* . . . as well as the Appendix. With postage and packing the cost was £6 8s. 0d. (© Linnean Society of London; reproduced by courtesy.)

Although Linnaeus cited plates from almost all of the published parts (the exception being the sixth, which contained no plants), detailed comparison of the plates with their citation in Linnaeus's publications and their identities shows a puzzling distribution.[35] While all but twelve of the seventy-seven plants depicted in the first four parts were cited by Linnaeus, parts 5 and 7 show a very different pattern. Although eighteen plants feature in these forty plates, only two of them were cited by Linnaeus (*NH* II: plate 26 under *Phyllanthus epiphyllanthus*, and plate 33 under *Amyris elemifera*). Plates from subsequent parts are cited much more heavily, with twenty-six out of forty-seven plants cited from parts 8 and 9, twenty-three out of twenty-eight from part 10, and fifteen out of eighteen from the Appendix (part 11).

The low incidence of citation from parts 5 and 7 is intriguing. Given that Linnaeus stated that these parts were present in Clifford's library, it seems unlikely that part 5 (from which no plate was cited) was unavailable to him in the Netherlands or later in Sweden. Plates depicting species in plant groups in which Linnaeus had scant interest (for example, *NH* I: plate 36—a fungus; *NH* II: plate 38—an aquatic grass-like plant) were perhaps understandably passed over, but the omitted plates are all of flowering plants. While at least some of them (for example, *NH* I: plates 82, 83, and 85) lack the key floral characters necessary to place a species within the Linnaean sexual system, perhaps it was simply chance that so many poorly characterized plant species appeared in these two parts, leading to their omission from *Species plantarum* in 1753. They were apparently never revisited and do not feature in any of Linnaeus's later accounts. This is in contrast to the reptiles, about which Linnaeus, though initially unconvinced of the status of a number of the snakes depicted by Catesby, later revised his opinion. He then coined binomial names for them in the light of subsequent observations on scale counts (regarded as crucial characters by Linnaeus) made by Alexander Garden.[36]

Catesby as a source of information on plants for Linnaeus

While the frequency with which Linnaeus cited Catesby's work confirms that he clearly found it valuable, study of the original text in which Linnaeus published his new binomial species names in 1753 shows a range of different situations. For some plants, it appears that Catesby was the sole source of information for Linnaeus, but, more commonly, a Catesby plate is just one of a number of sources used and cited. This should not be surprising, as Catesby himself cited, and sometimes adopted, the names coined by earlier authors for quite a number of the plants he featured. Of earlier authors, he cited the works of Leonard Plukenet most frequently, followed by those of Plumier and Sloane,[37] but he also adopted polynomial names already published by authors such as John Ray, Robert Morison (1620–1683), John Parkinson (1567–1650), Jan Commelin (1629–1692), and Joseph Pitton de Tournefort (1656–1708).[38]

John Clayton's Virginia specimens

Linnaeus, too, had access to these publications, and there were two significant collections of pressed plants that contributed important further information. The first of these was in the possession of his Dutch friend Johan Gronovius in Leiden, and it comprised the specimens collected by John Clayton, for many years the Clerk of Gloucester County in Virginia. Based in part on a manuscript prepared by Clayton and partly on the specimens themselves, Gronovius published his *Flora Virginica* in two volumes in 1739 and 1743, in which Clayton's specimens are explicitly cited. Linnaeus saw many of Clayton's specimens while he was working with Gronovius in Leiden. Some six hundred sheets still survive at the Natural History Museum, London, with a small number of duplicates in Linnaeus's herbarium at the Linnean Society of London.[39] Catesby himself had a hand in the transmission of some of these specimens to Gronovius.[40] Clayton was also in direct contact with Catesby, supplying him with material, which was acknowledged in *The natural history of Carolina. . . .*[41] Catesby reciprocated with a gift to Clayton of the published work.[42]

Pehr Kalm's North American collection

A further rich collection of North American specimens reached Linnaeus in 1751 with the return from Philadelphia of his compatriot Pehr (or Peter) Kalm (1716–1779). Kalm had been dispatched to North America with the primary purpose of obtaining stocks of economically useful plants (which it was anticipated would also thrive in the Swedish climate) that would serve to improve the impoverished Swedish economy. Having arrived in Philadelphia in September 1748, Kalm traveled to Delaware, New Jersey, New York, and southern Canada. In Linnaeus's possession in time to allow newly recognized species to be described in *Species plantarum* (1753), the specimens of six hundred or so species with which Kalm returned were a major source of new information for Linnaeus.[43]

Catesby Specimens in the Natural History Museum, London

Although Linnaeus never saw them, Catesby prepared many dried specimens that were sent to various of his supporters in England, notably William Sherard,[44] Sir Hans Sloane,[45] and Samuel Dale. Sloane's vast herbarium, now at the Natural History Museum, London, contains two bound volumes of specimens he received from Catesby. Some of the specimens carry Catesby's own handwritten labels. In one example (accompanying a specimen of the water tupelo, *Nyssa aquatica*), he noted that the plant "produces clusters of purplish colour'd berries of the shape of Spanish olives but somewhat less which are greedily eat by Bears," an observation that does not feature in Catesby's

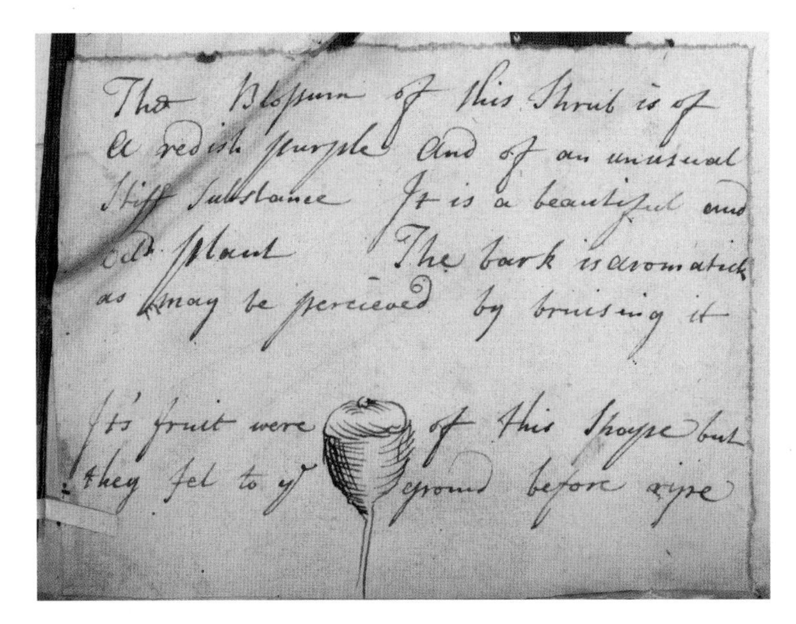
The Blossum of this Shrub is of a redish purple and of an unusual Stiff Substance It is a beautiful and odd plant The bark is aromatick as may be perceived by bruising it

Its fruit were of this shape but they fel to ye ground before ripe

FIGURE 14-6. One of Mark Catesby's labels with a pen-and-ink sketch of an unripe fruit of the Carolina allspice (*Calycanthus floridus*) (Sloane Herbarium 212, fol. 16). (© Natural History Museum, London; reproduced by courtesy.)

published text for this species. In another example, a specimen of the Carolina allspice (*Calycanthus floridus*) has a Catesby label that bears a pen-and-ink sketch of an unripe fruit (figure 14-6).[46] The herbarium of Samuel Dale, acquired by the Museum through donation by the Apothecaries' Company in 1862, also contains more than a hundred Catesby collections (including many from Virginia), frequently accompanied by detailed labels in the hands of both Catesby and Dale.

Catesby as a sole source of information for Linnaeus

Some Linnaean names were solely based on information supplied by Catesby, and the plant named in Catesby's honor, originally by Johan Gronovius and subsequently adopted by Linnaeus, is such an example. In Linnaeus's original publication of *Catesbaea spinosa* in 1753 (figure 14-7), the only source Linnaeus indicated was Catesby's description and plate. There are no specimens of this species in Linnaeus's own herbarium, and the fine hand-colored engraving (figure 14-8) serves as the type of the name.[47]

CATESBÆA.

1. CATESBÆA *ſpinoſa.*

Frutex ſpinoſus, buxi foliis plurimis ſimul naſcentibus, flore tetrapetaloide pendulo ſordide flavo tubo longiſſimo, fructu ovali croceo ſemina parva continente. *Catesb. carol.* 2. *p.* 100. *t.* 100.

Habitat in Providentia. ♄

FIGURE 14-7. C. Linnaeus, 1753, *Species plantarum*, volume I: 110. For *Catesbaea spinosa* (lily thorn), the only source Linnaeus indicated was Catesby's *The natural history of Carolina . . .*, volume 2, page and plate 100. (© Natural History Museum, London; reproduced by courtesy.)

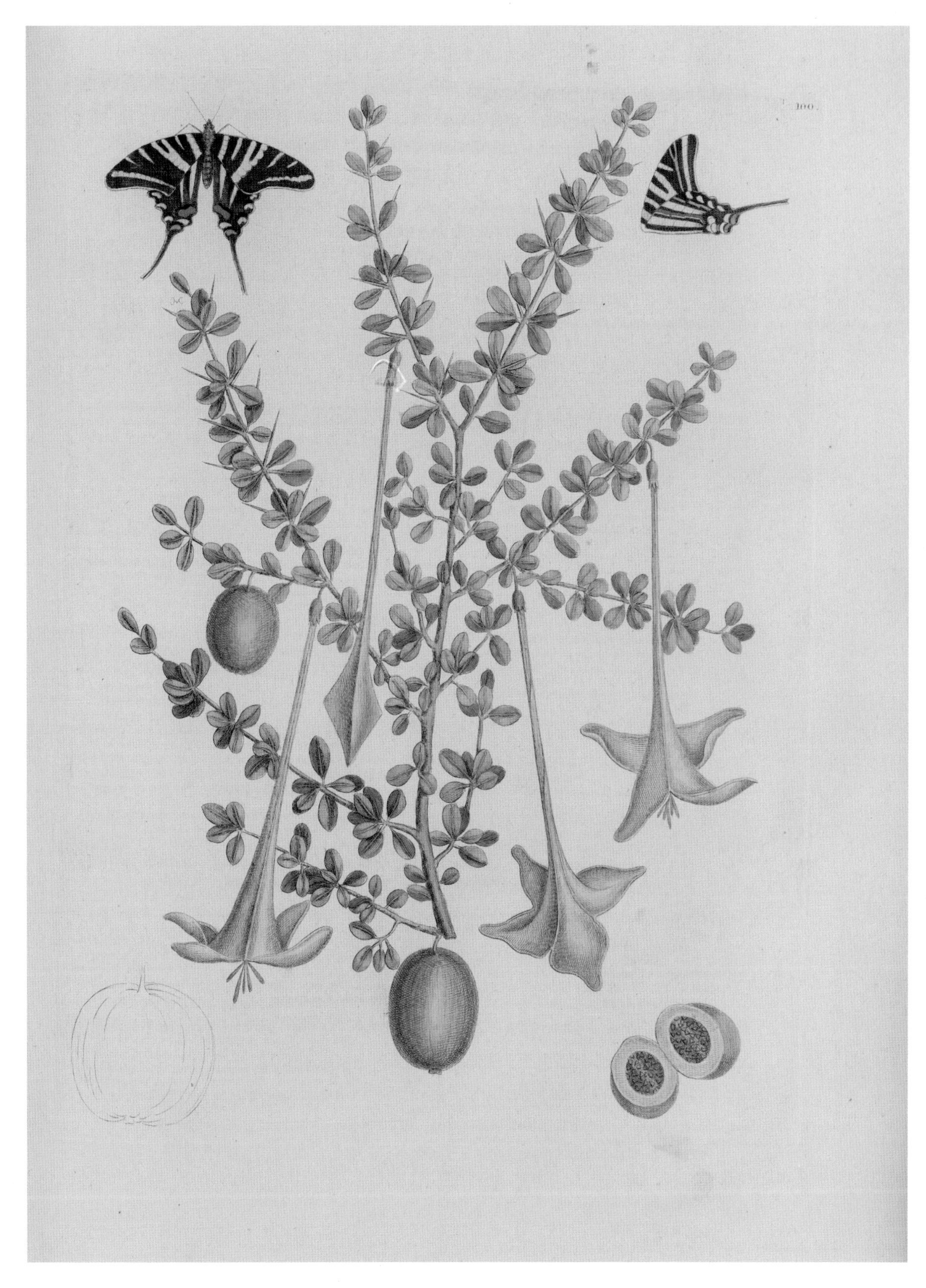

FIGURE 14-8. Lily thorn (*Catesbaea spinosa*); plate 100, M. Catesby, 1743, *The natural history of Carolina* . . . , volume II. (Digital realization of original etchings by Lucie Hey and Nigel Frith, DRPG England; courtesy of the Royal Society ©.)

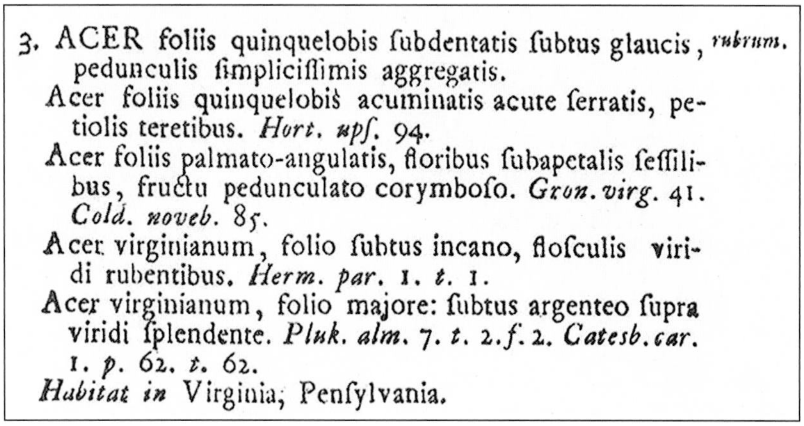

3. ACER foliis quinquelobis ſubdentatis ſubtus glaucis, pedunculis ſimpliciſſimis aggregatis. *rubrum.*
Acer foliis quinquelobis acuminatis acute ſerratis, petiolis teretibus. *Hort. upſ.* 94.
Acer foliis palmato-angulatis, floribus ſubapetalis ſeſſilibus, fructu pedunculato corymboſo. *Gron. virg.* 41. *Cold. noveb.* 85.
Acer virginianum, folio ſubtus incano, floſculis viridi rubentibus. *Herm. par.* 1. *t.* 1.
Acer virginianum, folio majore: ſubtus argenteo ſupra viridi ſplendente. *Pluk. alm.* 7. *t.* 2. *f.* 2. *Cateſb. car.* 1. *p.* 62. *t.* 62.
Habitat in Virginia, Penſylvania.

FIGURE 14-9. C. Linnaeus, 1753, *Species plantarum*, volume II: 1055. Linnaeus's original publication of the Latin name *Acer rubrum* (red maple) included reference to four synonyms, cited via six different authors including Catesby. (© Natural History Museum, London; reproduced by courtesy.)

Catesby and eastern North American plants

Many of the species illustrated by Catesby have extended distributions in the eastern United States, of which the red maple (*Acer rubrum*) is an example. Linnaeus's original publication of the Latin name (figure 14-9) included reference to four synonyms cited via six different authors.[48] The first is to an earlier account by Linnaeus of a plant growing in the Botanic Garden at Uppsala University in Sweden.[49] The second comes from Johan Gronovius's *Flora Virginica*,[50] based on a John Clayton specimen[51] that Linnaeus would have seen when he was in the Netherlands. A third is to another account of the plant in cultivation, this time in the Botanic Garden in Leiden, published in 1698 by the Dutch Professor of Botany Paul Hermann.[52] The fourth is to Catesby's plate. Catesby himself referred to an earlier description and illustration published by the English botanist Leonard Plukenet.[53] Lastly, there is a specimen in Linnaeus's own herbarium that was collected in Pennsylvania by Pehr Kalm and that would have reached Linnaeus in 1751, shortly before he finalized the text of *Species plantarum*. In this case, the information supplied by Catesby's publication would have been only contributory (rather than diagnostic) for Linnaeus because the species in question was already comparatively well known in Europe. It is Kalm's herbarium specimen that serves as the type of *Acer rubrum*.[54]

11. LAURUS foliis enerviis obovatis obtusis. *Mat. med.* 196. *Winterana.*
Winteranus. *Hort. cliff.* 488.
Cinnamomum s. Canella peruana. *Bauh. pin.* 409.
Cassia lignea jamaicensis, laureolæ foliis subcinereis, cortice piperis modo acri. *Pluk. alm.* 89. *t.* 160. *f.* 1.
Cassia cinnamomea s. Cinnamomum sylvestre barbadensium. *Pluk. alm.* 89. *t.* 160, *f.* 7.
Arbor baccifera laurifolia aromatica, fructu viridi calyculato racemoso. *Sloan. jam.* 165. *hist.* 1. *p.* 87. *t.* 191. *f.* 2. *Catesb. car.* 2. *p.* 50. *t.* 50.
Habitat in Jamaica, Barbados, Carolina. ♄
Flores ipse non vidi.

FIGURE 14-10. C. Linnaeus, 1753, *Species plantarum*, volume I: 371. When Linnaeus named *Laurus winterana* (now *Canella winterana*, cinnamon bark), he was aware that it had been described by earlier authors from places including Barbados and Jamaica. Information from Catesby supplemented what was already known to Linnaeus. (© Natural History Museum, London; reproduced by courtesy.)

Catesby and Caribbean plants

Other species depicted by Catesby and named by Linnaeus have distributions extending to the Caribbean and beyond. Linnaeus was evidently aware that the species he had named *Laurus winterana* (figure 14-10), now *Canella winterana* (cinnamon bark), had been described by a number of earlier authors from places such as Barbados[55] and Jamaica.[56] Once again, the information from Catesby supplemented what was already known to Linnaeus from other, earlier authors.

In summary, of the 187 plants figured by Catesby in the 220 published plates, 131 of them were cited by Linnaeus and given his new binomial names. Of these 131, 17 of Linnaeus's new species names were based solely upon Catesby's plates and descriptions, and the remaining 114 contributed, along with other sources, to Linnaeus's understanding of the species involved. A further 17 of these 114 names have Catesby plates as their nomenclatural types. This total of 34 of Catesby's plates of plants serving as types for Linnaean binomial names gives some measure of the enduring scientific and nomenclatural importance of Catesby's botanical work, despite Linnaeus's own reservations over its cost (and consequent exclusivity).

Contacts between Linnaeus and Catesby

Although Catesby was in England at the time of Linnaeus's brief visit there in 1736, it seems that Catesby and Linnaeus never met and had almost no direct contact. There are a few specimens in Linnaeus's herbarium that can be linked directly with Catesby, and these probably came via either Johan Gronovius or Peter Collinson.[57] Only a single letter, written in April 1745, from Catesby to Linnaeus is extant.[58] In it, Catesby announced that he was dispatching to Sweden, on behalf of their mutual friend, the Scottish physician Isaac Lawson (1704–1747), a selection of eighteen living American plants chosen on the basis of their comfortable survival of English winters in the hope that they would not find the Swedish climate too cold.[59] A few years later, via a letter sent to Linnaeus by the British physician John Mitchell (1690–1768), Catesby enquired whether the plants he had sent to Sweden were "alive and flourishing."[60]

While on his way to North America in 1748, Pehr Kalm spent some months in England, during which he recorded meeting Catesby, initially (21 April 1748) following a meeting of the Royal Society and again a month later (23 May 1748) at Catesby's home. Kalm praised the high quality of Catesby's illustrations: "incomparably well represented with lifelike colours . . . so that no one can see that they are not living where they stand with their natural colours on the paper" but, echoing Linnaeus's opinion, Kalm observed that the two volumes "both now cost in England twenty-two to twenty-four guineas, therefore not for a poor man to buy."[61]

Catesby and Collinson

Although Collinson had sent the last three parts of Catesby's work, destined for Charles De Geer, to Linnaeus in late 1748, he evidently had some difficulty in recouping the costs. On 1 May 1750, he wrote to Linnaeus, "[Y]ou have not given Mee a Line,—neither am I paid that money."[62] Collinson then sent a revised complaint that although De Geer had sent him £4, he was still owed £2 6s. 0d.—"a small sum," he wrote, "but to[o] much to loose."[63] In the same letter, Collinson reported the death, on 23 December 1749, of "our ingenious friend Mr Catesby at the age of 70—much lamented" and he went on to encourage Linnaeus to purchase a copy of Catesby's book, as Catesby's widow had reduced the price in order to encourage the sale of the remaining copies.

Several further comments on De Geer's failure to settle his bill are found in a series of letters sent to Linnaeus, with Collinson becoming increasingly irritated (25 August 1749—". . . no care is taken to send the money").[64] On 8 May 1753, he wrote to inform Linnaeus of the death of Sir Hans Sloane, adding, "My Dear Friend, no doubt well remembers, he wrote for 3 Books of Mr Catesbys

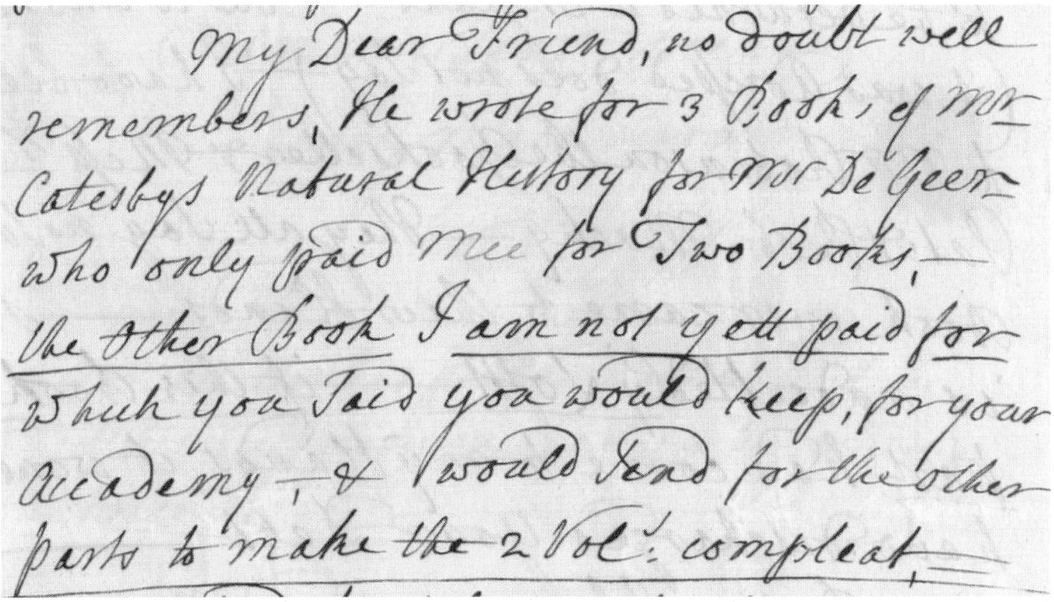

My Dear Friend, no doubt well
remembers, He wrote for 3 Books of Mr
Catesbys Natural History for Mr De Geer
who only paid Mee for Two Books,—
the Other Book I am not yett paid for
Which you said you would keep, for your
Academy, & would send for the other
parts to make the 2 Vol: compleat,

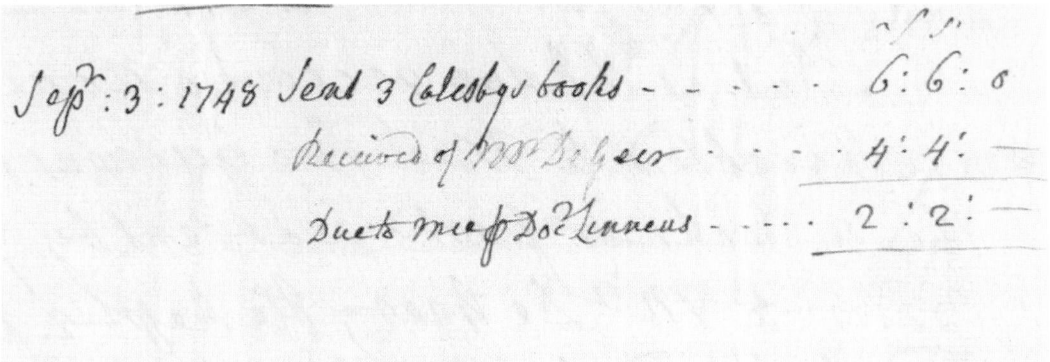

Sepr: 3 : 1748	Sent 3 Catesbys books - - - - - -	6 : 6 : 0
	Received of Mr De Geer - - - - -	4 : 4 : —
	Due to mee of Doc Linneaus - - - -	2 : 2 : —

FIGURE 14-11. Peter Collinson became increasingly frustrated by De Geer's failure to pay his bill (*upper*), noting that Linnaeus owed two guineas (*lower*). Five years later, however, Collinson was pleased to write, "I must not forget to tell you that I have received the Two Guineas, so now all accounts are ajusted between us." (© Linnean Society of London; reproduced by courtesy.)

Natural History for Mr De Geer who only paid Mee for Two Books,—<u>the Other Book I am not yett paid for</u>" (figure 14-11, *upper*).[65] However, in the well-known letter (20 April 1754)[66] in which Collinson criticized Linnaeus for perplexing "the delightful Science of Botany with Changing Names that have been well received, and adding new names quite unknown to us" and announced the Swede's election as a Fellow of the Royal Society, Collinson added, "I must not forget to tell you that I have received the Two Guineas, so now all accounts are ajusted between us."

As to Catesby, his botanical work was of considerable significance to Linnaeus. Many of the plates and descriptions were, for Linnaeus, his first glimpse of over twenty new species, including cascarilla (*Croton eluteria*), pond apple (*Annona glabra*), umbrella magnolia (*Magnolia tripetala*), frangipani (*Plumeria obtusa*), West Indian mahogany (*Swietenia mahagoni*), pawpaw (*Asimina triloba*), tievine (*Ipomoea carolina*), and, of course, the lily thorn, which, thanks to Carl Linnaeus, continues to commemorate Mark Catesby in its generic name, *Catesbaea*.

15

The economic botany and ethnobotany of Mark Catesby

W. HARDY ESHBAUGH

Mark Catesby's *The natural history of Carolina, Florida and the Bahama islands* is recognized as the first published account of the flora and fauna of North America. What set Catesby apart from his contemporaries and those who would follow him in the next century—for example, Alexander Wilson (1766–1813) and John James Audubon (1785–1851)—were the details he provided about the natural history that surrounded him. Indeed, he gave us a remarkable view into how plants were used by those who lived with them. In contemporary terms, he taught us about the economic botany and ethnobotany of that time.

Clearly, Mark Catesby was aware of his predecessors' work, particularly that of John White and John Lawson, and he borrowed freely from them.[1] As has been noted, some of Catesby's writing is strikingly similar to Lawson's, even to the detail of his descriptions, but "such borrowing was common in those days, and even egregious plagiarism did not carry the same stigma as it does today."[2]

Carl Linnaeus used many of the plates in *The natural history of Carolina, Florida and the Bahama islands* as the basis for naming many new plant species from North America and the Bahamas.[3] In all, 119 plates were the basis of his names.[4] Catesby is commemorated by the Linnaean binomial *Catesbaea spinosa* (figure 14-8, p. 199): he himself remarked, "it is not without reluctancy, that I here exhibit a plant with my name annexed to it. . . ."[5]

Mark Catesby was not only a naturalist curious about the New World and all the biological diversity it contained but also a highly practical man. His journey through a portion of eastern North America and the Bahamas, including a visit to Jamaica, had him looking for plants that would be of interest to his sponsors. The notes he included for each of the species he illustrated or discussed in his book are a collection of insights into the natural history, economic botany, and ethnobotany of those plants. The plants of eastern North America were better known to his sponsors in England, although he did make many new discoveries that represented new introductions into European gardens.

Catesby was in the Bahamas before the major arrival of settlers in the 1780s. At the time of his adventure only the islands of New Providence, Eleuthera, and Harbour Island had permanent settlements.[6] His discoveries in the Bahamas were by and large unknown to Europe. Of 171 plants depicted in *The natural history of Carolina . . .* , 65 were from the Bahama Islands. Six of Catesby's North American plants are also found in the Bahamas.[7] Another eight plants are described and discussed by Catesby but not illustrated, including four palms. It is hard to understand why the Caribbean pine (*Pinus caribaea* var. *bahamensis*) is neither illustrated nor discussed. It would have been the most abundant and among the largest plants that Catesby would have seen when he visited Andros Island.

Catesby's *The natural history of Carolina, Florida and the Bahama islands* remained the standard for the Bahamas flora until the twentieth century and the publication of Alice Northrop's (1864–1923) treatment of the flora of New Providence and Andros Island[8] and *The Bahama flora* by Nathaniel Lord Britton (1859–1934) and Charles Frederick Millspaugh (1854–1923).[9]

North American plants of interest

BALD CYPRESS (*Taxodium distichum*)

> The Cypress (except the Tulip-tree) is the tallest and largest in these parts of the world. Near the ground some of 'em measure 30 foot in circumference, rising pyramidally six foot, where it is about two thirds less; from which to the limbs, which is usually 60 or 70 foot, it grows in like proportion of other trees. Four or five foot round this Tree (in a singular manner) rise many Stumps, some a little above ground, and others from one to four foot high, of various shape and size, their tops round, cover'd with a smooth red Bark. These Stumps shoot from the roots of the Tree, yet they produce neither Leaf nor Branch, the Tree increasing only by seed, which in form are like the common Cypress and contain a balsamic consistence of a fragrant smell. The Timber this Tree affords, is excellent, and particularly for covering Houses with, it being light, of a free Grain, and resisting the Injuries of the weather better than any other here. It is an Aquatic, and usually grows from one, five and six foot deep in water; which secure situation seems to invite a great number of different Birds to breed in its lofty branches; amongst which this Parrot delights to make its Nest, and in October, (at which Time the Seed is ripe) to feed on their Kernels.[10]

Bald cypress is a prized ornamental tree. The odorless wood is valued for its water resistance. Many historic southern homes are paneled with the wood of bald cypress. Still-usable prehistoric wood is often found in swamps in the Southeast. Mineralized wood is recovered from some swamps in the Southeast and is prized for specialty uses. Pecky cypress is sometimes used in decorative wall paneling and is the result of the fungus *Stereum taxodii*.

> The usual Height of the Live Oak is about 40 foot; the Grain of the wood course, harder and tougher than any other Oak. Upon the edges of Salt-Marshes (where they usually grow) they arrive to a large size. Their Bodies are irregular, and generally lying along, occasioned by the looseness and moisture of the soil, and tides washing their roots bare. On higher lands they grow erect, with a regular pyramidal-shaped Head, retaining their leaves all the year. The Acorns are the sweetest of all others; of which the Indians usually lay up store, to thicken their venison-soop, and prepare them other ways. They likewise draw an oil, very pleasant and wholesom, little inferior to that of Almonds.[11]

We know that Native Americans extracted oil from the acorns and used it in cooking. The bark was the source of dye, while the leaves were used in making rugs. Southern live oak (figure 15-1) produces sweet, edible acorns that wild turkeys, deer, and other animals eat. The wood is hard, heavy, and difficult to work with, yet it is very strong and is often used for structural beams and posts. The USS *Constitution* got its nickname "Old Ironsides" because the southern live oak wood from which it was constructed was so strong and dense that the ship withstood a cannonade during the war of 1812.

FIGURE 15-1. Southern live oak was used in construction of "Old Ironsides," or the USS *Constitution*, famed for its successes in the War of 1812 and the world's oldest commissioned naval vessel afloat. (© Sylvia W. Bacon.)

WAX MYRTLE (*Morella cerifera*) (FIGURE 15-2)

These are usually but small Trees or Shrubs, aboat Twelve Foot high, with crooked Stems, branching forth near the Ground irregularly. The Leaves are long, narrow, and sharp-pointed. Some Trees have most of their Leaves serrated; others not. In May, the small Branches are alternately and thick set with oblong Tufts of very small Flowers resembling, in Form and Size, the Catkins of the Hazel-Tree, coloured with Red and Green. These are succeeded by small Clusters of blue Berries, close connected, like Bunches of Grapes. The Kernel is inclosed in an oblong hard Stone, incrustated over with an unctuous mealy Consistence; which is what yields the Wax; of which Candles are made in the following Manner.

In November and December, at which time the Berries are mature, a Man with his Family will remove from his Home to some Island or Sandbanks near the Sea, where these Trees most abound, taking with him Kettles to boil the Berries in. He builds a Hut with Palmeto-Leaves, for the Shelter of Himself and Family while they stay, which is commonly Three or Four Weeks.

The Man cuts down the Trees, while the Children strip off the Berries into a Porrige-Pot; and having put Water to 'em, they boil them 'till the Oil floats, which is skim'd off into another Vessel. This is repeated till there remains no more Oil. This, when cold, hardens to the Consistence of Wax, and is of a dirty green Colour. Then they boil it Again, and clarify it in brass Kettles; which gives it a transparent Greenness. These Candles burn a long Time, and yield a grateful Smell. They usually add a fourth Part of Tallow; which makes them burn clearer.[12]

HERCULES-CLUB (*Zanthoxylum clava-herculis*)

Catesby reported that Hercules-club known as "The Pellitory or Tooth-ach tree" has leaves that "smell like those of Orange; which, with the Seeds and Bark, is aromatic, very hot and astringent, and is used by the People inhabiting the Sea Coasts of *Virginia* and *Carolina* for the Tooth-ach, which has given it its name."[13] It is sometimes called tingle tongue because chewing the leaves or bark makes the mouth, teeth, and tongue numb. Numbness may be attributed to the chemical constituent berberine, which is a broad-spectrum bactericide. Native Americans and settlers were said to have used the plant for a wide range of problems; for example, gonorrhea was treated with a decoction of the bark.

SASSAFRAS (*Sassafras albidum*)

"The Virtue of this Tree is well known, as a great Sweetner of the Blood: I shall therefore only add, that in *Virginia* a strong Decoction of the Root has been sometimes given with good Success for an intermitting Feaver."[14] Little did Catesby know that sassafras (figure 15-3) had many other important applications with Native Americans. The oil, roots, and bark are reputed to have medicinal value as antiseptics and analgesics. The primary chemical constituent is safrole, which occurs in the leaves, bark, and roots. Sassafras is said to have

FIGURE 15-2. Colonial families used wax myrtle berries to produce candles; plate 69, M. Catesby, 1731, *The natural history of Carolina . . .* , volume I. (Digital realization of original etchings by Lucie Hey and Nigel Frith, DRPG England; courtesy of the Royal Society ©.)

FIGURE 15-3. Sassafras was used for a variety of medicinal purposes and also to produce root beer. Sassafras is now banned due to its carcinogenic properties; plate 55, M. Catesby, 1730, *The natural history of Carolina . . .* , volume I. (Digital realization of original etchings by Lucie Hey and Nigel Frith, DRPG England; courtesy of the Royal Society ©.)

a wide variety of medicinal uses, including treatment for scurvy, skin sores, kidney problems, toothaches, rheumatism, swelling, menstrual disorders and sexually transmitted diseases, bronchitis, hypertension, and dysentery.

Safrole was used as an additive to sassafras tea and root beer, among others, but was banned by the United States Food and Drug Administration after its demonstrated carcinogenicity in rats. It is now banned from being used in certain cosmetics, including perfumes and soaps.

BRISTLY GREENBRIER (*Smilax tamnoides*) (FIGURE 15-4)

> This Plant shoots forth with many pliant thorny Stems; which, when at full Bigness, are as big as a Walking Cane, and jointed; and rises to the Height usually of twenty Feet, climbing upon and spreading over the adjacent Trees and Shrubs, by the Assistance of its Tendrels. In Autumn it produces Clusters of black round Berries, hanging pendent to a Foot Stalk, above three Inches long, each Berry containing a very hard roundish Seed. The Roots of this Plant are tuberous divided into many Knots and Joints; and, when first dug out of the Ground, are soft and juicy, but harden in the Air to the Consistence of Wood. Of these Roots the Inhabitants of *Carolina* make a Diet-drink, attributing great Virtues to it in cleansing the Blood &c. They likewise in the Spring boil the tender Shoots and eat them prepared like Asparagus. 'Tis call'd there *China* Root.[15]

FIGURE 15-4. Bristly greenbrier was used by Carolinians to make a drink for cleansing the blood; plate 52, M. Catesby, 1730, *The natural history of Carolina . . .* , volume I. (Digital realization of original etchings by Lucie Hey and Nigel Frith, DRPG England; courtesy of the Royal Society ©.)

FIGURE 15-5. A tea made from yaupon leaves was enjoyed by American Indians as a vomit-inducing medicinal tonic. Catesby, however, preferred it as a beverage with milk and sugar. (© David J. Elliott.)

YAUPON (*Ilex vomitoria*)

The yaupon holly (figure 15-5) is North America's version of the South American yerba maté (*Ilex paraguariensis*). John Lawson wrote extensively about the use of this plant by the Indians of the Carolinas.[16] Catesby encountered it and commented as follows:

> But the great Esteem and Use the *American Indians* have for it, gives it a greater Character. They say, that from the earliest Times the Virtues of this Shrub has been known, and in Use among them, prepared in the Manner they now do it, which is after having dryed, or rather parched the Leaves in a Porrage-Pot over a slow Fire, they keep it for Use: Of this they prepare their beloved Liquor, making a strong Decoction of it, which they drink in large Quantities, as well for their Health as with great Gust and Pleasure, without any Sugar or other Mixture, yet they drink and disgorge it with Ease, repeating it very often, and swallowing many Quarts. . . .[17]

None of these observations were original, since Lawson had made the same ones much earlier. However, long after his return from his second trip, in a letter to Professor Johan Jacob Dillenius, Catesby told how the settlers had acquired a taste for the drink and how he had brought some of the yaupon "Tea" to London and shared it with William Sherard ("the Consul") and, presumably, Dillenius:

> The English inhabitants of the sea-coasts by living amongst the Indians, learn'd it of them, but they drink it by way of refreshment and not as a vomit except when their stomacks are foul, or they are otherwise out of order. The leaves which are never used without being parched are also drunk in the manner of Tea by some people amongst whome I have an hundred times made an agreeable breakfast of it but with milk and sugar; so used it is very pleasant & never provokes to vomit, whether it was from the opinion of was possessed, of its' virtues, but I thought while I drank it I had more spirits & was better in health than ordinary. I brought a large quantity of it over prepared, & I remember to have produced some of it at the Consuls.[18]

FIGURE 15-6. The roots of sweet leaf have been used in a tonic for the treatment of fever, and a yellow dye is obtained from the leaves, the bark, and the fruits. (Used by permission of the North Carolina Native Plant Society and photographer Tom Harville.)

SWEET LEAF (*Symplocos tinctoria*) (FIGURE 15-6)

> This Shrub has a slender Stem, and grows usually about 8 or 10 Feet high. Its Leaves are in Shape like those of a Pear, growing alternately on Foot-stalks of an Inch long; from between which proceeds small whitish Flowers, consisting of five Petals; in the Middle of which shoot forth many tall *Stamina*, headed with yellow *Apices*. The Roots of this Plant are made Use of in Decoctions, and are esteemed a good Stomachic and Cleanser of the Blood. The Fruit I have not seen. This Plant grows in moist and shady Woods, in the lower Parts of *Carolina*.[19]

The aromatic and bitter roots of sweet leaf have been used in a tonic and a decoction in the treatment of fever. A yellow dye is obtained from the leaves, the bark, and the fruits.

Other plants

Catesby collected hundreds of specimens in eastern North America, and while many of those that survive have no information with them, a few of the handwritten labels give brief particulars about the practical or medicinal uses of the plants. For example, on the labels attached to specimens of white basswood (*Tilia americana* var. *heterophylla*) (figure 15-7), Catesby noted a use for the bark; the specimen sent to Samuel Dale had this note: "The bark is exceeding tough & strong and is much used in Country affairs for binding &c as Ropes are."[20] Regarding the sorrel tree (*Oxydendrum arboreum*), his handwritten label included a note that "It is in great esteem here [Carolina] for feavers and other Malady's."[21] Catesby noted that various plants, including golden colicroot (*Aletris aurea*), were used as antidotes against snake venom: "The Root of this plant the Indians esteem good for the Bite of the RattSnake [*sic*]." Another example accompanies a specimen of a species of coneflower (*Rudbeckia*): "It is one of those roots made use of ag[ains]^{t} the bite of y^{e} Rattlesnake."[22] Other annotations concerned the potential use of the plant in gardens. In *The natural history of Carolina . . .* , Catesby referred to the wood of the loblolly bay

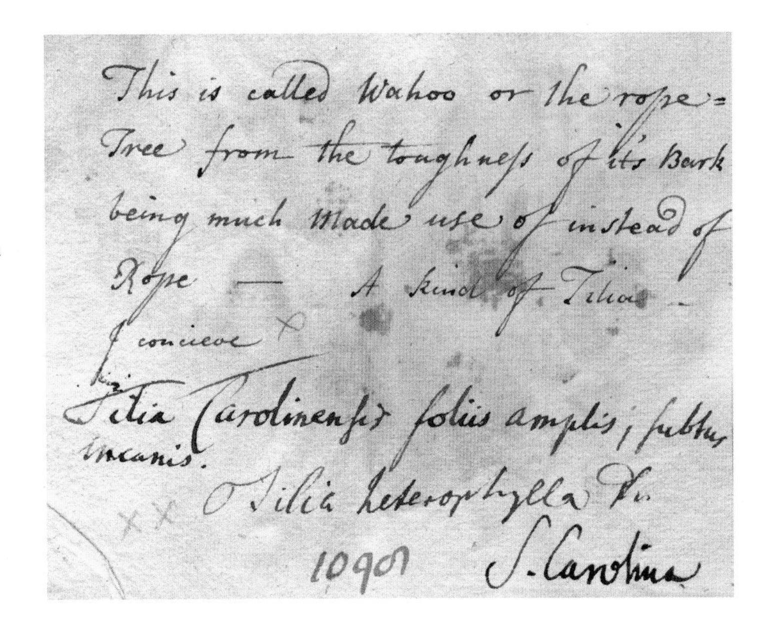
This is called Wahoo or the rope=
Tree from the toughness of its Bark
being much Made use of instead of
Rope — A kind of Tilia
I conceive
Tilia Carolinensis foliis amplis, subtus
incanis.
Tilia heterophylla
S. Carolina

FIGURE 15-7. Mark Catesby's label attached to a specimen of the white basswood (*Tilia americana* var. *heterophylla*) that he sent to William Sherard: the upper five lines are in Catesby's hand (Sher-1098; reproduced by permission of Oxford University Herbaria).

(*Gordonia lasianthus*), which he said was "somewhat soft; yet I have seen some beautiful Tables made of it." On a specimen label, he had added his opinion that the loblolly bay was "certainly one of the Most ornamental and proper Tree in Nature for Avenues."[23]

Bahamian plants of interest

MANCHINEEL (*Hippomane mancinella*)

Catesby provided a vivid picture of manchineel:

> The Wood of this Tree is close grained, very heavy and durable, beautifully shaded with dark, and lighter Streaks, for which it is in great Esteem for Tables and Cabinets, and other curious Works in Joynery; but the virulent and dangerous Properties of these Trees, causes a general Fear, or at least Caution, in felling them; this I was not sufficiently satisfyed of, 'till assisting in the cutting down a Tree of this Kind on *Andros* Island, I paid for my Incredulity, some of the milky poisonous Juice spurting in my Eyes, I was two Days totally deprived of Sight, and my Eyes, and Face, much swelled, and felt a violent pricking Pain, the first twenty-four Hours, which from that Time abated gradually with the Swelling, and went off without any Application, or Remedy, none in that uninhabited Island being to be had: It is no Wonder that the Sap of this Tree should be so virulent, when Rain or Dew, falling from its Leaves on the naked Flesh, causes Blisters on the Skin, and even the *Effluvia* of it are so noxious as to affect the Senses of those which stand any Time under its Shade.[24]

Manchineel is a tree up to forty-five feet (fifteen meters) tall. It has gray bark, shiny green leaves, and small greenish flowers. Its fruits when ripe are green to greenish-yellow. One current Spanish name is manzanilla de la muerte, meaning little apple of death. Throughout much of the Caribbean, this is the largest tree growing in the beach-strand zone. The fruit does smell somewhat like an apple, and since it is attractive, unsuspecting strollers along the beaches are too often tempted to take a bite to their regret.

GUMBO LIMBO (*Bursera simaruba*)

Although widely known as gumbo limbo throughout the Caribbean, in the Bahamas it is better known as gum elemi. Catesby noted: "This is a large tree; the bark remarkably red and smooth."[25] The tree is called tourist tree throughout the Bahamas since it has a red bark that peels, just like the tourists who spend too much time on the beach.

Catesby observed that "most of the Bahama Islands abound with these trees." Gum elemi has many uses. It is burned as incense. "This tree produces a large quantity of gum, of a brown colour, and of a consistence of turpentine." The gum is used throughout the islands as a medicinal, especially as an anti-inflammatory and for circulation problems and as a topical treatment for dermatitis caused by poison wood (*Metopium toxiferum*). Catesby stated that it "is esteemed a good vulnerary [used in the treatment and healing of wounds], and is much used for Horses." It is an ingredient of love potions or aphrodisiac teas.[26]

POISONWOOD (*Metopium toxiferum*)

> This is generally but a small Tree; has a light coloured smooth bark. Its leaves are winged, the middle rib seven or eight inches long, with pairs of *Pinnae* one against another on inch-long foot-stalks. The fruit hang in bunches; are shaped like a Pear, of a purple-Colour, covering an oblong hard Stone.
>
> From the trunc of this Tree distils a Liquid black as Ink, which the Inhabitants say is Poison. Birds feed on the Berries, particularly this *Gross-beak* on the mucilage that covers the Stone. It grows usually on Rocks in *Providence*, *Ilathera* and other of the *Bahama* Islands.[27]

The oleoresin urushiol, characteristic of many plants in the cashew family, Anacardiaceae, is found in all parts of the poison wood. Once in contact with air, it oxidizes and turns black. This induces dermatitis similar to but much harsher than that caused by poison ivy (*Toxicodendron radicans*). If poison wood resin gets on clothing, it will turn black when washed. The resin can be painted on wood surfaces, giving a black, lacquer-like appearance when it hardens and is polished.

WEST INDIAN MAHOGANY (*Swietenia mahagoni*)

> These Trees grow to great Height, and are usually four Foot Diameter; the Bark is of a brown Colour, the Leaves are pinnated, growing by Pairs on slender Stalks, the Ribs of the Leaves (like those of the *Tilia*) run on one Side, dividing the Leaf unequally. . . . The Excellency of this Wood for all Domestick Uses is now sufficiently known in *England*: And at the *Bahama* Islands, and other Countries, where it grows naturally, it is no less Esteem for Ship-building, having Properties for that Use excelling Oak, and all other Wood. . . .[28]

It did not take many years of exploitation to reduce this species everywhere throughout the Bahamas. Large trees can still be found in remote areas, especially on Andros Island. However, on San Salvador few original trees remain, and many extant trees represent recent introductions.

JOEWOOD (*Jacquinia keyensis*)

Soap was always in short supply in colonial times in the Bahamas, especially on the out islands. Thus, Catesby noted that

> The Bark and Leaves of this Tree being beat in a Morter produces a Lather; and is made Use of to wash Cloaths and Linnen, to which last it gives a yellowness. The Hunters who frequent the desolate Islands of *Bahama*, (where this Shrub grows on the Sea-Coast) are frequently necessitated to use this Sort of Soap to wash their Shirts, for want of better.[29]

BLACK MANGROVE (*Avicennia germinans*)

> This grows to the size of a small Tree; the leaves stand by pairs on foot-stalks about an inch long; they are long, thick and succulent: at the ends of the stalks grow in pairs, and sometimes singly, round flat Seed-vessels, about the breadth of a shilling: the fruit is of the substance of a Bean, and, like that, divides in the middle: it is cover'd with a thin membrane of a pale green colour. I had no opportunity of seeing the Blossoms, tho' I was told they were very small and white. The Bark of this Tree is used for tanning of Sole-leather.[30]

Black mangrove wood is strong, heavy, and hard. The wood is difficult to work because of its interlocked grain and somewhat difficult to finish due to its oily texture. It is used for posts, pilings, and fuel. The bark contains tannin and was previously used in tanning leather products. More recently, honey from black mangrove has become a marketable product in Florida.

SCOTCH ATTORNEY (*Clusia rosea*)

> These Trees grow on Rocks, and frequently on the Limbs, and Trunks of Trees; occasioned by Birds scattering or voiding the Seeds which being glutinous like those of Misletto, take Root and grow. But finding not sufficient Nutriment to increase in Growth, the Roots spread on the Bark or Superficies of the Tree till they find a decayed Hole or other Lodgment wherein is some small Portion of Soil, into which they enter and become a Tree: But the Fertility of this Second Plantation being exhausted, one or more of the Roots are discharged out of the Hole, and fall directly to the Ground, tho' at forty Feet distance, here again they take Root, and become a much larger Tree than before: The Rosin of this Tree is used for the Cure of Sores in Horses, and also instead of Tallow for Boats and other Vessels. They grow on the *Bahama* Islands, and on many other Islands of *America*, between the *Tropicks*.[31]

This impressive tree is now widely known as the "autograph tree" because when anyone scratches or writes on the surface of the thick leathery leaves, their names or initials will remain as a scar on the leaf for as long as it persists. The latex is orange and sometimes gathered for use as incense. It also may be applied to wounds to assist in healing and, as noted by Catesby, to caulk boats.

CASCARILLA (*Croton eluteria*)

> Shrubs grow plentifully on most of the *Bahama* Islands, seldom above ten Feet high, and rarely so big as a Man's Leg, tho''tis probable, that before these Islands were exhausted of so much of it, that it grew to a larger Size: The Leaves are long, narrow, and sharp-pointed, and of a very pale light green Colour; at the Ends of the smaller Branches grow Spikes of small hexapetalous white Flowers, with yellow *Apices*, which are succeeded by tricapsular pale green Berries, of the Size of Peas, each Berry containing three small black Seeds, one in every *Capsule*. The Bark of this Tree being burnt, yields a fine Perfume; infused in either Wine or Water, gives a fine aromatic Bitter.[32]

Catesby provided a good technical description of cascarilla but could not have envisioned how important it would become to the developing economy of the Bahamas. In the bush-medicine tradition of the islands, it is used to stimulate the appetite and to treat an unsettled stomach and intestinal gas. The bark has a significant fraction of volatile oils that contain a unique series of compounds known as cascarillins and so is used to flavor the liqueurs Campari and vermouth, as well as certain tobaccos.[33] The plant is still harvested and collected commercially in the Bahamas.

SILVER LEAF PALM (*Coccothrinax argentata*)

"The usual Height of these Trees, is about sixteen feet, the Leaves somewhat less than the precedent [Thatch palmeto, *Sabal palmetto*], but thicker set, and a shining Silver Colour. Of these Leaves of these Trees are made Ropes, Baskets, &c. The Berries are large and sweet, and yield a good Spirit."[34] Today this palm, commonly called "silver top," is harvested at the full moon and dried to make

FIGURE 15-8. The only place where baskets of this type are made in the Bahamas is in remote Red Bays on Andros Island. This African-derived craft was probably brought there by descendants of escaped slaves from the South Carolina Lowcountry, where the skill continues to be used extensively. The Bahamans use the silver leaf palm to make them, a different raw material to that used in the Lowcountry. (© W. Hardy Eshbaugh.)

a unique basket known only from Red Bays on the northeast end of Andros Island. This community was isolated until 1967, when a road finally reached it. The settlers of the original village were a mixture of Seminole Indians and escaped African slaves. The Seminole culture adopted their basket-making techniques from the dominant African slave culture of the community to give rise to this locally unique art form (figure 15-8).

HOG PALMETTO (*Pseudophoenix sargenti*)

> The Singularity of this Tree is remarkable, for as eatable Parts of all Plants is in their Fruit, Roots, or Leaves, the Trunks alone of these Trees is an excellent Food for Hogs; and many little desart Islands, that abound with them, are of great use to the *Bahamians* for the Support of their Swine. The exterior Bark of the Trunks of these Trees is somewhat hard, and in Appearance like those of the other *Palmettoes*, within which is contained that soft and pithy Substance of a luscious sweet Taste, which the Hogs are delighted with.[35]

The hog palmetto is widely dispersed in the Bahamas but is often found on the little off-shore cays and smaller islands. These served as ideal places to turn hogs loose because there was a plentiful natural food resource to meet the hogs' needs (figure 15-9).

FIGURE 15-9. The edible hog palmetto, a northern Caribbean palmetto species, is used by Bahamans to feed hogs, especially on remote islands and cays where the hogs can run wild. (© Ethan H. Freid.)

16

"Of birds of passage": Mark Catesby and contemporary theories on bird migration and torpor

SHEPARD KRECH III

Many who have taken measure of Mark Catesby's life and work, in particular the birds that he etched and described, have lauded not merely his magisterial *The natural history of Carolina, Florida and the Bahama islands* but also "Of birds of passage," a brief paper published in 1747 in the *Philosophical transactions of the Royal Society of London*.[1] For example, in his monograph *Catesby's birds of colonial America*, the biologist Alan Feduccia wrote that "Of birds of passage" was Catesby's "most acclaimed" paper, and in the foreword to that work, Russell Peterson, then President of the National Audubon Society, stated that Catesby was "years ahead of his time" in suggesting that birds absent after the end of summer had migrated away.[2] Feduccia and Peterson were not alone. Catesby's biographers George Frick and Raymond Stearns considered "Of birds of passage" not merely Catesby's "most important contribution to the deliberations of the [Royal] Society" but the most "advanced" of his ornithological thoughts. Given that they also linked Catesby's "lasting fame" to "his contributions as an ornithologist,"[3] it is somewhat surprising to find that "Of birds of passage" has not received sustained treatment in recent years.[4] The time to revisit it is overdue.

Where do birds go?

"Of birds of passage" addressed the puzzle of the periodic seasonal absence of birds especially after the end of summer. What happened to breeding birds when they left their summer grounds? Catesby had apparently been mulling migration in general for some time. In 1722, more than two decades before "Of birds of passage" was published, he was en route to America when an owl hovered over the ship, and (as he later recalled) he wondered how a bird with such short wings that often flew only quite short distances over land could fly so far over the ocean. Embedded in the question were assumptions linking specific

FIGURE 16-1. Bobolink and rice; plate 14, M. Catesby, 1729, *The natural history of Carolina . . .* , volume I. (Digital realization of original etchings by Lucie Hey and Nigel Frith, DRPG England; courtesy of the Royal Society ©.)

> "I and the company with me heard, three nights successively, flights of these birds passing over our heads northerly, which is their direct way from Cuba to Carolina. From whence I conceive, after partaking of the earlier crop of rice at Cuba, they travel over sea to Carolina for the same intent."

structural or physical attributes with flight over long distances and vast bodies of water.[5]

Catesby soon settled on food as a (if not *the*) major reason for migration and other movements of birds. In 1725, he was "lying upon the deck of a sloop" off Andros Island in the Bahamas, north of Cuba, when three nights in succession he heard birds whose calls he recognized. They were bobolinks (*Dolichonyx oryzivorus*) (figure 16-1) "passing over our heads northerly, which is in their direct way from Cuba to Carolina." Catesby, who called them "Rice Birds," mused on the reasons for their flight path: "From whence I conceive, after partaking of the earlier crop of rice at Cuba, they travel over sea to Carolina for the same intent, the rice there being at that time fit for them." Catesby's inferences were correct: rice planted in April in Carolina was ready for harvest in September, and bobolinks, which loved rice, went with regularity where they could find it (or other food).[6]

In the first volume of *The natural history of Carolina . . .* , issued between 1729 and 1732, Catesby reported on the comings and goings of nearly fifty species of birds. The following comments are typical. Where little blue herons (*Egretta caerulea*) came from, he remarked, "and where they breed, is to me a mystery."

He could not recall seeing green herons (*Butorides virescens*) in winter, "wherefore I believe, they retire from Virginia and Carolina more south." Swallow-tail kites (*Elanoides forficatus*), he thought, must be "birds of passage" because he saw none in winter. Blue-winged teal (*Anas discors*) arrived "in great plenty" in August and gorged on rice in Carolina and wild oats in Virginia marshes; many then went elsewhere in October. In winter, some birds like Canada geese (*Branta canadensis*) came to Carolina "from northern parts of America." Many "birds of passage," from the yellow-billed cuckoo (*Coccyzus americanus*) and red-eyed vireo (*Vireo olivaceus*) to the purple martin (*Progne subis*) (figure 16-2) and Baltimore oriole (*Icterus galbula*), arrived in spring and left in fall. Yet others like American kestrels (*Falco sparverius*) seemed to be year-round residents.[7]

In the second volume of *The natural history of Carolina . . .* , which was completed during 1743, Catesby expanded his interest in migration to speculation on how birds initially came to America. And he wrote what amounted to a précis of "Of birds of passage."[8] Thus by the time he tackled the question of seasonal absence in "Of birds of passage," Catesby had had the benefit of more than two decades of thought on migration in general as well as a first run at the thesis of "Of birds of passage." In fact, the latter would appear in three versions

FIGURE 16-2. Purple martin; plate 51, M. Catesby, 1730, *The natural history of Carolina . . .* , volume I. (Digital realization of original etchings by Lucie Hey and Nigel Frith, DRPG England; courtesy of the Royal Society ©.)

"They retire at the approach of winter, and return in the spring to Virginia and Carolina."

between 1743 and 1748: in the second volume of *The natural history of Carolina . . .*; in the essay in *Philosophical transactions*; and in a two-page extract in *The gentleman's magazine*, which had a large readership.[9]

The argument was fullest in the essay in *Philosophical transactions*. First, birds of passage, like all birds, went where they could find food. If their food was close and their bills were up to the task, they remained nearby. If not, they migrated or perished. The "Want of Food," Catesby remarked, "seems to be the chief if not the only Reason of their Migration." Second, birds went from one place to another "by the common natural Way of flying." Third, "the Places, to which [birds] retire," Catesby speculated, "lie probably in the same Latitude in the Southern Hemisphere as the Places from whence they depart." He admitted that this was "Conjecture."[10] Fourth, other explanations for the absence of birds, especially reports of birds "torpid in Caverns and hollow Trees" or "resting in the same State at the Bottom of deep Waters, are so ill attested, and absurd in themselves, that the bare Mention of them is more than they deserve." Moreover, Catesby remarked, "a late broach'd Hypothesis, which sends [birds] above our Atmosphere for a Passage to their Retreat . . . seems as remote from Reason, as the Ethereal Region is from the Aereal." In connection with this notion of flight at extreme altitudes, it needs to be mentioned that Catesby, according to Elsa Allen,[11] the historian of ornithology, held that birds, well fed, rested, and strong, ascended to a high altitude and then glided to a destination seen in the distant landscape below. Allen was mistaken. This was not Catesby's idea but that of the unidentified author "A. B.," who commented on the summary of Catesby's paper published in *The gentleman's magazine* in 1748.[12]

Catesby's final major point concerned the necessity (or at least ideal) of what he called "ocular testimony" to consider evidence as adequate. He was far from the first to insist on eyewitness or firsthand evidence. In 1708, for example, William Derham called for precise observations and records on the arrival and departure of birds of passage.[13] Nevertheless, the appraisal of Catesby's "Of birds of passage" can be summarized in the comment by his biographers that it represented "a rational view" that was "ahead of much of the best opinion of his time . . . even the great Linnaeus himself," and that Catesby's emphasis on "ocular testimony" was a critical element in his methodological approach.[14]

Before Catesby

Catesby was interested especially in "summer birds" (birds that appeared in summer but went elsewhere in winter) in general and insectivorous birds known today as house martins (*Delichon urbicum*), sand martins (*Riparia riparia*), and swallows or barn swallows (*Hirundo rustica*)—members of the family commonly known as the hirundines—as well as common swifts (*Apus apus*). His interest in these birds and in migration as an explanation for their absence did not spring from a vacuum. Torpidity (inertness that ranges widely from a brief mild condition to an extended state of hibernation) and migration as competing explanations for the seasonal absence of birds were ancient ideas, and swallows and martins as well as swifts seem often to have captured attention. For example, Herodotus (484–425 BCE) wrote that swallows remained on the Nile throughout the year, while Aristotle (384–322 BCE) noted that while some birds migrated away at the end of summer, others "hide themselves where they are." Among the latter were swallows, which "have often been found in holes, quite denuded" and, with others, in a torpid state.[15]

Other observers and theorizers, from Pliny the Elder in the first century CE to Olaus Magnus (1490–c. 1557) and Ulisse Aldrovandi (1522–1620), joined this particular discussion before Catesby came into the picture. Pliny was especially interested in the movements of birds: some remained all year and others migrated, but small birds seemed not to go very far. The swallow went to "sunny retreats" in adjacent lands where some lost their feathers (as did turtle doves). Magnus, a Swede who spent most of his life in Italy and wrote a history of northern people and lands that contained a dose of fancy, spoke not of migration but of torpidity and of fishermen on the edge of the sea pulling in nets bulging with both fish and torpid birds.[16]

The most important thinkers on these matters inasmuch as Catesby was concerned were John Ray (1627–1705) and Francis Willughby (1635–1672), whose *Ornithology* of 1678 had lasting influence on the structure and content of Catesby's *The natural history of Carolina. . . .* With regard to migration, they mused:

> What becomes of *Swallows* in winter time, whether they fly into other Countries, or lie torpid in hollow trees, and the like places, neither are natural historians agreed, nor can we certainly determine. To us it seems more probable that they fly away into hot Countries, *viz. Egypt, Aethiopia*, &c. then that either they lurk in hollow trees, or holes of Rocks and ancient buildings, or lie in water under the Ice in Northern Countries, as *Olaus Magnus* reports. For as *Herodotus* witnesseth, they abide all the year in Egypt, understand it of those that are bred there (saith *Aldrovandus*) for those that are bred with us only fly thither to winter. I am assured of my own knowledge (saith *Peter Martyr*) that *Swallows, Kites*, and other Fowl fly over Sea out of *Europe* to *Alexandria* to winter.[17]

In his travels in America, Catesby carried the *Ornithology*; and it is clear that *The natural history of Carolina* . . . owes much to it. If further concrete evidence supports a closer link between Catesby and the great naturalist, Ray, either from propinquity, as George Edwards, a friend and colleague, wrote in a retrospective comment, or from Catesby's assistance to the botanist Samuel Dale, who in turn apparently collected for Ray, then Catesby's ties to the *Ornithology* would have run even deeper.[18] Regardless, given the obvious influence of the *Ornithology* on him and his knowledge of its contents, it is reasonable to suggest that Catesby was familiar with Willughby's and Ray's opinions on migration and that his own ideas on migration were formed in knowledge of their

FIGURE 16-3. Chimney swift; plate 8, M. Catesby, 1747, *The natural history of Carolina . . .* , volume II: Appendix. (Digital realization of original etchings by Lucie Hey and Nigel Frith, DRPG England; courtesy of the Royal Society ©.)

> "Their periodical retiring from, and returning to Virginia and Carolina, is at the same seasons as our swallows do in England: therefore the place they retire to from Carolina is, I think, most probably Brazil, some part of which is in the same latitude in the Southern Hemisphere, as Carolina is in the northern."

positions in the *Ornithology*. That said, Catesby's repeated emphasis on evidence based on personal eyewitness accounts, while of a piece with the emergent scientific method that was a mark of his day, represents a progression from the time of Ray and Willughby.

After Catesby

Catesby's impact on the debate over the fate of summer birds is unclear. He died in 1749, two years after the appearance of his essay in *Philosophical transactions*. In the decades that followed—the 1750s through the 1780s—opinions in the debate continued to be divided, with influential people on both sides. On the one hand were those who held to the now-familiar notion that birds went into caverns, crevices and fissures in rocks, or old nesting holes in banks and cliffs, where they entered a state of torpor or hibernation. Others went further, claiming that some birds of passage sank beneath the surface of the sea or of fresh water, or went into mud on the edges or bottoms of ponds, there to enter an inert state until the next season of warmth. Such was not only so-called vulgar opinion but a belief at all levels of society.

In March 1758, Peter Collinson, the merchant-botanist, Fellow of the Royal Society, and one of Catesby's patrons, weighed in on the same side of the debate as had his late friend. Collinson's position basically was that swallows, martins, and other birds migrated south where temperatures were warmer and food was available for birds that ate insects on the wing. Given their relationship and membership of the Royal Society, Collinson was surely familiar with Catesby's "Of birds of passage," and it is possible that it influenced his thoughts on migration. Yet Collinson evidently modified his thoughts through time. In the late 1750s, he apparently failed to persuade Linnaeus that migration, not hibernation, accounted for the disappearance of certain summer birds. In 1758, he argued strenuously against the idea that swallows could hibernate under water—a biological impossibility, he said, given the role of respiration in the circulation of blood. Yet three years later, he seemed open to the idea of torpor above the surface of waters, wondering if "swallows or martins" (they were probably sand martins) taken annually in late March from nesting holes in banks of the Rhine near Basel, some apparently inert, "stay behind in their dormitories all the winter." He evidently did not consider that, on the cusp of April, they were early spring migrants.[19]

Also in the 1760s, Thomas Pennant (1726–1798), the naturalist and antiquarian, published *British zoology*, which within a short time went through four editions, when Pennant concluded that nearly every "known kind [of the "swallow tribe"] observe a periodical migration, or retreat." He dismissed theories that placed swallows torpid beneath the waters, which "must provoke a smile" for the physiological impossibility, or in crevices of rocks. He accepted without doubt that swallows went where they could find aerial insects—Catesby's position as well—and cited with approval anecdotes about these birds massing in preparation for migration, resting in the rigging of ships at sea, or feeding on late hatches of insects—and inadvertently trapped by the onset of cold weather and at its mercy if it lingered.[20]

Daines Barrington (1727–1800), a judge, amateur naturalist, and Fellow of the Royal Society, begged to disagree, publishing, in 1772, a stiff rebuttal of migration. Confining migration to the movement across a vast sea of an entire species, he submitted not only that it did not occur—birds only "flitted" from one island, headland, or place to another—but that there was ample evidence from people in all stations of life for torpidity and submersion, under water and in mud, of swallows and martins.[21]

Between 1767 and 1774, both Pennant and Barrington were in correspondence with the Reverend Gilbert White (1720–1793), who, in 1788, published his side of the exchange in *The natural history and antiquities of Selborne*.[22] In time, this celebrated work on the natural history of a small village in Hampshire took its place at the heart of a natural history "craze" and became the "fourth most popular book in the English language."[23] Here it is important because it arguably amounted to an extended meditation on the coming and going of birds, on the debate over torpidity or migration.

White's letters to Pennant reveal their author as an advocate of migration and skeptic on torpidity—except for a few birds that might remain behind and "hide." White seemed uncertain if birds still present in late fall (autumn) or early winter left for another clime or found rest in a local "hybernaculum." His letters to Barrington, "no great friend to migration," reveal the depth of his lack of confidence on this issue. In 1771–1772, White remarked that Barrington's "well attested accounts" justified his sense that many of "the swallow kind" become torpid until sun and insects return. At one point White wrote, "Many of the swallow kind [figure 16-4] . . . lay themselves up in holes and caverns," coming out for mild spells and then returning. Then he expressed doubt, noting that sand martin holes excavated in winter failed to turn up torpid or hibernating birds. Then he slid back to the belief that entire species of swallows and martins, "or at least many individuals" of each species, "never leave this island at all, but partake of the same benumbed state." White, it seems, never entirely abandoned his faith in "secret dormitories" for wintering martins and swallows and never could decide if swallows and martins undertook migration or entered torpidity during that season.[24]

The nineteenth century spelled the end of doubt over what happened to birds in winter. As global knowledge of birds expanded, methodological principles set in motion long before took hold, quite simple experiments proved that swallows could not survive submersion, and "the swallow tribe" were seen more and more at, near, or en route to wintering grounds far to the south, where they fed and thrived. The idea, once considered unlikely by many, took hold that birds disappeared from their familiar breeding and summer haunts

FIGURE 16-4. "The Swallows and Martins clustering on the chimneys of the neighboring cottages," by George Edward Collins (1880–1968), from Gilbert White, *The natural history and antiquities of Selborne*. (1911 edition by Macmillan & Company; courtesy of University of California Libraries and Biodiversity Heritage Library.)

because they migrated to more hospitable climes where the insects on which they fed were available. This alteration occurred sooner in England than in America, where the layperson's opinion on a state like torpidity or submersion was often considered as the equal of the scientist's—a democratizing impulse that delayed the final acceptance of migration as the reason for disappearance.[25]

Local knowledge

In England, proponents on all sides of the great debate over the disappearance of "the swallow tribe" often coupled a person's opinion on migration or torpidity with his status, reputation, or morality. They deemed a report more worthy not only if it was firsthand rather than at farther remove but if it came, as they said, from a "very honest man," a person "worthy of credit," an "observing or respectable gentleman," or one of "veracity and ability," a "naturalist," or a "clergyman, of an inquisitive turn." Far less worthy of attention were the reports of the incurious, the poor, those lacking in respect or ability, or "ignorant peasants and credulous people."[26]

Thomas Pennant and Gilbert White offer a useful way to think about those who claimed that "the swallow tribe" entered a state of torpidity or hibernation rather than migrated. Pennant dismissed the possibility of torpidity or hibernation during submersion beneath the surface of water, or of life after submersion, yet he was far less certain about the truthfulness of accounts of torpidity in birds found in the walls of houses, hollow trees, or rock crevices. White, who was (and is) often seen as a "migrationist," in reality waffled. He could not make up his mind, and he ended up thinking that if some swallows or martins left, that did not mean that all did (eliding their state if they remained behind).

In essence, both White and Pennant cautioned not to throw the baby out with the bathwater. It makes sense to consider three rather than two alternatives—submersion, concealment on land, or migration—that is, to keep separate the two major alternatives to migration involving torpidity or hibernation: first, submersion beneath the surface of water or mud; second, inertness in dense copses, rock crevices, nesting holes, hollow trees, or holes in the walls of houses. By this time, Ray and Willughby and others, including Catesby, had firmly rejected submersion. Pennant, Edward Jenner (1749–1823), and others sounded its death knell in the increasingly rational, scientific, and empirical data-driven and hypothesis-testing atmosphere of the late eighteenth and early nineteenth centuries.[27] But the idea of inertness on land proved far more resistant. Perhaps not all accounts were false. Perhaps White waffled because his and others' knowledge of the local "patch" of Selborne was so fine-grained that he was aware of different possibilities for hirundine adaptations.

For insight on these last matters, we turn finally to what is known today about torpidity or hibernation in what was loosely called "the swallow tribe," that is, in house martins, sand martins, and barn swallows—the hirundines—and in common swifts.

Torpidity and hibernation

Torpidity is far from uncommon in birds. Facultative hypothermic responses to energy fluctuations and thermoregulatory demands, of which torpor and hibernation are two kinds, are widespread. Today torpidity is known in about a hundred species of birds.[28] The boundaries of torpidity and hibernation are much discussed. For some biologists, torpor, which is marked by inactivity and reduced responsiveness, is a shallow and daily inertness with body temperatures between 10° and 25°C (50° and 77°F); in contrast, hibernation consists of prolonged bouts of torpor lasting from days to a season, with body temperatures below 10°C (50°F). The data, however, do not always fit the two categories neatly, for which reason other biologists define torpidity by length of inertness: torpor if less than twenty-two hours in duration (average eleven hours), hibernation if longer than ninety-six hours in duration.[29] Among birds, much attention on torpidity has been focused on hummingbirds, sunbirds, mousebirds, manikins, swifts, and goatsuckers. Many hummingbirds, for example, regularly enter torpor at night, swifts have extremely labile temperatures, and the poorwill, a caprimulgid or goatsucker, is known to hibernate. Even though knowledge is not yet adequate for generalization, food deprivation—actual or perceived—is an important precondition for entering a state of torpor, as are extremes of temperature and weather, and the facultative response is arguably one that conserves energy for varying periods of time.[30]

The hirundines and swifts

The hirundines—house martins, sand martins, and swallows—which, with swifts, were a particular focus of Catesby, White, and others who mused on their disappearance in fall, often become visible on their breeding and summer grounds early in spring and tend to have second (third, for the house martin) broods; thus they are often on the same grounds well into fall. These habits and the birds' exclusive dependence on insects caught on the wing mean that they are often in the wrong place at the wrong time. That is, a cold spell with heavy rain that damped their prey could also put them in dire jeopardy and kill many that are unable to take refuge or find insects to consume. In 1931, a great late-September storm killed thousands of martins and swallows in central Europe, and many others were found numbed and inert in houses, stables, and barns. Remarkably, nearly ninety thousand were taken by train to Venice, where they flew off. In 1974, another September storm in central Europe killed vast numbers of hirundines—estimates range from hundreds of thousands to millions. During very heavy rain in Estonia in August 1967, sand martins came into the attic of a building, clustered together, and were hypothermic, with body temperatures 4–8°C (7–14°F) below normal (warmer in the clusters) for sixty hours. There is now firm evidence that during such events, hirundines huddle for warmth, succumb to hypothermia, and seek and take man-made or natural shelter where they can find it until the return of their aerial insect prey. Furthermore, the events need not be cataclysmic for facultative hypothermic responses by the hirundines to fluctuations in temperature, precipitation, and other weather-related phenomena.[31] When starved, healthy house martins can enter torpor at eleven days of age; so can young swifts. Moreover, adult house martins are also capable of this state and can decrease their body temperature by 13°C (23°F) for energetic efficiency under conditions in which their prey is also stressed.[32] Hirundines on other continents show signs of similar proclivities and capacities.[33]

Finally, that swallows spend the winter in South Africa and elsewhere in sub-Saharan Africa has been firmly established for more than half a century. House martins also winter in Africa, south of 20°N, and an analysis of isotope composition of primary feathers indicates that house martins that breed in the Netherlands spend the northern winter in West Africa.[34]

Mark Catesby, in ways that neither he nor others of his day could have anticipated, was correct in his brief paper on the central fact that many birds, including hirundines and swifts, migrated to and from England in spring and fall, as well as on his insistence on empirical evidence, on eyewitness testimony. His position on migration surely developed in part from pervasive curiosity about where birds in America (and the Bahamas) came from and went to in the seasons when they were absent. Yet his thoughts also evolved, as in much of his writings on birds, from Ray's and Willughby's *Ornithology*, and in his insistence on firsthand observation, he was very much an empiricist of his day (and in stating the need for precise observations of the coming and going of birds, like Derham in particular). As for the "conjecture" that some birds went to the same latitude south of the equator as they left in the north, Catesby was wrong: for example, London is 51°N and Cape Town 33°S. From one standpoint it is unfortunate that Catesby was myopic on evidence of torpor that today can be attested to physiologically and linked to diet, annual cycle, and growth and development, but it is surely too much to expect him to have thought otherwise. In not giving credence to accounts of torpor that were not simply anecdotal but "ill attested," Catesby was neither prescient nor anachronic but merely the well-placed naturalist of the times.

17

Catesby's animals (other than birds) in *The natural history of Carolina, Florida and the Bahama islands*

AARON M. BAUER

Mark Catesby's *The natural history of Carolina, Florida and the Bahama islands* stands as a landmark of both eighteenth-century natural history illustration and colonial American botany and zoology. Although Catesby's chief interests were botanical, the work is rightly known to zoologists for its treatment of birds of the American Southeast.[1] However, Catesby also considered, in greater or lesser detail, other animals he encountered, most notably other vertebrates, and indeed 109 of the plates in *The natural history of Carolina* . . . depicted one or more non-avian zoological subjects (nine in volume I, ninety in volume II, and ten in the Appendix). Collectively, 148 images of animals are presented, not counting two moth cocoons and several anatomical details (for example, a cutaway view of the rattlesnake's rattle). Of these, fish dominate with forty-nine images, followed by reptiles and insects with twenty-nine images each. Of remaining vertebrates, mammals are represented by ten images and amphibians by seven. Eight crabs are depicted, as are seven corals, four mollusks, two starfish, one spider, and one centipede.

In their 1961 biography of Catesby, George Frick and Raymond Stearns provided a balanced assessment of Catesby's zoology, focusing on the accuracy (or not) of his depictions but also touching upon his natural history. Their assessment was that his observations and imagery were both perceptive and flawed but that they represented an amazing accomplishment for their era.

Organization

Most of the images and text devoted to non-avian zoology are in the second volume of *The natural history of Carolina*. . . . Undoubtedly, Catesby's decision to concentrate on birds over other animals in the first volume reflected his own interests and competency. It also allowed him to treat in a single volume the one animal group that he believed he had surveyed comprehensively while in Carolina. This was probably also a calculated business decision as Catesby needed to generate income through subscriptions and including the plates of birds, the most popular subjects of study and general interest at the time, in the first parts to be issued would have best served this aim. Indeed, only eleven animals other than birds are represented on nine different plates in the first volume. Only two of these, both depicting corals (figure 17-1)—considered plants by Catesby—are described. The northern mole cricket (*NH* I: plate 8) is discussed briefly (see below), but two other insects, the giant leopard moth (*Hypercompe scribonia*, *NH* I: plate 35) and an unidentifiable wasp or fly (*NH* I: plate 55), are not mentioned. In the remaining images, animals are depicted merely as secondary elements in what are otherwise portraits of birds. Fish are depicted being captured or consumed by birds (*NH* I: plates 1, 2, and 69), and two starfish (*NH* I: plates 73 and 83) and one bivalve (*NH* I: plate 83) are minor decorative elements.[2]

Other zoological observations are scattered throughout the first volume. The texts of twenty-three bird accounts, for example, mention animal prey. This is usually a generalization (for example, "lizards" or "insects"), but in one case Catesby mentions having found the remains of mole crickets (probably *Neocurtilla hexadactyla*) in the stomach of a dissected bird. While most of Catesby's assertions regarding diet have been substantiated, his report that turkey vultures (*Cathartes aura*, *NH* I: plate 6) capture live snakes may be based in part on observations of birds carrying or eating snake carrion.[3] His botanical accounts also contain information relevant to animals, such as the possible efficacy of saltmarsh morning-glory (*Ipomoea sagittata*) against snake venom (*NH* I: p. 35), or the use of black tupelo (*Nyssa sylvatica*) berries by raccoons and other mammals (*NH* I: p. 41).

In the second volume, animals are arranged mostly by group, starting with the fish.[4] The only disruptions in the pattern result from intervening botanical plates with no animal subjects and from the incongruous "Pol-Cat" (skunk) between the snakes and lizards. For the most part, the animals depicted in the Appendix are haphazardly arranged.[5]

Classification and taxonomy

The greatest influence on Mark Catesby, with respect to animal classification, was John Ray (figure 17-2). Ray's classification system[6] was in popular use by naturalists of the early eighteenth century and largely held sway until it was replaced by that of Carl Linnaeus, which gradually supplanted it in the decades following the publication of the first edition of *Systema naturae* in 1735. For the identification of species and the application of scientific names, Catesby also consulted the works of James Petiver, Martin Lister, and especially Hans Sloane, one of his major patrons. For the most part, Catesby's major groupings—mammals, birds, serpents, and so on—were not controversial. However,

FIGURE 17-1. Bent sea-rod (*Plexaura flexuosa*), one of several octocorals illustrated by Catesby, was considered to be a marine plant until the 1750s; plate 74, M. Catesby, 1731, *The natural history of Carolina* . . . , volume I. (Digital realization of original etchings by Lucie Hey and Nigel Frith, DRPG England; courtesy of the Royal Society ©.)

FIGURE 17-2. John Ray was one of the most influential pre-Linnaean systematists and a source for Mark Catesby's zoological writings in *The natural history of Carolina, Florida and the Bahama islands*. Portrait from Ray 1693. (By courtesy of Kraig Adler.)

in his accounts of fishes of both Carolina and the Bahamas, he included whales and other cetaceans, as well as crabs, among the fishes. Ray had recognized that whales were live-bearing lung-breathers like other mammals; nonetheless, he included them among the fishes, although he included crabs, along with other invertebrates, in a group he called the Exanguia.[7] Corals appear in seven of Catesby's plates, but, like his contemporaries, he considered sea fans and their relatives to be marine plants. It was not until the 1750s that the animal nature of these organisms was generally recognized by John Ellis (1710–1776).[8]

Catesby's organization of his plates probably reflects his perceived completeness of coverage rather than any classification scheme. He considered his treatment of Bahamian fishes to be second only to that of Carolinian birds, followed by that of "reptiles" (including amphibians), then mammals, and finally insects, of which he "was not able to delineate a great Number."[9] Reptiles were, in turn, also organized largely by completeness: serpents, then lizards, then frogs. Sea turtles, clearly reptiles but aquatic and hard-shelled, were placed in a transitional position, on the three plates between the crabs and the terrestrial reptiles.

Ichthyology

Although Catesby discussed sharks and rays in his notes on the fishes of Carolina and of the Bahamas and listed the lamprey as a Carolinian sea fish, all of his illustrations are of bony fishes. Of these, all but one are teleosts (an infraclass of ray-finned fishes), the exception being the Florida gar (*Lepisosteus platyrhincus*, *NH* II: plate 30). In addition, sturgeons (*Acipenser oxyrinchus* and *Acipenser brevirostrum*), which are chondrostean (largely cartilaginous) fish, are discussed at great length in "Some Observations concerning the Fish on the Coasts of Carolina and Virginia"[10] Most of Catesby's fish accounts are merely descriptions of external morphology, and he was often inattentive to details, occasionally omitting some fins altogether.[11] The "Cat Fish" was copied from a sixteenth-century painting attributed to John White and is certainly the least accurate rendering of a fish in *The natural history of Carolina . . .* (figure 17-3).[12] The rather crudely depicted "green Gar-Fish" (Florida gar; see above), as well as several other illustrations, were also taken from White's drawings,[13] copies of which were in the collection of Sir Hans Sloane. Catesby's description of the "Cat Fish" was apparently written from memory and was based on more than one species.[14] He erroneously stated that the barbed pectoral spines were bones borne on the sides of the jaws and also incorrectly stated that they were lacking in the species he purported to figure (they are present in all North American catfish). Nonetheless, the vast majority of the fish images are unambiguously identifiable usually based on color pattern or body form.[15]

Nearly half of the fish illustrated are assessed with respect to their palatability. As might be expected, given that many of the Bahamian fish would have come to Catesby via fish markets and fishermen, the majority were reported to

FIGURE 17-3. "The Cat Fish," Catesby's least accurate fish illustration, was drawn from memory and was one of the few fishes of Carolina figured; plate 23, M. Catesby, 1736, *The natural history of Carolina . . .*, volume II. (Digital realization of original etchings by Lucie Hey and Nigel Frith, DRPG England; courtesy of the Royal Society ©.)

be edible, with the mutton fish (*Lutjanus analis*, *NH* II: plate 25) reported to be the best eating, and only a few species considered poisonous or bad tasting (barracuda, *Sphyraena barracuda*, *NH* II: plate 1; yellowfin grouper, *Mycteroperca venenosa*, *NH* II: plate 5; scrawled filefish, *Aluterus scriptus*, *NH* II: plate 19; spotted muray, *Gymnothorax moringa*, *NH* II: plate 21). Catesby was correct that these and other predatory reef fish can cause ciguatera, a food-borne illness resulting from the accumulation of toxins ultimately derived from certain dinoflagellates in tropical waters.[16] He was also correct that certain fish could be edible or not, depending on where and when they were caught.

Catesby's treatment of the fishes of Carolina was much less extensive than that of the Bahamian species, but at least some of those he described were also noteworthy for their usefulness as food. Sturgeons, although not illustrated, were discussed in greatest detail. Catesby recounted catching 9-foot-long (2.7 meters) sturgeons in the Savannah River, and he published recipes for pickled sturgeon and caviar provided by "his Excellency Mr. Johnson, late Governor of South Carolina, which . . . he got translated from the Original in High Dutch, which was wrote in Gold Letters, and fixed in the Town Hall at Hambourg."[17] These must have been Atlantic sturgeon (*Acipenser oxyrinchus*), which achieve a maximum length of about 8 feet (2.5 meters) and still spawn in many areas of the Savannah River, where they have historically been known to occur.[18]

Other biological information about fishes in *The natural history of Carolina* . . . is limited, although Catesby did discuss defense in the tang (*Acanthurus coeruleus*, *NH* II: plate 10) and diet of the scrawled filefish (*NH* II: plate 19), and he proposed a hypothesis about the unique morphology of the "Old Wife" (queen triggerfish, *Balistes vetula*, *NH* II: plate 22). The flying fish (*NH* II: plate 8), being an oddity of special interest to naturalists, however, was discussed in more detail by Catesby, and information about not only its locomotion but also its distribution and role in the economy of Barbados was considered. Although at least some of the details provided were based on his own observations, Catesby clearly drew on other unidentified sources as well. Likewise, his account of the "Sucking-Fish" (remora, *Remora remora*, *NH* II: plate 26) drew chiefly on Sloane's discussion of the species.[19] The illustration was taken from John White's drawing in Sloane's collection.[20]

Herpetology

Catesby's contributions to herpetology were substantial. Many of the species he illustrated were subsequently commented upon by William Bartram[21] and others, but his were the first descriptions of many of the common species of the southeastern United States. Sea turtles follow the same principle as fish: they could be eaten and were thus of practical value and so received extensive comments. At least some of the information presented was derived from Sloane's work on Jamaica[22] but it appears that Catesby added his own observations, both on biology and on the turtle fisheries. His images of green (*Chelonia mydas*, *NH* II: plate 38), hawksbill (*Eretmochelys imbricata*, *NH* II: plate 39), and loggerhead (*Caretta caretta*, *NH* II: plate 40) sea turtles are all unambiguously recognizable. He reported accurately on aspects of mating and egg laying, correctly noting that sea turtles lay multiple clutches in a season, with about a two-week inter-nesting period. However, he underestimated incubation to take only three weeks instead of the actual period of about two months.

Although Catesby was observant and correct about certain aspects of snake biology, such as the live birth of rattlesnakes, he was wrong on others, suggesting that the forked tongue of snakes was used to catch insects; in fact, it is used to chemically sense the environment. He rejected many folktales, such as the story that the coachwhip could cut a man in half with a snap of its tail, but he reserved judgment on others, describing the ability of the rattlesnake to charm its prey but noting that he had not observed this himself. Although rattlesnakes were well known in Europe, Catesby considered them of sufficient interest to devote some space to little-known aspects of their biology. He also reported on the methods used by the Native Americans who lived in Carolina to treat bites, including sucking venom from the wound and the use of botanical remedies. Despite clear attention to some detail in his illustrations of the vipers, as witnessed by the inset illustrations of the rattle and fang (*NH* II: plate 41; figure 9-7, p. 118) of the timber rattlesnake (*Crotalus horridus*), neither the "Small Rattle-Snake" (pygmy rattlesnake, *Sistrurus miliarius*, *NH* II: plate 42) nor the "Water Viper" (eastern cottonmouth, *Agkistrodon piscivorus*, *NH* II: plate 43), both of which were shown with open mouths, illustrates the typical enlarged, hinged front fangs of viperids, and the latter species lacks the thermosensory pit organ between the eye and nostril, a feature characteristic of all American vipers that is shown in the two rattlesnakes.

The "black Viper" and "Brown Viper" depicted by Catesby (*NH* II: plates 44 and 45) have both been identified as *Heterodon platirhinos*,[23] but these are not, or are only in part, based on the eastern hog-nosed snake. The illustration of the "Brown Viper" (figure 17-4) clearly shows enlarged front fangs, yet in his account accompanying plate 56, which is unambiguously a hog-nosed snake, Catesby explicitly used the absence of enlarged front fangs as evidence that the snake was not a viper, despite the fact that he believed the body to be very viper-like. The "black Viper," although similar in its stout body form to the hog-nosed snake, which is often melanistic (darkly pigmented), lacks the characteristic upturned snout of this species. Further, the short, secondhand accounts of the "black Viper" and "Brown Viper" are in sharp contrast to the more confident and detailed accounts of the cottonmouth and timber rattlesnake, snakes certainly seen alive by Catesby, and suggest that the former species were drawn from memory or from the descriptions of others.

Most non-venomous snakes were treated in less detail than vipers, and these vary in their accuracy of depiction. The illustration of the plain-bellied watersnake (*Nerodia erythrogaster*, *NH* II: plate 46) is one of the most accurate (figure 17-5), beautifully showing the strongly keeled scales of the species, and Catesby

FIGURE 17-4. Catesby's depiction of "The Brown Viper"; plate 45, M. Catesby, 1737, *The natural history of Carolina . . .*, volume II. The snake depicted by Catesby remains unidentified. Although previously regarded as the eastern hog-nosed snake (*Heterodon platirhinos*), this seems unlikely, given both the accuracy of Catesby's illustration of this species and the clear presence of front fangs, a feature he knew was lacking in the non-venomous hog-nosed snake. (Digital realization of original etchings by Lucie Hey and Nigel Frith, DRPG England; courtesy of the Royal Society ©.)

correctly reported that it was not venomous, despite its superficial resemblance to the cottonmouth. The "Chain-Snake" (eastern kingsnake, *Lampropeltis getula*, *NH* II: plate 52) was also well rendered, although he had seen only a single specimen. The common garter snake (*Thamnophis sirtalis*, *NH* II: plates 51 and 53) and eastern ribbonsnake (*Thamnophis sauritus*, *NH* II: plate 50) are also well depicted, although reports that the former raided henhouses perhaps reflect mistaken identity, as the species typically feeds on relatively small prey, often near water, such as fish and frogs. The majority of Catesby's snakes are easily identifiable, but there are exceptions. The image of *Opheodrys aestivus* (rough greensnake, *NH* II: plate 47) shows an exaggerated, elongate, upturned snout, perhaps suggesting that Catesby's illustration was influenced by depictions of tropical Asian vine snakes (*Ahaetulla* spp.) that were well known to eighteenth-century naturalists.[24] Catesby's "little brown Bead Snake" (*NH* II: plate 49) has been identified as *Storeria dekayi*, although it is a very poor likeness.[25]

The identity of the "Wampum Snake" (*NH* II: plate 58) has long been problematic.[26] The blue-on-blue color pattern ascribed by Catesby to the species (figure 17-6) is not consistent with any known snake in the Carolinas. However,

FIGURE 17-5. "The Copperbelly Snake" is unambiguously referable to the plain-bellied watersnake (*Nerodia erythrogaster*) and is one of Catesby's finest reptile images; plate 46, M. Catesby, 1737, *The natural history of Carolina* . . . , volume II. (Digital realization of original etchings by Lucie Hey and Nigel Frith, DRPG England; courtesy of the Royal Society ©.)

FIGURE 17-6. "The Wampum Snake" has been variously interpreted as a water snake (*Nerodia* sp.) or the eastern mudsnake (*Farancia abacura*). Although its color does not match any American snake, its smooth scales and bluish dorsum are consistent with the latter species; plate 58, M. Catesby, 1737, *The natural history of Carolina* . . . , volume II. (Digital realization of original etchings by Lucie Hey and Nigel Frith, DRPG England; courtesy of the Royal Society ©.)

at least the dorsum of the mudsnake is an iridescent blue-black, and its scales are smooth, as depicted by Catesby.

Included among Catesby's snakes was one legless lizard, the "Glass Snake" (*Ophisaurus ventralis*, *NH* II: plate 59). The likeness is good, and the ventrolateral fold that characterizes the lizard family Anguidae, to which this species belongs, is clearly shown, although the movable eyelid—typical of lizards but not snakes—is not. Catesby recognized that the "Glass Snake" was not a typical snake, commenting on its unique small juxtaposed scales. Significantly, he also described and illustrated its fleshy tongue and contrasted it with the slender forked tongues of typical snakes. Tongue morphology was subsequently to become an important character in establishing phylogenetic relationships among lizards and snakes.[27] The skink (*Plestiodon fasciatus*), which is also quite well illustrated (*NH* II: plate 67), is noteworthy, as Catesby mentions that it was regarded as venomous by some people in Carolina. This is still true in rural areas of the southeastern United States, where skinks are called "scorpions" by people who believe that the blue tail can be used as a "stinger."

Several of the other lizards illustrated in Catesby's *The natural history of Carolina . . .* have hitherto been misidentified. The "*Green Lizard* of Jamaica" (*NH* II: plate 66) is clearly an anole, and it has been considered to be *Anolis* (or *Norops*) *garmani*.[28] The members of this species are, however, "crown giant" anoles (one of several species that share certain similarities in niche and body form), with males reaching more than 14 inches (350 millimeters) long. Males also bear a serrated dorsal crest and have a greenish-yellow dewlap with an orange center.[29] This is a poor match for Catesby's description and illustration: 6 inches (150 millimeters) long, no crest, and a scarlet dewlap. This is a much better match to another Jamaican lizard, *Anolis grahami aquarum*, although its dewlap is typically more orange than scarlet.[30] Catesby correctly hypothesized that dewlap displays by male "green lizards" in Jamaica (*NH* II: plate 66) were a signaling mechanism. Another misidentified animal is the "Lyon Lizard" (*NH* II: plate 68), which had been associated with *Cnemidophorus sexlineatus* (now *Aspidoscelis sexlineata*). This species does not occur in the West Indies, but the name "lion lizard" is still in use in the Bahamas for a related species, the blue-tailed lizard (*Ameiva auberi*). Catesby appears to have applied this local name to a lizard of a different family—a curly-tailed lizard of the genus *Leiocephalus*. Although the rendering is not very good (figure 17-7), its description leaves no doubt as to its identity: "It cocks its Tail with a round twirl . . . frequenting the Rocks on the Sea Shores of *Cuba*, *Hispaniola*. &c."[31] Catesby would have undoubtedly seen the common species of the Bahamas (*Leiocephalus carinatus*) and perhaps others elsewhere in the Greater Antilles. He may have also had access to Jamaican specimens in Sloane's collection. Although nothing resembling a curly-tailed lizard is mentioned in Sloane's book on Jamaica, published in 1725, earlier voyagers, such as Charles de Rochefort, clearly figured *Leiocephalus* in its characteristic posture.[32]

FIGURE 17-7. "The Lyon Lizard" is poorly rendered, but its distinctive tail posture and Catesby's description of its habitat are unmistakably those of the West Indian curly-tailed lizards (*Leiocephalus* spp.); plate 68, M. Catesby, 1739, *The natural history of Carolina . . .*, volume II. (Digital realization of original etchings by Lucie Hey and Nigel Frith, DRPG England; courtesy of the Royal Society ©.)

FIGURE 17-8. Among the most iconic images in *The natural history of Carolina . . .* is the bullfrog, named *Lithobates catesbeianus* in Catesby's honor; plate 72, M. Catesby, 1737, *The natural history of Carolina . . .* , volume II. (Digital realization of original etchings by Lucie Hey and Nigel Frith, DRPG England; courtesy of the Royal Society ©.)

The alligator (*NH* II: plate 63; figure 10-11, p. 139) is represented by an image stated by Catesby to have the "Size and Figure of an Alligator, soon after breaking out of the Shell," but is most likely based on one or more adult models.[33] Catesby noted that his account did not need to repeat details of this "formidable Animal" published by earlier authors (although many actually confounded information relating to alligators and crocodiles). Nonetheless, the text accompanying the plate is extensive and provides mostly accurate data about the alligator, describing winter torpor, reproduction, and hatching, even giving the Neuse River in North Carolina as the northern extent of the species in America.[34]

In contrast to reptiles, only a few amphibians were noted by Catesby. Not surprisingly, salamanders, which are generally inconspicuous to the casual observer, despite their sometimes astounding densities, were largely missed by him. Only two species of mole salamanders (*Ambystoma*) appear in *The natural history of Carolina . . .* , one as prey of a great blue heron (*NH* II: Appendix, plate 10), and the other as a nondescript and unmentioned figure (*NH* II: plate 45). Frogs, on the other hand, would have been evident by their calls throughout the Carolina landscape. Catesby's namesake, the bullfrog (*Lithobates catesbeianus*, formerly *Rana catesbeiana*, *NH* II: plate 72) (figure 17-8), with its bellowing call and huge size, was an obvious choice to figure, as were the green tree frog (*Hyla cinerea*, *NH* II: plate 71), the southern leopard frog (*Lithobates sphenocephalus*, *NH* II: plate 70), and the southern toad (*Anaxyrus terrestris*, *NH* II: plate 69), all common species throughout the Carolinas. It is somewhat surprising that the ubiquitous green or banjo frog (*Lithobates clamitans*) was not recognized by Catesby, but he may have taken it for a juvenile of the bullfrog.

Mammalogy

Catesby often erroneously considered North American species to be conspecific with their European counterparts and as such did not illustrate them or provide information about their biology. This was most evident for mammals, where such ubiquitous species as white-tailed deer (*Odocoileus virginianus*) were not illustrated, and bats, constituting more than a fifth of the non-marine mammal fauna of South Carolina, were not even mentioned. Catesby's "Of Beasts" section divided native North American mammals into those identical to European forms, those belonging to the same genus, and those unique to the American continent.[35] Only three of thirty-six mammals listed fall into the latter category, so the dearth of mammal illustrations is perhaps not surprising. For example, he considered the American black bear identical with the European brown bear. Despite what modern zoologists would recognize as fairly obvious differences between the two, Catesby may be forgiven his error. Although he saw bears in South Carolina, it is likely that his familiarity with European bears would have been limited to the literature or perhaps imported animals used in bear-baiting, which persisted in Britain until the early nineteenth century, as wild bears had been extirpated from Britain by the tenth century.[36]

The three mammals that Catesby considered unique to the New World were the raccoon, the opossum, and the "Quickhatch" (wolverine), the last of which actually does occur in northern Europe, unbeknownst to Catesby, who probably did not realize that this was the same animal as "the gulo" or "glutton," as it was known to European naturalists. These novelties were discussed but were not among the ten mammals illustrated in *The natural history*

FIGURE 17-9. Southern flying squirrel, based on an original drawing by Everhard Kickius (see figure 11-9), is perhaps the most lifelike of all the zoological images in *The natural history of Carolina . . .*; plate 76, M. Catesby, 1737, *The natural history of Carolina . . .* , volume II. (Digital realization of original etchings by Lucie Hey and Nigel Frith, DRPG England; courtesy of the Royal Society ©.)

of Carolina. . . . Some of the other North American mammals that were discussed but not figured, such as the "White Bear" (polar bear), moose (*Alces alces*), and "Greenland Deer" (reindeer, *Rangifer tarandus*), were northern species, living beyond Catesby's area of activity and thus unavailable to him to draw. Oddly, however, the "Java Hare" (*NH* II: Appendix, plate 18; figure 21-3, p. 300), believed by Catesby to live in Java and Sumatra but actually referable to the agouti (*Dasyprocta leporina*) of northern South America and the Lesser Antilles, was illustrated based on a specimen in the collection of the Duke of Richmond. These animals were apparently popular as pets during the mid-1700s, as Mary Delany (née Granville) saw a tame one in the menagerie of the Duchess of Portland at Bulstrode Hall, Buckinghamshire, in 1749 and made a sketch of what was probably the same animal in 1755.[37]

Several of the mammals that Catesby chose to portray were noteworthy for being crop pests (squirrels) or predators on poultry (gray fox), and some (eastern fox squirrel, *Sciurus niger*, *NH* II: plate 73; Bahamian hutia, *Geocapromys ingrahami*, *NH* II: plate 79; eastern spotted skunk, *Spilogale putorius*, *NH* II: plate 62) were edible, carrying on the emphasis on animals of direct concern to humans. Likewise, the methods of hunting were discussed for many of the beasts, and some basic anatomy was also outlined. Usually this was superficial, although the shape of the baculum (os penis) was described in the raccoon.

The flying squirrel was unique in being figured on two plates and having a concomitantly extensive associated text. In addition to being odd and, therefore, interesting, this was a species that Catesby would have had ample opportunity to observe. His claims of groups of ten or a dozen traveling together are consistent with the known group nesting of the southern flying squirrel (*Glaucomys volans*), and his estimates of glides of up to 80 yards (about 73 meters) are within the known capabilities of the species.[38] The image of the animal in gliding posture (*NH* II: plate 77) is the more dynamic of the plates, but the animal sitting on a branch (*NH* II: plate 76) is truly striking and may be the most lifelike of any of the vertebrates in *The natural history of Carolina* . . . (figure 17-9).

Entomology and invertebrate zoology

Catesby's treatment of insects in the second volume of *The natural history of Carolina* . . . was usually purely descriptive, and there is no mention at all of the green lynx spider, the single arachnid figured (*NH* II: plate 71). His list of insects of Carolina included twenty-eight entries, among them true insects but also spiders, centipedes, earthworms, mollusks, and even nematodes. For the most part, the largest and most conspicuous moths and butterflies were shown (for example, the monarch butterfly, luna moth, and swallowtail butterflies), and

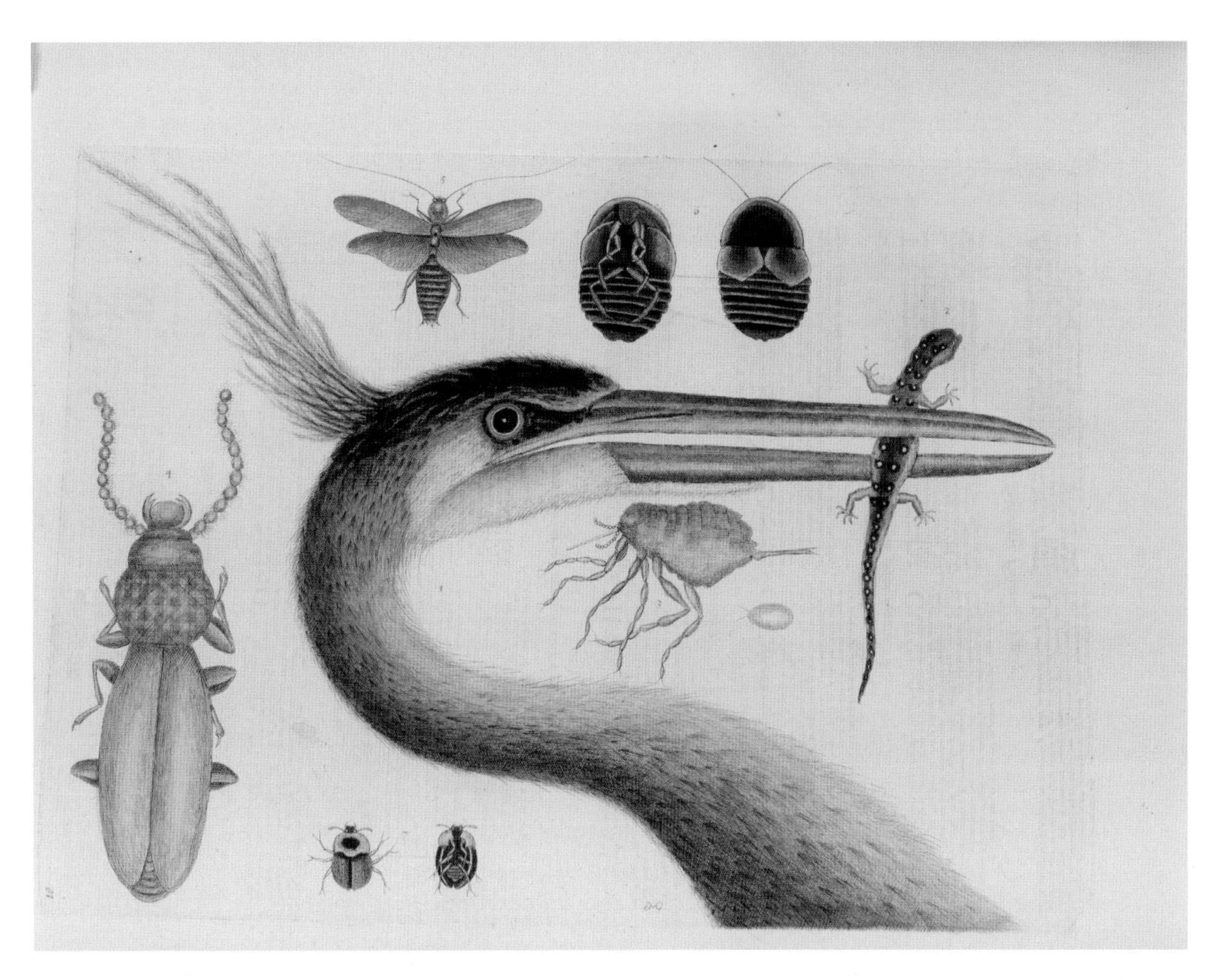

FIGURE 17-10. Mark Catesby combined the head of a great blue heron with some of his insect images, including the parasitic chigoe flea (*center*) and an American carrion beetle (collected in Pennsylvania), as well as one of only two salamanders in *The natural history of Carolina* . . . , to create this startling engraving; plate 10, M. Catesby, 1747, *The natural history of Carolina* . . . , volume II: Appendix. (Digital realization of original etchings by Lucie Hey and Nigel Frith, DRPG England; courtesy of the Royal Society ©.)

the cocoons were also figured for the cecropia and luna moths. In the "Account of Carolina" the use of wasp larvae as food by Indians was noted.[39] Most of the insects that are explicitly discussed were seen alive by Catesby in Carolina or the Bahamas, but two were sent from Pennsylvania, a giant ichneumon wasp (*Megarhyssa atrata*, *NH* II: Appendix, plate 4) and an American carrion beetle (*Necrophila americana*, *NH* II: Appendix, plate 10, figure 7) (figure 17-10), the former collected by John Bartram (1699–1777), the famous Philadelphia botanist, horticulturist, and traveler. In contrast to the insect accounts in the second volume, those in the Appendix provided more biological detail. Among those discussed at length were "tumble-turds" (dung beetle, *Canthon pilularius*, *NH* II: Appendix, plate 11), the parasitic chigoe flea of the tropics (*Tunga penetrans*, *NH* II: Appendix, plate 10, figure 3) (figure 17-10), and the blue mud wasp (*Chalybion californicum*, *NH* II: Appendix, plate 5), known for its spider-hunting techniques.

Crabs also received some attention from Catesby, with eight species being depicted. As for fishes, the palatability of crabs was noted, the purple land crab (*Gecarcinus ruricola*, *NH* II: plate 32) and the land hermit crab (*Coenobita clypeatus*, *NH* II: plate 33) both being mentioned in this regard. The migration of the former species was noted in particular, as were the types of shells used by the giant hermit crab (*Petrochirus diogenes*, *NH* II: plate 34). Indeed, uniquely among invertebrates, natural history rather than description dominates in most of Catesby's crab accounts.

Although corals were treated as plants, and mollusks and echinoderms (sea stars) were incidental to the few plates in which they appeared, Catesby did discuss the difference in gastropod shell abundance as it related to island orientation and commented on the specificity of certain species to particular habitats or water depths, mentioning five species by name. As for some other groups, this brevity was to avoid redundancy with information and images that were available in comprehensive conchological works at the time, such as that by Martin Lister upon which many of Catesby's identifications of mollusks were based.[40]

Catesby's choice of zoological subjects

Even a cursory review of Catesby's subjects reveals that only a tiny fraction of even the most conspicuous animals of Carolina and the Bahamas were illustrated by him. Some of the reasons for this were laid out explicitly by Catesby himself.[41] He noted that his coverage of land birds and perhaps snakes was near comprehensive but that for some groups, most notably insects and other invertebrates, the impossibility of providing a comprehensive treatment led him to provide only representative illustrations. Thus, in Carolina it was birds that were the object of his zoological focus, whereas in the Bahamas it was fish.

Moreover, like many naturalists of his time and indeed of the early nineteenth century as well, Catesby believed that it was not particularly important to provide information about animals already known to his chiefly northern European audience unless there were significant new facts to report. Exceptions were made for oddities such as flying fish and exotic American endemics like rattlesnakes, but just as Catesby's patrons prized new and different specimens for their collections, so subscribers to *The natural history of Carolina* . . . wanted to see illustrations of novelties. A similar attitude was expressed a century later by Charles Darwin who, on the voyage of HMS *Beagle*, collected few specimens and commented little on animals he encountered in places such as Australia and South Africa, believing them to be sufficiently well known as not to be worth the effort.[42]

However, even for groups that Catesby thought he had sampled well, many species were missed. For example, his statement "of serpents very few I believe have escaped me" was rather optimistic.[43] Discounting the two dubious "vipers" and the snake-like glass lizard, Catesby only depicted sixteen species (the garter snake was figured twice). All but one of these occur in either the coastal plain or Piedmont of South Carolina, but his "Green Snake" (*Opheodrys vernalis*, *NH* II: plate 57) reaches only as far south as the higher elevations of Virginia, where he may have encountered it during his trip to the "Apalatchian mountains" in 1714.[44] However, at least twenty-three other snakes occur in these areas of South Carolina.[45] A convenient comparison may be made with the snake fauna of the Savannah River Site (SRS) in Aiken, Barnwell, and Allendale Counties, South Carolina, which has been as well surveyed herpetologically as any place in North America.[46] It lies less than a dozen miles (20 kilometers), as the crow flies, downstream from the site of Fort Moore, Catesby's main center of activity in the Piedmont. Fifteen of Catesby's snakes occur at the SRS, but so do an additional twenty, including numerous conspicuous species, including the black racer (*Coluber constrictor*) and the pine snake (*Pituophis melanoleucus*). To be fair, however, many of the snakes missed by Catesby are either inconspicuous (for example, the worm snake, *Carphophis amoenus*) or similar in appearance to species he did illustrate (several other *Nerodia* species superficially resemble the plain-bellied watersnake).

Catesby's animal specimens

According to Catesby, in order to capture the most accurate likenesses of his subjects, fishes were painted shortly after capture, and reptiles were drawn from life.[47] At least some of the latter were preserved by Catesby and, along with birds and shells, were dispatched to Catesby's patrons in England, most notably Sir Hans Sloane. However, of the zoological specimens illustrated, the particular whereabouts of only the two fish illustrated on plate 19 of the Appendix (neither collected by Catesby himself) were explicitly stated.

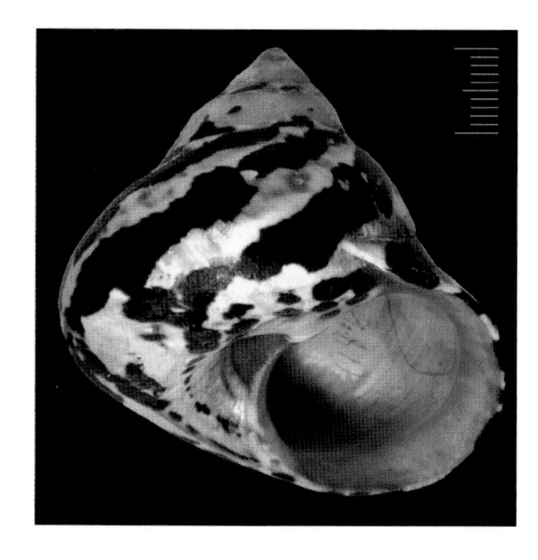
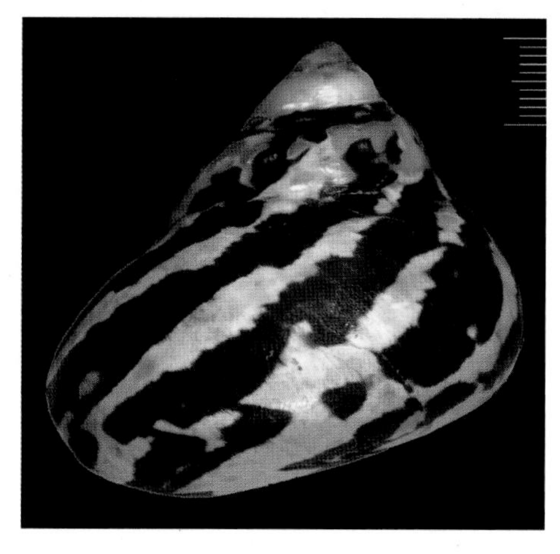

FIGURE 17-11. *Cittarium pica* (West Indian topsnail); two views of a specimen sent by Mark Catesby to Sir Hans Sloane (catalog no. 1182) preserved in the Natural History Museum, London. (Photography by Eddie Hardy, by courtesy of Kathie Way 2013 © NHMUK.)

Both were in Sloane's private collection, the viper fish having been sent from Gibraltar by Mark's youngest brother, John, and the armored catfish from "the coast of *New England*."[48] Although many of Sloane's herbarium specimens are extant,[49] there appear to be no surviving bird specimens from Sloane's original bequest.[50] Some of Sloane's shells from Catesby are extant.[51] Remarkably, these include an example of the West Indian topsnail, the only identifiable mollusk illustrated by Catesby (figure 17-11).[52]

The identity of the images of plants and animals figured by Mark Catesby is of both historical and taxonomic relevance. An index, usually credited to George Edwards and added in the third edition of *The natural history of Carolina, Florida and the Bahama islands*, published in 1771, was one of the first attempts to apply Linnaean binominal names to the biota. Subsequently, the zoological names have been re-evaluated and updated by a succession of authorities.[53] Catesby's *The natural history of Carolina* . . . pre-dates the publication in 1758 of the tenth edition of Carl Linnaeus's *Systema naturae*, the starting point for animal binominal nomenclature. As such, Catesby's names have no standing as formal zoological nomina. However, his images portray many specimens that were referenced by Linnaeus and other zoologists of the eighteenth and early nineteenth centuries in their original descriptions. Indeed, as the author of one of the relatively few early illustrated works featuring North American fauna, Linnaeus relied heavily on Catesby's illustrations as the basis of his descriptions.[54]

Mark Catesby's treatment of the American fauna was incomplete and his choice of subjects idiosyncratic, reflecting not only his own access to specimens but also the broader goals of his American sojourn, the interests of his patrons in

England, and his perception of the novelty and utility of his observations. His images and observations were imperfect, sometimes lacking important details or (rarely) confounding fact and fiction, or confusing the attributes of one species for those of another. Nonetheless, for a time when no formal training in zoology existed and in a place virtually unknown to European naturalists, Catesby's achievements were monumental, and *The natural history of Carolina* . . . stands as the first and most important faunal work dating from before the American War of Independence (1775–1783).

18

Catesby's fundamental contributions to Linnaeus's binomial catalog of North American animals

KRAIG ADLER

Mark Catesby's *The natural history of Carolina, Florida and the Bahama islands* became the key reference for North America in the first systematic catalog of the animals of the world. Carl Linnaeus (1707–1778), the Swedish inventor of binomial scientific nomenclature—typically formed as Latinized names—for the world's plants and animals, was first and foremost a botanist (figure 18-1). His *Genera plantarum*, first published in 1737, with six editions through 1764, and *Species plantarum*, issued in 1753, with three editions through 1764, are the cornerstones of systematic botany.[1] His other major treatise—a series of catalogs of animals and plants (and sometimes also minerals) entitled *Systema naturae*, which was issued in twelve editions from 1735 to 1766—displayed his considerable knowledge of zoology. Because these books were globally comprehensive, coincidentally they provided the first summary of North American animals according to Linnaeus's system of nomenclature.

Through later agreement by the international community of zoologists, the tenth edition of *Systema naturae*, published in 1758 (figure 18-2), was designated as the starting point for zoological scientific nomenclature. In this work, Linnaeus named and described nearly 4,400 species in a mere 823 pages. Excluding the introductory pages, this corresponds to an average of about eighteen species per page. Needless to say, the descriptions are very brief; they range from one to no more than a half-dozen lines of text. *Homo sapiens*, for example, is described in only eight words.

Linnaeus's sources of information and specimens

Linnaeus's descriptions were based on two sources of information: his own examination of actual specimens and the descriptions and illustrations published by others. Catesby was one of these published sources. Except for visits to examine collections in Hamburg, London, Paris, and several cities in

FIGURE 18-1. Carl Linnaeus by Per Krafft (1724–1793) and Jean Gustaf Haagen-Nillson, 1938. (Reproduced by permission of the Linnean Society of London.)

CAROLI LINNÆI
EQUITIS DE STELLA POLARI,
ARCHIATRI REGII, MED. & BOTAN. PROFESS. UPSAL.;
ACAD. UPSAL. HOLMENS. PETROPOL. BEROL. IMPER.
LOND. MONSPEL. TOLOS. FLORENT. SOC.

SYSTEMA
NATURÆ
PER
REGNA TRIA NATURÆ,
SECUNDUM
CLASSES, ORDINES,
GENERA, SPECIES,
CUM
CHARACTERIBUS, DIFFERENTIIS,
SYNONYMIS, LOCIS.

TOMUS I.

EDITIO DECIMA, REFORMATA.

Cum Privilegio S:æ R:æ M:tis Sveciæ.

HOLMIÆ,
IMPENSIS DIRECT. LAURENTII SALVII,
1758.

FIGURE 18-2. Title-page of *Systema naturae*. (Kraig Adler.)

the Netherlands between 1735 and 1738, Linnaeus never left Sweden again. Fortunately, in his homeland, he had available to him a large number of zoological collections that had been acquired by King Adolf Fredrik, by Queen Lovisa Ulrika, and by Swedish aristocrats and other wealthy citizens.[2] In those days before the establishment of public museums, private collections (generally called "cabinets") of plants, animals, and other natural products such as minerals were *de rigueur* ornaments of high society. Linnaeus and his students made careful studies of virtually all of the zoological cabinets in Sweden and published extensive catalogs of them in various works, including the students' doctoral dissertations.

Linnaeus also amassed zoological collections that he had made during his numerous excursions in Scandinavia. However, his most important sources of fresh animal specimens were his students, specifically one group of seventeen former pupils whom he called his "apostles." Between 1742 and 1778, the "apostles" collectively visited six continents (figure 18-3).[3] The seventh continent, Antarctica, was closely approached by one of them, and another made it to Spitsbergen Island within the Arctic Circle. Eight of the "apostles" died during

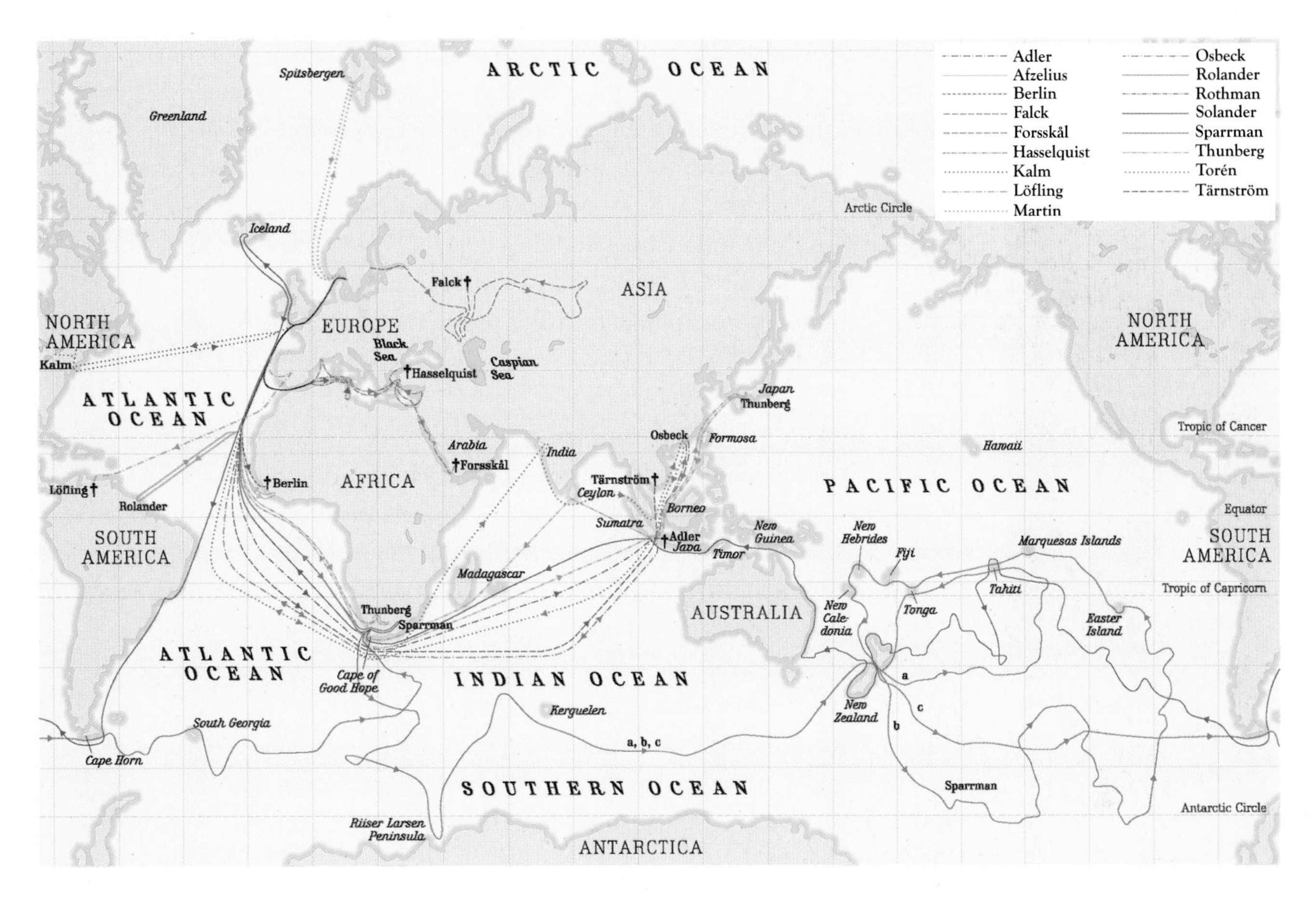

FIGURE 18-3. Expedition routes for Linnaeus's "Apostles"; daggers (†) indicate places of death for those who died overseas. Peter Kalm was the only former student of Linnaeus who reached North America. (Adapted from Broberg 2006; reproduced with permission of Stig Söderlind.)

their travels, but even then most of their collections were returned for Linnaeus to study and publish. Linnaeus himself was averse to such long-distance journeys. In 1738, he had declined opportunities to visit Dutch colonies in South America and southern Africa, but he exhorted his students to travel overseas in search of specimens for him. As reported by one of his departing "apostles," Carl Peter Thunberg, Linnaeus said to him: "I avoid long voyages. Now you are on your own. You'll manage well."[4]

Pehr Kalm: pupil of Linnaeus and his collector in northeastern North America

Only one of Linnaeus's "apostles" visited North America: Pehr (or Peter) Kalm, an ethnic Finn who was born in northeastern Sweden (figure 18-4).[5] Kalm studied with Linnaeus at Uppsala University during 1741 and 1742, after which he explored western Sweden, Russia, and Ukraine through 1746. At that time, Sweden wanted to develop a domestic silk industry, which would require the leaves of mulberry trees on which the moth caterpillars could feed. Kalm was

FIGURE 18-4. Pehr (or Peter) Kalm. (Kraig Adler.)

sent to North America in 1748 primarily to obtain the American red mulberry (*Morus rubra*) for introduction into Sweden. Linnaeus took advantage of this opportunity to obtain animal and plant specimens for himself. He specifically instructed Kalm to make "observations on Birds and Fishes, on Snakes and Insects. . . ." On his return, Kalm published a popular book on his travels and several articles about rattlesnakes and the cures for their bites.

En route to the British colonies in North America, Kalm spent approximately six months in London to make preparations for his trip. While in England, he visited Mark Catesby, who was then aged sixty-five and had completed *The natural history of Carolina. . . .*

In North America, Kalm's base was in the Swedish-Finnish community of Raccoon (now Swedesboro), New Jersey. He made excursions as far north as Quebec and west to Niagara Falls, as well as to Delaware and Pennsylvania, a region that does not overlap the area that had been visited by Catesby. Kalm returned to Sweden in 1751 after nearly two and a half years in North America. Of some seven hundred North American plant species named by Linnaeus, Kalm collected sixty, including two species of *Kalmia* (mountain laurel), a genus of shrubs that Linnaeus named for him.

In the tenth edition of *Systema naturae*, there are twenty-nine references to Kalm (table 18-1). Almost all of these refer to animal specimens that Kalm provided to Linnaeus. In comparison, Linnaeus referenced Catesby's *The natural*

TABLE 18-1. References to Catesby's text and illustrations in Linnaeus's tenth edition of *Systema naturae* (1758)

References to Peter Kalm are provided for comparison.

Mammals	7
(Kalm 0)	
Birds	81
(Kalm 2)	
Amphibians and reptiles	7
(Kalm 6, all snakes)	
Fishes	23
(Kalm 0)	
Insects, etc.	20
(Kalm 20)	
Worms, etc.	1
(Kalm 1)	
TOTAL CITATIONS	
Catesby	139
Kalm	29

history of Carolina . . . 139 times, but the taxonomic distribution of Linnaeus's citations to Kalm and Catesby is quite different. Whereas Kalm was cited for "Amphibia," all of which were snakes,[6] and for insects and other invertebrates about as frequently as was Catesby, the latter was cited much more frequently for fishes and especially for birds.

J. R. Forster's book on North American animals

The first book on the animals of North America to incorporate Linnaean binomial nomenclature—entitled *A catalogue of the animals of North America*—was published in 1771 by John Reinhold Forster (1729–1798), a Prussian of Scottish ancestry then living in London. His book is strictly a compilation based on the literature and information from his correspondents, for he never set foot in North America. Catesby was his most frequent authority for fishes, amphibians, reptiles, and birds.[7] Forster's was the only book covering North America's animals that William Bartram took with him on his celebrated travels in southeastern North America from 1773 to 1776.[8]

Linnaeus's principal source for information on North American animals

Linnaeus's principal authority for North American animals was Mark Catesby. Of the 139 references to Catesby in Linnaeus's species entries, for 52 of them Catesby was his sole reference (table 18-1). From this number alone, one can see how indebted Linnaeus was to Catesby. For the other eighty-seven new species, Catesby's book provided illustrations and information that helped Linnaeus to decide the status of the species.

Focusing on the Catesby species included in the tenth (1758) and twelfth (1766) editions of *Systema naturae*, the last that were personally revised by Linnaeus, it is instructive to compare Linnaeus's references to Catesby's birds with those to his amphibians and reptiles and to note what seems at first to be an inexplicable difference in Linnaeus's acceptance of Catesby's animals in these two groups (table 18-2). Linnaeus accepted most of Catesby's birds in his 1758 book (thirty-three of the total of eighty-one were based solely on Catesby's plates and text) and added only a few more species in his twelfth edition. In contrast, only two of Catesby's amphibians and reptiles—both lizards—were accepted by Linnaeus in 1758 based solely on his plates and text. Yet just eight years later he was willing to accept eight more, and most of these additional species were snakes.

Recall that in his tenth edition Linnaeus accepted six of Kalm's reptiles as new species, all of them snakes. How can we account for Linnaeus's apparent unwillingness to credit any of Catesby's twenty kinds of snakes in his tenth edition when he was willing to accept most of his birds and two lizards (and all of Kalm's snakes)?

TABLE 18-2. Summary of references to Catesby in Linnaeus's descriptions of birds compared to amphibians and reptiles

	Tenth edition (1758)	*Twelfth edition (1766)*
BIRDS		
Total	81[a]	+18[a]
Catesby only	33[b]	1[b]
AMPHIBIANS AND REPTILES		
Total	7	+8
Catesby only	2[c]	7[d]

[a] Total number of new species described by Linnaeus in which Catesby *and* other authors are referenced.

[b] Number of new species described by Linnaeus in which Catesby's text and plates are the sole reference.

[c] Both lizards.

[d] Mostly snakes.

Accuracy of Catesby's illustrations and observations

Was Linnaeus's reluctance about Catesby's snakes because his illustrations and descriptions were not generally accurate? I think not. Like John James Audubon and other outstanding artist-naturalists, Catesby was a careful observer of nature and a worthy artist. But the many myths about reptiles biased contemporary observers—for example, lizards that were thought to have venomous bites or a snake that supposedly could cut a person in half with a snap of its tail—and made them fearful to capture or handle reptiles for close observation. When compared to illustrations drawn by other artists of the eighteenth century, Catesby's were among the best in terms of their accuracy, although some of them were rather two-dimensional and not very lifelike.

Several examples from among Catesby's reptiles can be employed to make this point. His illustration of what he called "The Copper-belly Snake" (plain-bellied watersnake, *Nerodia erythrogaster*, *NH* II: plate 46) is quite accurate down to the keeling, or longitudinal ridging, of the animal's dorsal scales. Catesby was the first naturalist to separate two similar species of gartersnake, the eastern gartersnake (*Thamnophis sirtalis*), which he called "The Green Spotted Snake" and "The Spotted Ribbon-Snake," and the eastern ribbonsnake (*Thamnophis sauritus*), his "Ribbon Snake" (figure 18-5). These two snakes are often confused today, for the quantifiable differences are subtle. In the former, the lateral stripe is on scale rows 2 and 3, whereas in the other, it is on rows 3 and 4. The eastern ribbonsnake has a longer and more slender tail, and there are some pattern differences, but these require experience to detect reliably in every specimen.

FIGURE 18-5a. Common gartersnake (*Thamnophis sirtalis*). (© David M. Dennis.)

FIGURE 18-5b. Eastern ribbonsnake (*Thamnophis sauritus*). (© David M. Dennis.)

Catesby was also the first to distinguish North America's two species of greensnakes, the rough greensnake (*Opheodrys aestivus*), his "Blueish Green Snake," and the smooth greensnake (*Opheodrys vernalis*), which he simply called "Green Snake" (figure 18-6). Like the two gartersnakes, these greensnakes are superficially similar, but the rough greensnake has keeled dorsal scales, giving them a rough appearance, and a more slender body and tail. Catesby was apparently more impressed by differences in the shape of the snout: "The Nose of this Species [*Opheodrys aestivus*] turning up, sufficiently distinguishes it from another green Snake I shall hereafter describe."[9] He did, however, overemphasize the degree of "turning up" in his image. Although Linnaeus gave a scientific name to the rough greensnake in 1766, the other species was not given one until 1827 (see table 18-3), nearly a century after Catesby correctly reported that there were two species of greensnakes in eastern North America. Catesby had a keen eye for differences.

Perhaps the best example of Catesby's attention to detail among the reptiles was his illustration and text about the legless eastern glass lizard (*Ophisaurus ventralis*), which he called "Glass Snake" (figure 18-7). In Catesby's day, all legless lizards were thought to be snakes. It was only in 1810 that dissections revealed the presence of a sternum, or breastbone, in legless lizards, which were thenceforth classified as lizards. Catesby's observations, published in 1737, nearly seventy-five years earlier, were prescient. He noted the short and wide lizard-like tongue and the lateral folds, absent in all snakes, and he placed the vent properly. He missed the eyelids (no snakes have them, but legless lizards do) and probably the external ear (lacking in all snakes, but a close examination of his plate is equivocal on this point). He specifically wrote: "Their Skin is very smooth, and shining with smaller Scales, more closely connected, and of a different Structure from other Serpents."[10]

FIGURE 18-6a. Smooth greensnake (*Opheodrys vernalis*); compare with Figure 18-6c. (© David M. Dennis.)

FIGURE 18-6b. Rough greensnake (*Opheodrys aestivus*); compare with Figure 18-6d. (© David M. Dennis.)

FIGURE 18-6c. Smooth greensnake; plate 57, M. Catesby, 1737, *The natural history of Carolina . . .*, volume II. (Digital realization of original etchings by Lucie Hey and Nigel Frith, DRPG England; courtesy of the Royal Society ©.)

FIGURE 18-6d. Rough greensnake; plate 47, M. Catesby, 1737, *The natural history of Carolina . . .* , volume II. (Digital realization of original etchings by Lucie Hey and Nigel Frith, DRPG England; courtesy of the Royal Society ©.)

TABLE 18-3. Amphibians and reptiles described and illustrated by Mark Catesby in *The natural history of Carolina . . .*

Identifications that are questionable at the generic or specific levels are marked accordingly with ?. When Alexander Garden sent a specimen to Linnaeus, an asterisk is in the fifth column.

Volume and plate numbers	*Catesby's name*	*Current scientific name*	*Describer and date*	*Garden*
II: 38	green Turtle	*Chelonia mydas*	Linnaeus 1758	
II: 39	Hawks-bill Turtle	*Eretmochelys imbricata*	Linnaeus 1766	
II: 40	Loggerhead Turtle	*Caretta caretta*	Linnaeus 1758	
II: 41	Rattle-Snake	*Crotalus horridus*	Linnaeus 1758	
II: 42	Small Rattle-Snake	*Sistrurus miliarius*	Linnaeus 1766	*
II: 43	Water Viper	*Agkistrodon piscivorus*	Lacepède 1789	
II: 44	black Viper	? *Heterodon platirhinos*	Latreille 1801	
II: 45	Brown Viper	? *Heterodon platirhinos*	Latreille 1801	
II: 45	(no name given)	*Ambystoma* ? *talpoideum*	Holbrook 1838	
II: 46	Copper-belly Snake	*Nerodia erythrogaster*	Forster 1771	
II: 47	blueish green Snake	*Opheodrys aestivus*	Linnaeus 1766	*
II: 48	Black Snake[a]	*Pantherophis alleghaniensis*	Holbrook 1836	
II: 49	little brown Bead Snake	?*Storeria dekayi*	Holbrook 1839	
II: 50	Ribbon-Snake	*Thamnophis sauritus*	Linnaeus 1766	*
II: 51	Spotted Ribbon-Snake	*Thamnophis sirtalis*	Linnaeus 1758	
II: 52	Chain-Snake	*Lampropeltis getula*	Linnaeus 1766	*
II: 53	Green Spotted Snake	*Thamnophis sirtalis*	Linnaeus 1758	
II: 54	Coach-whip Snake	*Coluber flagellum*	Shaw 1802	*
II: 55	Corn Snake	*Pantherophis guttata*	Linnaeus 1766	*
II: 56	Hog-nose Snake	*Heterodon platirhinos*[b]	Latreille 1801	see text
II: 57	Green Snake	*Opheodrys vernalis*	Harlan 1827	
II: 58	Wampum Snake	*Farancia abacura*[c]	Holbrook 1836	see text
II: 59	Glass Snake	*Ophisaurus ventralis*	Linnaeus 1766	*
II: 60	Bead Snake	? *Cemophora coccinea*	Blumenbach 1788	
II: 63	Alligator	*Alligator mississippiensis*	Daudin 1802	*
II: 64	Guana	*Cyclura sp.*[d]	—	
II: 65	Green Lizard of Carolina	*Anolis carolinensis*	Voigt 1832	
II: 66	Green Lizard of Jamaica[e]	*Anolis grahami*	Gray 1845	
II: 67	Blue-Tail Lizard	*Plestiodon fasciatus*	Linnaeus 1758	
II: 68	Lyon Lizard	*Leiocephalus sp.*[f]	—	see text
II: 69	Land Frog	*Anaxyrus terrestris*	Bonnaterre 1789	
II: 70	Water Frog	*Lithobates sphenocephalus*	Cope 1886	
II: 71	Green Tree Frog	*Hyla cinerea*	Schneider 1799	
II: 72	Bull Frog	*Lithobates catesbeianus*	Shaw 1802	
Appendix, 10	Spotted Eft	*Ambystoma maculatum*	Shaw 1802	*

[a] This illustration has formerly been allocated to *Elaphe* (now *Pantherophis*) *obsoletus* (Say 1823), now considered to be a western species.

[b] Catesby's plate is clearly *Heterodon platirhinos*, but the specimen sent to Linnaeus by Garden was *Agkistrodon contortrix*, the copperhead.

[c] This snake has usually been allocated to *Nerodia fasciata* (Linnaeus 1766), but Catesby's illustration has been identified by James Reveal (2013: note 9) as *Farancia abacura*. *Nerodia fasciata* is based on a Garden specimen sent to Linnaeus.

[d] This illustration has usually been allocated to *Cyclura cornuta* (Bonnaterre 1789), but Aaron Bauer (this volume) notes that Catesby's animal lacks the rostral horn characteristic of *C. cornuta*.

[e] This illustration has often been allocated to *Anolis garmani* Stejneger 1899, but Bauer (this volume) has reidentified it as *A. grahami*.

[f] Catesby's plate has usually been allocated to *Cnemidophorus* (now *Aspidoscelis*) *sexlineatus* (Linnaeus 1766), but Bauer (this volume) has reidentified it as a curly-tailed lizard (*Leiocephalus*). Linnaeus's description of *A. sexlineata* is therefore based on Garden's specimen.

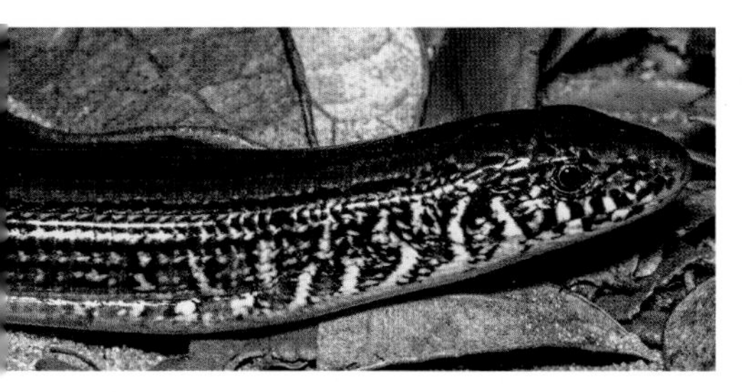

FIGURE 18-7a. Eastern glass lizard (*Ophisarus ventralis*). (© David M. Dennis.)

FIGURE 18-7b. Eastern glass lizard (*Ophisaurus ventralis*). (© David M. Dennis.)

FIGURE 18-7c. Eastern glass lizard; plate 59, M. Catesby, 1737, *The natural history of Carolina . . .*, volume II. (Digital realization of original etchings by Lucie Hey and Nigel Frith, DRPG England; courtesy of the Royal Society ©.)

Alexander Garden: source of Linnaeus's specimens from southeastern North America

At this point Linnaeus's other major source of information on North American animals must be introduced: Alexander Garden (1730–1791), a Scottish physician who moved to the warmer climate of Charleston, South Carolina, to alleviate his chronic tuberculosis. He arrived in 1752 and was soon corresponding with Linnaeus and sending him specimens. He collected in South Carolina, Georgia, Spanish East Florida, and westward to the Cherokee Indian lands in western North Carolina. As a Loyalist, after the Revolutionary War he lost all of his possessions, and he left Charleston as the last British troops departed in 1782 after having worked in the colonies for thirty years. The generic name *Gardenia* was published by John Ellis, a British linen merchant and naturalist, to honor him, although the type species was a Chinese plant, not one that Garden had ever collected. No portrait of Garden, as for Catesby, is known to exist.

Among the amphibians and reptiles illustrated by Catesby were all of the major taxonomic groups: turtles, snakes, lizards, alligator, frogs, and salamanders (table 18-3). With one exception—the timber rattlesnake (*Crotalus horridus*), which Linnaeus knew from the literature and from Kalm's specimens—all of the snakes illustrated by Catesby that Linnaeus eventually accepted and gave scientific names in 1766 were the six species for which he also obtained specimens from Garden: *Sistrurus miliarius* (pygmy rattlesnake), *Opheodrys aestivus* (rough greensnake), *Thamnophis sauritus* (eastern ribbonsnake), *Lampropeltis getula* (eastern kingsnake), *Pantherophis guttata* (red cornsnake), and the eastern glass lizard (*Ophisaurus ventralis*—both Catesby and Linnaeus regarded the last of these to be a snake). To understand why Linnaeus needed to have actual specimens when describing snakes but not other animals, a fundamental discovery about the identification of snake species that was announced in 1752 by Linnaeus himself has to be considered.

Linnaeus and the modern identification of snake species

Linnaeus was not fond of amphibians and reptiles, referring to them as "Pessima tetraque Animalis" (these most hideous animals).[11] Nevertheless, to make his catalogs comprehensive he was required to study them. Snakes, because they lack limbs and many other anatomical features found in other reptiles, became a subject of special interest to him. In 1752, he published a two-page note[12] on distinguishing characteristics between snake species.[13] Linnaeus discovered that the numbers of scutes[14] on the underside of snakes are relatively constant in each species of snake but often differ between species. In *Systema naturae*, he provided these data for snakes (and legless lizards) but not for other reptiles.[15]

Linnaeus gave Latin binomial names to the snakes collected by Kalm because he could count the scutes on the underside of his specimens. None of Catesby's plates illustrated snakes in a way that allowed Linnaeus accurately to count the scutes on their undersides. Thus Linnaeus probably withheld naming Catesby's snakes until he had actual specimens on which he could make the necessary counts. When Garden provided them, Linnaeus could then associate Garden's specimens with Catesby's illustrations, report the number of abdominal and subcaudal scutes, and finally describe and name the new species.[16] This also may explain why Linnaeus named the rough greensnake but not the smooth greensnake. He only had a specimen of the former, from Garden; the other species has a more northerly distribution but apparently was never encountered by Kalm.

In making these connections between illustrations and specimens, Linnaeus sometimes erred. Here are three examples of such mistakes. Linnaeus associated Catesby's illustration (*NH* II: plate 56) of the "Hog-nose Snake" (*Heterodon platyrhinos*) with Garden's specimen of the copperhead (*Agkistrodon contortrix*), which still exists in the collection of the Linnean Society of London.[17] Thus Linnaeus's written description and the scientific name apply to the copperhead, while the hog-nosed snake remained without a name until 1801 (see table 18-3). Similarly, Linnaeus associated Catesby's illustration of the "Wampum Snake" (*NH* II: plate 58) with Garden's specimen of the southern watersnake (*Nerodia fasciata*). Linnaeus's description and scientific name apply to the watersnake, and the red-bellied mudsnake that Catesby depicted was not given a Latinized name until 1836 (*Farancia abacura*; see table 18-3).

A non-snake example is afforded by Catesby's "Lyon Lizard," now identified as a curly-tailed lizard (*Leiocephalus* sp.), which Linnaeus incorrectly associated with Garden's specimen of the six-lined racerunner (*Aspidoscelis sexlineatus*). Thus, Linnaeus's scientific name for the racerunner is based on Garden's specimen, not on Catesby's plate.

Catesby's ability to distinguish species

Catesby described and illustrated thirty-five different kinds of amphibians and reptiles in his book. Thirty-two of these are recognized today as distinct species (table 18-3), but only twelve of them were included in Linnaeus's *Systema naturae*, if we exclude those named from Garden's specimens that he mistakenly associated with Catesby's illustrations. The remaining twenty species were given Latin binomial names by other naturalists, one as late as 1886. Linnaeus, therefore, recognized only about one-third of the true herpetological species diversity illustrated by Catesby.[18]

Among the species illustrated by Catesby but overlooked by Linnaeus are some of the most distinctive vertebrates found on the North American continent. Perhaps the most surprising is the American alligator (*Alligator mississippiensis*) (figure 18-8), which was not distinguished by Linnaeus despite Catesby's excellent account and figure of a young animal (*NH* II: plate 63; see figure 10-11, p. 139): the animal depicted by Catesby is not a hatchling, as indicated in his text, because of the long snout and lack of banding on the body. (Garden had even sent Linnaeus a small specimen in 1761.) Linnaeus regarded all of the world's crocodilians to be a single species, which he named *Lacerta*

FIGURE 18-8. American alligator (*Alligator mississippiensis*). (© David M. Dennis.)

FIGURE 18-9. American bullfrog (*Lithobates catesbeianus*). (© David M. Dennis.)

crocodilus in the tenth edition of *Systema naturae* and in which account he cited Catesby. In fact, Linnaeus examined specimens of at least four different crocodilian species in the collections he examined.

The American alligator, in fact, is one of the largest and most distinctive species of crocodilians.[19] The longest specimen measured authoritatively—a freshly killed animal that was shot in Louisiana in 1890 and not a skin, which can easily be stretched—was 19 feet 2 inches long (5.8 meters).[20] Thus, Catesby's statement that "They are found above twenty Foot in length" is not the exaggeration it might seem.[21] He also gave a detailed account of this reptile so extensive that the point size of the printer's type had to be greatly reduced in order for Catesby's text to fit on one page. The American alligator was finally given a scientific name in 1802 (table 18-3).

Another distinctive species apparently unrecognized by Kalm, Garden, and Linnaeus was the American bullfrog (*Lithobates catesbeianus*), called simply "Bull Frog" by Catesby, the largest frog native to North America and one of the largest species of frogs in the world (figure 18-9).[22] His drawing (*NH* II: plate 72; see figure 17-8, p. 242) is easily recognizable as this species and not the green frog (*Lithobates clamitans*), which often lives in the same habitats and with which the bullfrog is commonly confused even today. Fortunately, when the American bullfrog was finally named in 1802, Catesby's name was associated with it (see table 18-3). This species has a large range, from Nova Scotia south to Florida and west to the Great Plains. It has also been introduced to California, Mexico, Cuba, China, Japan, and other places around the world, mostly as a source of food for humans. It is, therefore, the most widely known of the five reptiles and one amphibian that carry Catesby's name.[23]

Mark Catesby's ability to distinguish different species of animals was exemplary. He rarely illustrated or gave different names to animals that have not been recognized by later specialists to be valid species. Among amphibian and reptile species, for example, thirty-two of the thirty-five that he illustrated are recognized scientifically today. Statistically, this is a far better record than almost every other naturalist who has worked in North America up to the present day. Catesby was indeed a gifted and careful observer of nature.

19

Mark Catesby's plant introductions and English gardens of the eighteenth century

MARK LAIRD

North American plants were avidly sought by gardeners in early eighteenth-century England, and Mark Catesby's role in stimulating this horticultural frenzy was twofold. He collected suitable plants in the American colonies during his visits, sending seeds and tubs of living plants to England, and later he promoted some of these plants in his publications, including *The natural history of Carolina, Florida and the Bahama islands*.

One of his discoveries was *Catalpa bignonioides* (figure 19-1), which, Catesby commented,

> was unknown to the inhabited Parts of *Carolina* till I brought the Seeds from the remoter Parts of the Country. And tho' the Inhabitants are little curious in Gard'ning, yet the uncommon Beauty of the Tree has induc'd them to propogate it; and 'tis become an Ornament to many of their Gardens, and probably will be the same to ours in *England*, it being as hardy as most of our *American* Plants; many of them now at Mr. *Bacon's* at *Hoxton*, having stood out several Winters without any Protection except the first Year.[1]

Catalpa quickly attracted the attention of the nursery trade. A thirteen-foot (four-meter) specimen (costing three guineas, equivalent to about £400 ($650 in 2014), was delivered by the nurseryman Robert Furber in the autumn of 1734 to Carlton House, the home of the Prince of Wales, where the garden was being remodeled by William Kent (1685–1748).[2] Between its arrival as a seed in London in the early 1720s and its taking root in a royal garden, this *Catalpa* had inched up to over twice the height of a man. It had benefited from the warm seasons that followed 1725. While in the Bahamas, Catesby had missed the summer of 1725, which proved the coldest ever recorded in England.[3]

The significance at Carlton House of Kent's breakthrough in picturesque design—the "new taste in gardening . . . without either level or line"—is more compelling when seen as a milestone in "horticultural culture."[4] That culture originated within a loosely affiliated group of men who gathered at the Temple

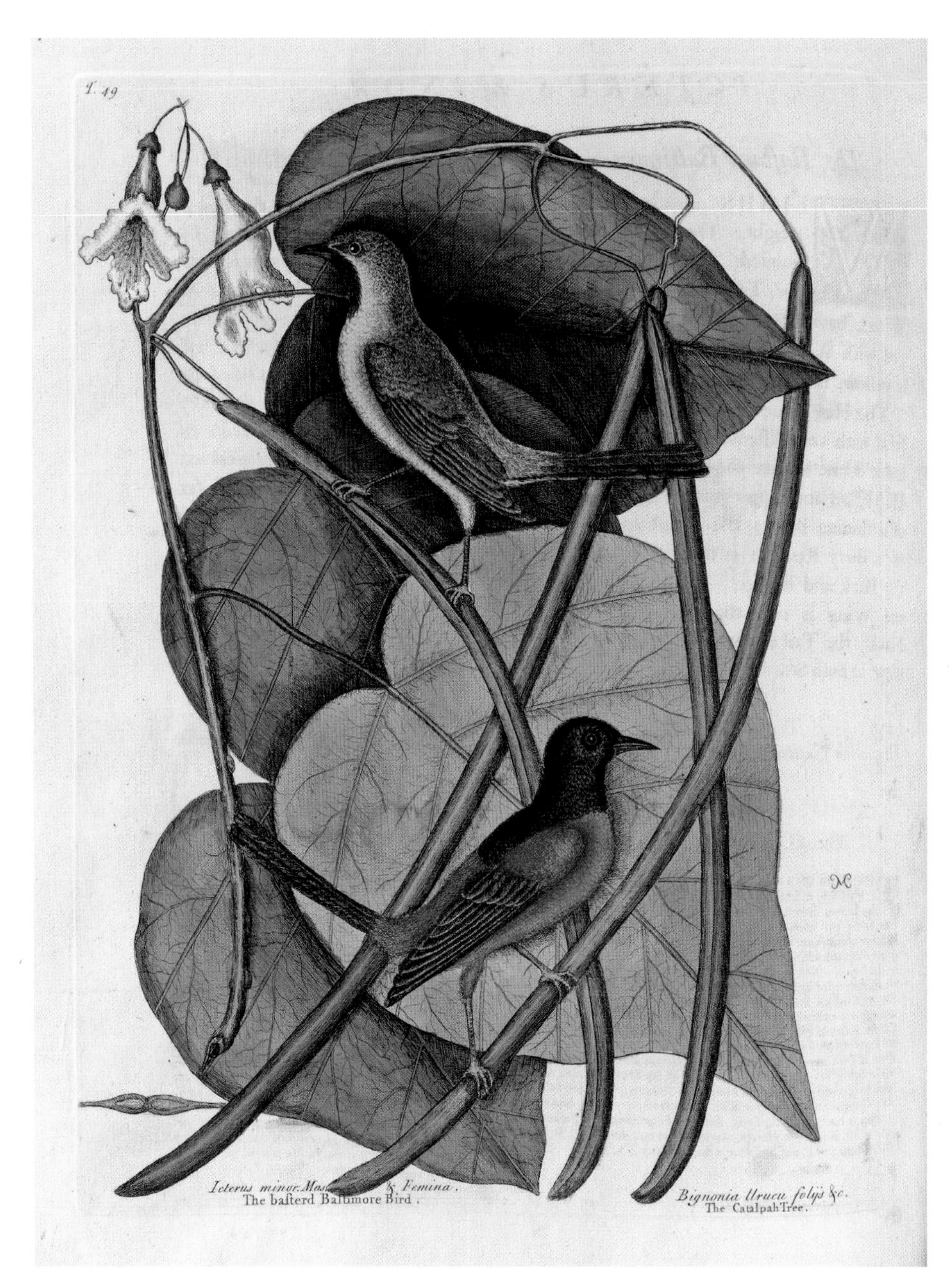

FIGURE 19-1.
Southern catalpa (*Catalpa bignonioides*); plate 49, M. Catesby, 1730, *The natural history of Carolina . . .* , volume I. (Digital realization of original etchings by Lucie Hey and Nigel Frith, DRPG England; courtesy of the Royal Society ©.)

Coffee-house in London from the 1690s, including two of the twelve sponsors of Catesby's 1722–1726 trip: William Sherard and Charles du Bois.[5] Having procured many North American plants, du Bois liked to show off candles made from boiling up berries of the common American candle-berry tree (or wax myrtle, *Morella cerifera*).[6]

Patronage involves subscribers, and subscription lists of the kind appearing at the front of Mark Catesby's *The natural history of Carolina* . . . are particularly revealing, as are prestigious dedications such as those to Queen Caroline and Princess Augusta. Yet of the 155 people named by Catesby, only 10 were women.[7] When compared to the women subscribers to Robert Furber's *Twelve months of flowers*, issued in 1730, who represented almost one-third of the patrons, the subscription list for Catesby's *The natural history of Carolina* . . . reflects a more masculine alliance of colonial enterprise, science, and medicine as opposed to the appeal of the decorative.[8]

There exists a distinction in the subscription makeup of eighteenth-century books featuring birds as opposed to flowers or butterflies.[9] Yet, gender stereotyping simplifies particularities,[10] and men who appear typical plant collectors turn out to be singularities.[11] It is thus a useful corrective to generalizations to single out one of Catesby's subscribers and look at that person through two North American species introduced to England by 1725. The first plant is the silver maple (*Acer saccharinum*), which was called Wager's maple from the 1730s to 1770s. Though not assumed to be a Catesby introduction, it introduces the subscriber, Sir Charles Wager. It also raises a question about Catesby's *Acer rubrum*—the only maple to appear in *The natural history of Carolina* . . . (figure 19-2). The second species is Catesby's *Magnolia grandiflora* (figure 19-3), which

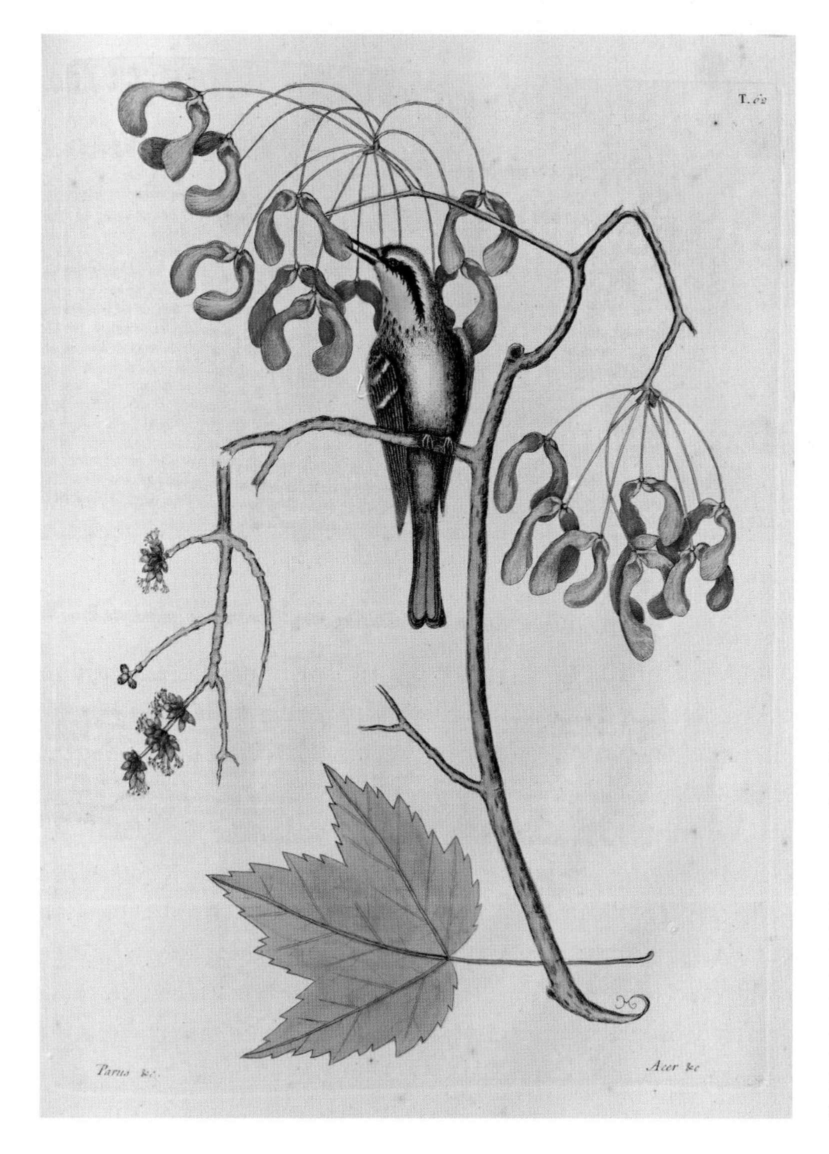

FIGURE 19-2. Red maple (*Acer rubrum*); plate 62, M. Catesby, 1731, *The natural history of Carolina* . . . , volume I. (Digital realization of original etchings by Lucie Hey and Nigel Frith, DRPG England; courtesy of the Royal Society ©.)

FIGURE 19-3. Southern magnolia (*Magnolia grandiflora*); plate 61, M. Catesby, 1739, *The natural history of Carolina* . . . , volume II. (Digital realization of original etchings by Lucie Hey and Nigel Frith, DRPG England; courtesy of the Royal Society ©.)

outdoes even *Catalpa* as an extraordinary plant with an extraordinary history in cultivation, and its early life in England is intimately linked to Sir Charles and his garden in Parsons Green in Fulham to the west of London.

Sir Charles Wager was among the "curious Gentlemen . . . in the Business of Gardening and Vegetation" mentioned in the preface to the Society of Gardeners' *Catalogue of trees, shrubs, plants, and flowers, both exotic and domestic* of 1730.[12] He was in the company of ten other "encouragers." These included Peter Collinson, celebrated for introducing new plants and developing new cultivation methods. The Society of Gardeners' *Catalogue* cited Catesby on occasion: for example, under "Acacia," two "Sorts" were sent as seed in 1723.

The puzzle of Sir Charles Wager's maple

During much of the eighteenth century, silver maple was the tree associated with Sir Charles. From the first to the eighth editions of Philip Miller's *Gardeners dictionary*, published between 1731 and 1768, a maple with large bunches of flowers was called very specifically "Sir Charles Wager's Flowering Maple." It was listed as a variety of the red maple (*Acer rubrum*).[13] In *The natural history of Carolina* . . . Catesby featured only the red maple, writing about the little red blossoms, which come out in February—the earliest of any forest tree—and about the succeeding red keys (see figure 19-2). He added: "They endure our English Climate as well as they do their native one; as appears by many large Ones in the Garden of Mr. *Bacon* at *Hoxton*."[14] Giving Leonard Plukenet as his authority, Catesby used the polynomial *Acer Virginianum, folio majore, subtus argenteo, supra viridi splendente*. In the eighth edition of his *Gardeners dictionary*, published in 1768, Miller supported the new binomial "Acer (Rubrum)" by citing that polynomial. He claimed, at the same time, that *Acer saccharinum* was more popular than *Acer rubrum*.

The red maple had been introduced to cultivation in England by 1656. It is not surprising, then, that Stephen Bacon's Hoxton nursery contained some large trees of that species. By the time "Index Horti Chelseiani" was compiled in 1772–1773, specimens of "Acer rubrum" were growing in both the "Wood" and the "Shrubbery" in Chelsea Physic Garden, but it is not clear which of the two species that meant.[15] Uncertainty also applies to Joseph Spence's account of touring the Duke of Argyll's estate, Whitton Park, near Richmond in 1760, where he saw "Sir Charles Wager's Maple, with red flowers, commonly called the Scarlet-Flowering Maple."[16]

At Chelsea Physic Garden, by the 1770s, as Stanesby Alchorne oversaw the transition from Philip Miller's long stewardship to William Curtis's six-year stint as *Praefectus Horti*, the silver maple had consolidated its position as the more ornamental of the two maples because of its "largest aggregate of flowers." Indeed, as late as 1778, William Malcolm's nursery catalog and *The universal gardener and botanist* by Thomas Mawe and James Abercrombie both sustained the name "Wager's Maple," distinguishing it from the red maple by its flowers in "numerous large clustering heads."[17]

While the association between Sir Charles and his maple began with Philip Miller in 1731, the distinction "large flowering" may be traced further back, possibly to Robert Furber's 1727 catalog and certainly to the Society of Gardeners' *Catalogue* of 1730:

> There is also another Sort of the *Virginian-flowering Maple*, which hath lately been brought into *England*, that produces large Bunches of Flowers at the Joints of the young Shoots, so that at a small Distance, all the Branches seem to be cover'd with Flowers, but this is at present very rare in *England*.[18]

FIGURE 19-4.
Silver maple (*Acer saccharinum*), previously determined a red maple (*Acer rubrum*), Sloane Herbarium 212. (© Natural History Museum, London.)

Philip Miller wrote in 1731 of *Acer saccharinum* being sent to Sir Charles from America, but no date of introduction is given. He commented: "This Tree is at present very rare in *Europe*, but as it produces ripe Seeds in *England*, so it is to be hoped it will in Time be more common in the Gardens of the Curious."[19] Whether a ten-year-old tree would set seed is perhaps doubtful. So did silver maple come as a sapling in or before 1725? Certainly Wager would have seen the mature tree during his early years in New England, and he might have brought a living plant to London on a return voyage. Or is there another way that a sapling or seeds could have reached Wager by 1725? Specimens in Sloane's and Sherard's herbaria (figure 19-4) indicate that Mark Catesby collected several maples in Virginia as well as in Carolina, apparently including the silver maple in Carolina in 1723.[20] He may also have gathered seed of the silver maple, although clear evidence is wanting.

One further puzzle is the lack of contemporary representation of the silver maple from any English garden. The "Acer Virginianum" that Georg Ehret painted in 1741 is *Acer rubrum* (figure 19-5).[21] Ehret's undated "Acer Americanum" is ambiguous and could be a variant of *Acer rubrum*, possibly *A. rubrum* var. *trilobum* (figure 19-6).[22] Mark Catesby's *Hortus britanno-americanus*, which was published posthumously in 1763, contains no likenesses of the four *Acer* species listed. Only "*Acer Virginianum* . . . The Red flowering Maple," which was illustrated in *The natural history of Carolina* . . . , is given a full description. The text mentioned the difficulty of getting seeds from red maple trees growing in

FIGURE 19-5. "Acer Virginianum" by Georg Ehret. (Courtesy of the Hunt Institute for Botanical Documentation, Carnegie Mellon University, Pittsburgh, Pennsylvania.)

FIGURE 19-6. "Acer Americanum" by Georg Ehret. (Courtesy of the Hunt Institute for Botanical Documentation, Carnegie Mellon University, Pittsburgh, Pennsylvania.)

England and thus the need to propagate "by laying, or possibly by inarching on our native maple."[23] The three other species listed in cultivation included "Acer Americanum" ("The American flowering Maple, with larger bunches of flowers"), which is surely the silver maple.

In short, then, Ehret apparently did not paint "Sir Charles Wager's Maple" on his visits to Wager's Parsons Green home in or after 1737 (when drawing *Magnolia grandiflora*). The confusion over identities among the various maples, including the sugar maple (*Acer saccharum*), introduced by Peter Collinson about 1735, was possibly a factor. The drawings of maples the young William Bartram

FIGURE 19-7. "The Great Silver Leafed River Maple" by William Bartram. (© Lord Derby, Knowsley Hall, Liverpool, England.)

sent to Collinson in 1755 might have ended all the confusions in Europe (figure 19-7). Yet only Collinson ever saw these, and even he was still confused by one that resembled the Norway maple: "I have a fine sort of Mapple thous formerly Sent Mee—I can't find by Billey's Drawings." In writing to John Bartram on 18 February 1756, Collinson confirmed that he had his own living specimen of silver maple:

> Since the Striped Bark Mapple [*Acer pensylvanicum*] will afford us none of its Seeds I wish Thou would gently bend down thy 10 foot Tree & Layer it on the Ground to Strike Root since it is Like to bear no seed—I have served my fine Large Silver Leafed Mapple So with great Success.[24]

Interestingly, then, neither Collinson nor Bartram is known to have used the name "Wager's Maple" in their correspondence. "Silver Leafed Maple" was their standard name, appearing in four of their lists that span the period from about 1757 to about 1769.[25]

The "Laurel-leaved Tulip Tree"

Both Mark Catesby and Georg Dionysius Ehret made representations of a plant of *Magnolia grandiflora*, "The Laurel Tree of Carolina . . . which in the Year 1737 produced Blossoms at the Right Honourable S^r^. Charles Wager's at Parsons green" near Fulham.[26]

Charles Wager was born in 1666 into a family with navy forebears on "both sides."[27] At a young age, and through his Quaker stepfather, Wager was apprenticed to a Quaker merchant captain of New England, John Hull. By 1693, in command of the *Samuel and Henry*, he helped convoy the New England trade. He was in charge of the *Hampton Court* when the Great Storm of November 1703 blew his tattered ship to port in Torbay. Knighted in 1709, he was by then a very rich man. His wealth came largely from silver on board a captured galleon off Cartagena (estimated at over £60,000), but also from his flag share of other prizes in the West Indies. Thus, like William Sherard and Charles du Bois, the considerable fortune he put into gardening and natural history came from new wealth rather than from landed wealth. In other words, his is a bourgeois history.

What marks him out as distinct, however, is his naval career. Service in the Royal Navy culminated in his being appointed First Lord of the Admiralty in 1733. He held this position until his appointment as Treasurer of the Navy in 1742. This meant that, until his later sedentary years, Wager, unlike Sherard and Du Bois, was personally on the high seas, leaving his wife ashore for as many months as he flew his flag between 1726 and 1731. Perhaps, on occasions, he had direct access to new exotic plants in distant lands. But in the very cold summer of 1725, when the first *Acer saccharinum* is supposed to have taken root in English soil, we should picture him at home in the stately brick house, Hollybush, which was situated at the southeast corner of Parsons Green, Fulham.

In 1691 Wager had married Martha Earning, and in 1720 they moved into Hollybush, which Sir Charles leased for twenty-two years, until 1742. George Edwards recorded that Lady Wager, like the small number of female subscribers to his four-volume *A natural history of uncommon birds*, shared her husband's interest in keeping exotic birds, from "The Transverse Striped or Bared Dove" to the "Schomburger."[28] She is known to have given a "She Posham with three young ones" to Sir Hans Sloane around 1730.[29] This makes her distinct from fellow gentlewomen who were more interested in the decorative arts than in natural history. (The fact that Lady Sarah Featherstonhaugh, wife to Sir Matthew, painted a capriccio around 1750 by copying animals from George Edwards and Benjamin Wilkes also points to how a woman who was not a subscriber transformed ornithological science into the decorative arts by her accomplishments.)[30]

Wager's official appointment as First Lord of the Admiralty was dated 21 June 1733, and hence the period leading up to when Ehret walked around three miles from Chelsea to Parsons Green several times each week to draw Wager's *Magnolia grandiflora* (called southern magnolia or bull bay) coincides with Wager's retirement from active naval service. Wager became MP for Westminster in 1734. He was closely allied to Prime Minister Robert Walpole as a "Parliament man." Yet he retained as a senior admiral an unusual grasp of maritime geography in association with commerce and colonial rule. Horace Walpole praised his character as the "fairest."[31]

Given their Quaker upbringing, and with common interests in natural history and horticulture, it was scarcely surprising that he and Peter Collinson became closely allied. Wager was a "Worthy Man" *par excellence* for Collinson, to whom Wager entrusted care of his greenhouses during many absences at sea. Indeed, in 1764, recalling all who helped in his effort to plant English gardens with exotic trees, Collinson wrote: "That Gentle Tree so like a Cypress looks uncommon, thats the Syrian Cedar. The Seed was gave Mee by Sir Charles Wager first Lord of the Admiralty, gather'd in the Isle of Iona."[32]

Mark Catesby's connection with Wager is in evidence, for example, with the tender "Cassia Bahamensis" (*Cassia ligustrina*), which Catesby introduced from the Bahamas in 1726.[33] This species soon flowered in Wager's garden as well as in Chelsea Physic Garden. Yet it is *Magnolia grandiflora* that provides the most compelling link between the two, and one involving the celebrated Georg Dionysius Ehret.

As with *Acer saccharinum*, the particulars of *Magnolia grandiflora*'s introduction to cultivation in England are confused.[34] The account in the ninth part of Catesby's *The natural history of Carolina . . .* , published in June 1739, however, is revealing:

> What much adds to the value of this Tree is, that it is so far naturalized and become a Denizon to our Country and Climate, as to adorn first the Garden of that worthy and curious Baronet, Sir *John Colliton*, of *Exmouth* in *Devonshire*,

> where, for these three Years past, it has produced Plenty of Blossoms, since that and in the year 1737, one of them blossom'd at *Parsons Green*, in the Garden of the Right Honourable Sir *Charles Wager*; one of which Blossoms expanded, measured eleven inches over.[35]

This suggests that the first flowering in Devon was perhaps as early as 1735, with the flowering in London two years later. It is clear, however, from Philip Miller's appendix to the *Gardeners dictionary*, published in 1735, that plants had been growing in England well before 1737 and perhaps, like *Catalpa*, before Catesby's return in 1726. Indicating that at the time of writing (1734 or 1735) these plants were yet to flower, Miller added his expectation of seeing "Flowers in a few Years; there being several Trees planted in the Gardens of some curious Persons near *London*, where they have endured the Cold of four or five Winters without Shelter."[36] Those four or five winters only take Miller's chronology back to around 1730. Catesby had sent seed several times in 1723 and 1724, which would mean a more plausible spurt from seedling to flowering within a dozen seasons in England. The mild seasons that followed the "winter in summer" of 1725 surely hastened that celebrated flowering on the coast of Devon. Eight of the winters of the 1730s proved among the mildest on record.[37] The average temperature in December 1733 to February 1734, for example, was 6.1°C (43°F), making it the sixth warmest winter on record.[38] The summers of the late 1720s and 1730s were very warm.

The confidence engendered by those seasons is reflected in Christopher Gray's *A catalogue of American trees and shrubs that will endure the climate of England*, published after 1737. It included "Acer Americanum floribus multis. Red Maple with larger bunches" (distinguished from "Acer Virginianum floribus coccineis. Red Flowering Maple"), and this was surely "Sir Charles Wager's maple." Gray's Fulham nursery lay close to Parsons Green. The image of *Magnolia grandiflora* (called "Magnolia Altissima") that dominates the broadsheet (see figure 7-6, p. 94) was drawn "in its exact Dimensions" from Wager's tree. Mark Catesby's monogram is visible at the base of the main stem, but the engraving is based on a watercolor done by Ehret. There are at least four other paintings of the same magnolia flower, all of which are indisputably by Ehret, and all can be linked to the representation of *Magnolia grandiflora* in Catesby's *The natural history of Carolina* . . . , which was signed not by Catesby but by Ehret (see figure 19-3).

The story of these portraits of Wager's *Magnolia grandiflora* is intriguing and complicated. In the Natural History Museum, London, a sketch by Ehret of a flower bud (figure 19-8), though undated, conveys expectation before the opening of the flower in a sequence of August days. That opening was recorded by Ehret in studies that became the basis of a signed engraving.[39] The handsome Oak Spring portrait,[40] which is on a dark brown ground, appears to be Ehret's clever combining of all the stages of a single bloom over several days (figure 19-9). The bud, for example, is refashioned at the top. The fruiting head

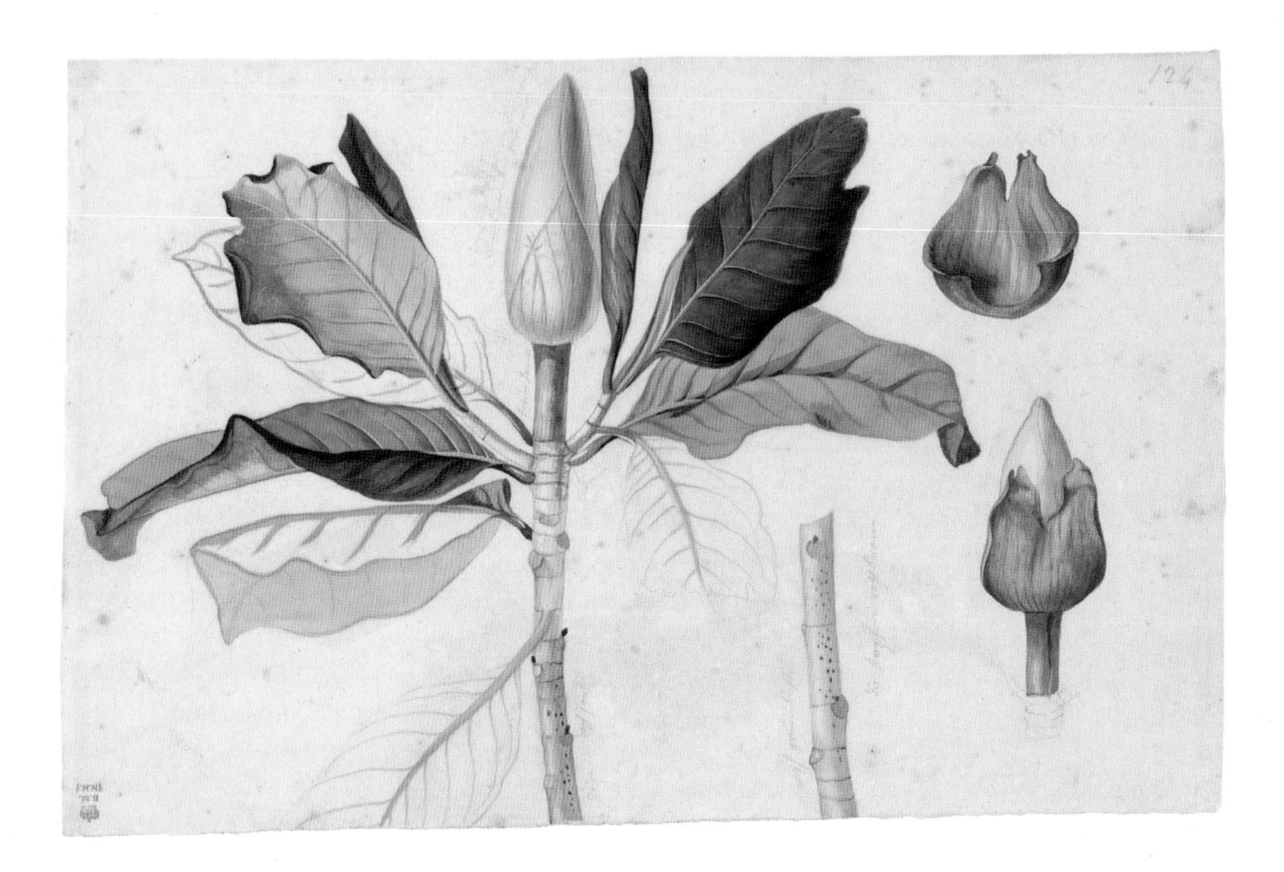

FIGURE 19-8. Sketch of southern magnolia (*Magnolia grandiflora*) bud by Georg Ehret. (© Botany Library, Natural History Museum, London.)

indicates that Ehret returned to Parsons Green well after the blossom faded. The Oak Spring vellum thus represents a sequence, much as one might record an opening blossom using time-lapse photography today.

Though undated, that resplendent vellum probably coincides with, or follows just after, the bitter winter of 1739–1740, when many evergreen magnolias that had been growing outdoors in English gardens for well over a decade were killed.[41] Peter Collinson has left an account of how that winter came as a profound shock to gardeners lulled into a false sense of security by the favorable seasons of the 1730s.

Writing to John Custis on 31 January 1740, Collinson launched into a lament:

> Wee have att this Juncture an Extreame of Cold. A sharp Frost began the 26 December and has continued Ever since. . . . This is a Trying Time to Our Gardens & south Country plants. I have been Obliged to keep Constant fires In My stove & Green house by which Means I am in Hopes I shall be but a small sufferer, but those that would not take the pains are quite Demolish'd. The frost was so very sharp & severe. The Like has not been known since the 1715/16 or 1709.[42]

Indeed, it was not just the bitter winter alone. 1740 proved the coldest calendar year in England since records began in 1659, averaging only 6.8°C (44°F) in central England. Even before the shock of that winter, when the daily mean temperature for December, January, and February averaged –0.4°C (31°F), there had been one severe warning. The cold summer of Catesby's Bahamas sojourn,

FIGURE 19-9. Southern magnolia (*Magnolia grandiflora*) combination by Georg Ehret. (Collection of Rachel Lambert Mellon, Oak Spring Garden Library, Upperville, Virginia.)

1725, averaged only 13.1°C (56°F) for the months of June, July, and August. As the winters either side of 1725 were not cold, the gap between Frost Fairs, when the River Thames froze over and fairs were held on the solid ice, amounted to twenty-four years (from 1715–1716 to 1739–1740), just enough time to let a memory lapse.

It should be noted that Miller's 1754 account of magnolias dying off also stated how Sir John Colleton's *Magnolia grandiflora* was the largest in England.[43] Presumably Exmouth's situation on the milder southwest coast of England led to his tree being spared. In December 1834, Robert Glendinning, then gardener to Lord Rolle, gave an account of that original *Magnolia grandiflora* var. *exoniensis*.[44] Apparently it was mistakenly cut down about 1794 instead of an old apple tree. However, layers had been taken that perpetuate the cultivar, which is assumed to be today's 'Exmouth'.

Peter Collinson, who had had several *Magnolia grandiflora* in 1737, grew others after 1740. Within twenty years (presumably withstanding his move from Peckham to Mill Hill in 1748), one flowered in 1760. He recorded: "Planted them against the piers of the Green House, in my Garden at Ridgeway House . . . and this 6th of August 1760, one of Them flowered for the First Time, the tree is 15 feet High."[45] Prior to this, he reported that a "Large Tree that Stands in the Open Air" at Christopher Gray's nursery was in flower in September 1758.[46]

Gardeners had become cautious. Dr John Hope of Edinburgh, touring English gardens in 1766, noted at Syon House: "Great Magnolia covered in winter with poles & mats open to South."[47] In 1778, Mawe and Abercrombie,[48] calling it "the finest ornamental ever-green yet known," advised the use of poles and mats from November onward, with reserve plants in pots. They had just lived through the cold winter of 1775–1776, when the Frost Fair returned to the Thames for the first time since 1739–1740.

Others were keeping *Magnolia grandiflora* under glass as an extra precaution. Mary Delany's collage of *Magnolia grandiflora*, dated 26 August 1776, was executed from a greenhouse plant just after that bitter winter of 1775–1776 (figure 19-10). The greenhouse was at the Countess Dowager Gower's residence, Bill Hill, near Reading. Despite living through three decades of warmer seasons, Delany and her close circle of female friends were perhaps wise enough to recall what had happened in their younger years to such promising exotics.

Developing her initial enthusiasm of 1772 for *Magnolia*, Lady Gower extended her offspring to Mrs. Boscawen. On 9 November 1773, she wrote of transporting it in her "chaize." She arrived with it just as Boscawen was throwing "*a spud*" at a tortoiseshell cat that threatened her almond-fed robin.[49] That *Magnolia* "child" went into the Glan Villa greenhouse at Colney Hatch. On 23 June 1774, Lady Gower wrote to Mrs. Delany: "Mag. y^e great had two blooms almost ready to blow before I came away. I shall be much disappointed if her children do not inherit her blooming charms."[50]

FIGURE 19-10.
Southern magnolia (*Magnolia grandiflora*) collage by Mary Delany. (© Trustees of The British Museum, London.)

On 30 June 1776, Lady Gower wrote again to Mrs. Delany: "*Mother Magnolia* has more buds y^{n} I can count, her daughters are all fruitfull."[51] On 2 August, after Mrs. Boscawen visited Bill Hill and found the countess as cool as a "Salamander" in the heat of the summer, she reported further blooming: "Magnolias innumerable perfume the air and delight the eye at the green-house; 23 have blown."[52] By the time Delany arrived with black paper, knife, scissors, and glue, it was more than three weeks later. No wonder some of the flowers were full-blown. Mary Delany chose to depict, or was obliged to depict, then, a flower that had opened at least two days before.

Ehret's Oak Spring vellum shows a similar flower, over two days old and with its stamens shed. Although Ehret's drawings and paintings of *Magnolia grandiflora* thus serve as the gold standard in Georgian botanical illustration, Delany's collage is a sterling amateur work. It was undertaken just weeks after the American Declaration of Independence, rounding off that period of colonial collecting to which Mark Catesby contributed so much.

Sir Charles Wager, convoying the New England trade and gathering the spoils of the West Indies trade, had used his captured colonial silver to help underwrite Catesby's endeavors and to facilitate Ehret's finest portraits of *Magnolia grandiflora*. While Ehret engraved the two-day-old flower for Mark Catesby's *The natural history of Carolina* . . . , it was Catesby who engraved the newly opened bloom for Christopher Gray's *Catalogue* (figure 7-6, p. 94). Catesby's engraving then became a template for the plate published in his posthumous *Hortus britanno-americanus* (figure 22-10, p. 328). The fate of *Acer saccharinum*, or so it would appear in the tangled web of naming and picturing, was to remain without depiction in England. Nevertheless, given that some of the silver worth £60, 000 went into importing and cultivating the species in Parsons Green (perhaps from seed supplied by Catesby), it seems appropriate that a tree once known as Sir Charles Wager's maple is now called simply silver maple. Every time a breeze blows in England, the silvery undersides of the leaves show off wealth turned into horticultural commonwealth.

20

Following in the footsteps of Mark Catesby

JUDITH MAGEE

Many significant artists and naturalists followed on the heels of Mark Catesby. In Britain, the outstanding ones were George Edwards, Georg Dionysius Ehret, and John Latham (1740–1837). In North America, these men were William Bartram (1739–1823); Alexander Wilson (1766–1813), the author of *American ornithology*; and John James Audubon (1785–1851), the creator of *Birds of America*. Like Catesby, the three Americans were both naturalists and artists rather than solely collectors or scientists. They spanned a period of a century from when Catesby was working, and their contribution to science can be identified as rooted in his pioneering work. Although each man had much in common with Catesby, each one also brought new and different contributions to science and art. While what these three men brought to these fields may have differed, a continuity can be traced from Catesby to Audubon in how each of them viewed nature, a view that was immersed in observing and understanding the relationships between species and comprehending nature as a whole, or, as it is now called, an ecological view.

However, it is Alexander von Humboldt (1769–1859), born in Berlin twenty years after Mark Catesby died, who is today considered the father of ecology and environmentalism. Humboldt laid the foundations for the study of climatology, oceanography, meteorology, and plant geography. If Humboldt is stripped of all his endless measuring, sampling, quantifying, calculating, and precision equipment, then his understanding of the natural world was not that much different from that of Catesby and his successors. When Humboldt argued that there was an underlying unity of nature and all things were connected, nothing in isolation, he was echoing the views of Mark Catesby, William Bartram, and others.[1] When he visited Philadelphia on his return from South America in 1804 and dined with the eminent scientists there, including Alexander Wilson and William Bartram, Humboldt probably found in his fellow diners kindred spirits in their shared view of nature. This view of the natural world, as a

FIGURE 20-1. Scarletsnake (*Cemophora coccinea*) and sweet potato (*Ipomoea batatas*); plate 60, M. Catesby, 1737, *The natural history of Carolina* . . . , volume II. Catesby was primarily a botanist but today is considered a pioneering ornithologist in his depiction and observations of birds. However, there are, surprisingly, thirty-one plates of fish and twenty plates that include snakes in *The natural history of Carolina*. . . . Many of the snakes are figured entwined in vines, demonstrating Catesby's artistic motivation in placing different subjects on the page together that do not always have a relationship in nature. In this plate, Catesby believed there was a connection between the two species, as the scarlet snake was "frequently found dug up with Potatoes." (© Natural History Museum, London; reproduced by courtesy.)

living unity of diverse and interdependent life forms, an organic whole, is what distinguishes these men from those who understood and interpreted nature purely through classification.

Mark Catesby's work preceded the revolution that came in the natural sciences with Linnaeus's classification system and binomial nomenclature. These new ideas and methods came to dominate and had an impact on the naming of species and ways of describing and identifying plants and animals, but they also had a significant bearing on natural history art. By the second half of the eighteenth century, strict conventions for scientific illustration had been put in place and were referred to as the Linnaean or scientific style of drawing (figures 20-2 and 20-3).

FIGURE 20-2. Taro (*Colocasia esculenta*); Georg Ehret, watercolor, c. 1740s. Ehret worked with Linnaeus as a young man and was one of the first artists to adhere to the strict conventions of the Linnaean style of botanical illustration. He was not only one of the foremost botanical artists of his day but also an excellent botanist. (© Natural History Museum, London; reproduced by courtesy.)

FIGURE 20-3. Red silky oak (*Grevillea banksii*); Ferdinand Bauer, watercolor, c. 1803. Considered one of the finest botanical artists of all times, Bauer sailed on HMS *Investigator* with Captain Matthew Flinders, circumnavigating Australia between 1801 and 1803. This drawing of *Grevillea* is a perfect example of a Linnaean-style botanical illustration. It details all the features needed by a botanist to describe and classify the plant. (© Natural History Museum, London; reproduced by courtesy.)

A drawing of a specimen is often of greater worth than the specimen itself, as it can record details not seen in a dried plant or skin or an animal preserved in spirits. All important in conveying information about the subject are the morphological features—the structure, form, color, and pattern. The aim of both botanical and zoological art is to produce an image of the organism that is typical of the species and to which all others of its kind can be compared. This usually means eliminating any distracting elements, so the subject is conventionally placed on the page in isolation and without context. The Linnaean style of illustration redefined some of the traditional ways of depicting a bird, animal, or plant; there was no background either of flora or fauna accompanying the subject. Birds were typically drawn motionless, possibly on a branch and in profile. Insects were rarely given any attention at all. What the Linnaean style introduced, and rigorously so, was the inclusion of features for identification of a species. For botanical art, a standard practice was to include

FIGURE 20-4. Bald eagle (*Haliaeetus leucocephalus*); plate 1, M. Catesby, 1729, *The natural history of Carolina . . .* , volume I. Catesby was one of the first artists to depict his subjects in active poses. Viewing the bird in flight not only brings life to the illustration but also reveals aspects of the species not normally seen in the traditional static profile of a bird on a branch. (© Natural History Museum, London; reproduced by courtesy.)

FIGURE 20-5. Bluejay (*Cyanocitta cristata*); plate 15, M. Catesby, 1729, *The natural history of Carolina . . .* , volume I. The plant depicted here with the bluejay is laurelleaf greenbrier (*Smilax laurifolia*), a vine native to the southeastern United States. The black berries of the plant ripen in October, and the bluejay is one of several birds that feed upon them.
(© Natural History Museum, London; reproduced by courtesy.)

the number and structure of the male and female parts of the flower, preferably depicted at greater magnification. In zoological illustration, certain anatomical features were also often included, again preferably magnified. Thus, the drawings of Maria Sibylla Merian, Mark Catesby, and, later, William Bartram, which to varying degrees gave a more rounded view of nature by placing the species in some context either with the prey of the main subject (figure 20-4) or with plants (figure 20-5), were no longer regarded as scientific.

Of course, not everyone adopted Linnaeus's system immediately and without question. And what can be seen in the second half of the eighteenth century is the development of two trends in the study of natural history. One was the Linnaean trend, with the domination of classification and taxonomy; the other trend was that of observing and describing living species, from which conclusions about plant habitat, animal behavior, and the relationship between living organisms could be drawn. The latter trend was influenced by important scientists such as the French naturalist Georges-Louis Leclerc, Comte de Buffon (1707–1788), the author of thirty-six volumes of *Histoire naturelle* and one of the most influential scientists of his time. However, it was not until the late nineteenth century that the ideas associated with this second trend were absorbed into the study of ecology and evolution.

Field naturalists

Catesby, Bartram, Wilson, and Audubon can all be identified with this second trend. They were self-taught artists and naturalists. None was privy to the academic discipline and training that a university education provided. Instead they spent a great deal of time observing the natural world. For the naturalist, drawing is a valuable activity in learning about the subject. As well as recording the diagnostic features for identifying the species, it is also important to capture other features such as motion, behavior, and character, and this requires observing the subject in its living habitat. The greatest natural history artists had excellent observational skills and were fine field naturalists. Humboldt claimed that the conclusions drawn from his research were possible only through the rigorous collection of primary data, recorded accurately and directly from observations of nature. Catesby's drawings and text are steeped in his fieldwork and observations. The ecological artistry that dominated his work is also present in the writings and drawings of Bartram, Wilson, and Audubon. It is no coincidence that these four men were, during their lifetimes, the naturalists who had the greatest understanding of bird migration. Their ideas on the subject were based on their long-term observations over the seasons and over the years. These men learned much of their science through experience, traveling through the wild and observing nature in its living environment. Each identified himself as a field naturalist. Bartram stated that he was no systematist and that he rejected the obsessive collecting habits that filled cabinets with curiosities devoid of life. Wilson claimed that he turned a thousand times from the "barren musty records" of the dry systematic writers with a "delight bordering on adoration to the magnificent repository of the woods and fields."[2] And Audubon, who described himself as a practical ornithologist, reiterates throughout the seven-volume version of *Birds of America*, published between 1840 and 1844, how his time spent in the wild observing birds proved to be so much more fulfilling and his conclusions about them more accurate than those of the "closet naturalists"[3] (figure 20-6).

Each of these artists broke with the conventions of natural history illustration. Catesby's illustrations introduced a move away from the standard lifeless depiction of a species. He presented visual associations between species that suggested an environmental relationship he had observed in his fieldwork. He often portrayed his species with background vegetation, sometimes with the plants and insects a species fed upon. He gave some context to his subjects

FIGURE 20-6. Northern mockingbird (*Mimus polyglottos*) and timber rattlesnake (*Crotalus horridus*); plate 21, J. J. Audubon, 1827, *Birds of America*. Critics raged against this illustration when it was first published, claiming that rattlesnakes did not climb trees and that the shape of the fangs was incorrect. Audubon was proved right on both counts, demonstrating that his field observations were more worthy and accurate than those of the "armchair naturalists." (© Natural History Museum, London; reproduced by courtesy.)

and introduced movement to many of them. This approach was evident in the work of each of the three naturalists who followed Catesby, culminating in Audubon's magnificent portrayal of birds.

Patriotism

The men who came after Catesby all introduced something new in their work, and their vision of the New World was different from that of Catesby. They spanned the period from the Enlightenment through the Romantic Movement and could not help but reflect the moods and fashion of their day. Catesby had a group of patrons he needed to satisfy by providing plants and seeds they could grow in their gardens and nurseries in England. His observations of plant habitat and the relationship between plants and animals were important sources of information for the success of transplanting. By the time Bartram, Wilson, and Audubon were writing and drawing, America had undergone considerable change. One of the driving forces behind these men's work was the concept of an independent scientific community in the new American republic, which was no longer reliant on Europe for the intellectual interpretation of American natural products. William Bartram was a distinguished figure in this developing milieu and a pivotal influence on a whole generation of young American scientists and artists who were fervent patriots. Neither Wilson nor Audubon was born in America, but they both became American citizens in adult life, and Wilson became more patriotic than many who were native-born. His form of patriotism spurred on his desire to do something new for American science and became the motivation in his work, the publication of *American ornithology*.

William Bartram had drawn nature from an early age. He spent countless hours poring over the volumes of Catesby and George Edwards. Many of his early drawings, which he sent to Peter Collinson in London, depicting a bird or animal were annotated with the words "Not in Catesby." Collinson had been one of Mark Catesby's principal patrons, and for the last thirty-four years of his life he was friend and correspondent of John Bartram, William's father. Collinson greatly admired William Bartram's drawings and compared Bartram to Georg Ehret. Collinson wrote to Linnaeus that "He will soone be another *Ehret*—His performances are So Elegant."[4] Later, Bartram's influence was widespread and significant due to his *Travels through North and South Carolina, Georgia and Florida*, published in 1791. Scientists and statesmen from Europe and across the United States, including Thomas Jefferson, went out of their way to visit Bartram and his garden in Kingsessing, Philadelphia. They read his work and consulted him on many aspects of natural history.

Bartram's approach to natural history was to study plants and animals in relation to their environment. He expressed his commitment to this ecological view of nature throughout his work, explaining how the interplay of underlying natural forces was integral to the renewal of the living earth. His writings contain a wealth of wonderful observations about the relationship between species. His work on American birds surpassed anything written previously, describing not only bird migration and nesting habits but also their songs. It was this subject that Charles Darwin found interesting, referencing it in his notebooks.

Unfortunately, few of Bartram's drawings were published during his lifetime. His illustrations went beyond mere surface characteristics and depicted the balance in nature, its underlying harmony, as well as the violent and all-consuming side of nature. In so doing, he brought life to his subjects. Of course, he could draw a perfect Linnaean-style image, such as his illustration of *Franklinia* (figure 20-7), but he preferred to portray nature in a more descriptive manner, capturing the complete cycle of nature in its unsentimental "progressive operations."[5] The two drawings of the American lotus (*Nelumbo lutea*) are perfect examples, demonstrating Bartram's understanding of the natural world (figure 20-8). Each stage of the development of the plant is depicted, and the cycle is completed with the portrayal of the seed vessel, hinting at the potential regeneration of the plant.

Meeting Alexander Wilson in 1802, Bartram soon took him under his influence, instructing him in bird identification, providing lessons in how to draw, and advising Wilson on the coloring of his plates for his books. (Wilson ended up coloring 4,500 of these plates himself.) As a protégé of Bartram, Wilson set out to travel the land to record and describe all the birds of America (figure 20-9). When he stayed at the Bartram house he had access to the volumes of Catesby and Edwards and, from the Philadelphia Library, John Latham's *General synopsis of birds*. Wilson's *American ornithology* ran to nine volumes, the last two completed and published under the supervision of George Ord (1781–1866) after Wilson's untimely death at the age of forty-seven in 1813.

Wilson's artistry was dwarfed by the magnificent life-size drawings produced by Audubon, but the quality of Wilson's writing and his major contribution to ornithology make up for his artistic limitations. His descriptions and narratives are engaging; they are both scientific and poetic and demonstrate his excellent observational skills.

Unfortunately, because Wilson was forced to economize, he depicted several birds on one page, thus losing background and context (figure 20-10). He was unable to depict the larger birds at full size but was careful to reduce those figured on the same page by the same scale. Wilson identified thirty-nine new species, refuted many claims made by others who had never visited America, and corrected many of the previously poor descriptions of birds. His observations on the habit, manners, and plumage of birds were insightful, and his comments from his travels and experiences were both humorous and entertaining. Wilson can be seen today as one of the great bird-men of his time, and had he lived longer he would no doubt have built a reputation for himself as the foremost ornithologist of the early nineteenth century.

Throughout Wilson's whole work runs the theme of American identity in the arts and sciences developed by and for Americans. He constantly referred to the United States as "my country" and his subjects as "our native birds." In

FIGURE 20-7. Franklinia (*Franklinia alatamaha*); William Bartram, watercolor, 1788. *Franklinia* was named by Bartram for Benjamin Franklin, a family friend, and after the Altamaha River, which runs through Georgia. Bartram and his father first discovered the plant in Georgia in 1765. William returned to the spot on his travels some ten years later and collected seed. In less than thirty years, the plant had become extinct in the wild. All plants in cultivation today owe their existence to William Bartram. (© Natural History Museum, London; reproduced by courtesy.)

FIGURE 20-8. American lotus and seed vessel (*Nelumbo lutea*); William Bartram, ink and watercolor, 1767. Bartram intended both drawings to be viewed together, demonstrating the complete life cycle of the plant. In the lower left corner of "Tab. I" is the first depiction of the Venus flytrap (*Dionaea muscipula*), described by Bartram as a "sportive vegetable" in reference to its carnivorous habit. (© Natural History Museum, London; reproduced by courtesy.)

FIGURE 20-9. Bald eagle; plate 36, A. Wilson, 1811, *American ornithology*, volume IV. This is one of only five plates in which Wilson depicted a single species. He has also placed the bird in a setting that includes Niagara Falls in the background and the eagle's main food source, fish, in its claw. The bird is drawn one-third its natural size. (© Natural History Museum, London; reproduced by courtesy.)

FIGURE 20-10. Western tanager (*Piranga ludoviciana*), Clark's nutcracker (*Nucifraga columbiana*), and Lewis's woodpecker (*Melanerpes lewis*); plate 20, A. Wilson, *American ornithology*,, volume III. Wilson was the first to illustrate and describe the birds that Lewis and Clark brought back from their expedition of 1804–1806. The birds had been donated to Charles Wilson Peale's Museum in Philadelphia, where Wilson regularly had access to the collections. (© Natural History Museum, London; reproduced by courtesy.)

the preface to the fifth volume, he praised the engravers, printers, and paper-makers, all of whom were American, and asked that the "generous hand of patriotism be stretched forth to assist and cherish the rising arts and literature of our country."[6]

Splendid though Wilson's volumes were at the time, within twenty years they had been superseded by the work of John James Audubon and his *Birds of America*, published in London between 1827 and 1838. Audubon demonstrated his excellent observational skills through his magnificent drawings. Although his text *Ornithological biography*, published in 1831 to accompany the plates, does not come close to Wilson's, it does contain some first-class descriptions of bird behavior. In both his drawings and his writings he revealed his intimate knowledge of the habits of birds and their environment. He was not overtly patriotic like Wilson, but he did refer to America as his "native land." He expressed the view that the time spent wandering through his "native land" resulted in his knowing and understanding it better than most. His image of an "American woodsman" was very much about the individual, but an individual who was a true American.

Audubon revolutionized the visual portrayal of birds. The birds he depicted were all drawn life-size, in active poses, and often with local habitat as a backdrop. Having shot his bird, Audubon would then attach it to wires to create animated poses, displaying habit and character he had observed in nature. The intention was to introduce the viewer, through the composition of Audubon's plates, to the beauty of these birds, as well as their dramatic behavior and a vibrant display of form. Audubon's contribution to ornithology was in his observations of bird behavior, which were fresh and new. His depictions of the form, coloring, and plumage of birds were accurate. And for those birds that are now extinct, Audubon's drawings have become valuable scientific references and important records of ornithological history (figure 20-11). He managed to change the way ornithological artists considered the picturing of birds. Audubon possessed copies of Catesby's and Wilson's works, and when he visited Florida in 1832 it was Bartram's *Travels*, a book published some forty years earlier, that he used as a guide.

FIGURE 20-11. (*Opposite*) Carolina parakeet; plate 26, J. J. Audubon, 1827, *Birds of America*. The Carolina parakeet was the only native parrot of the United States and is now extinct. This magnificent portrait almost brings this species back to life with its naturalistic setting and dynamic positioning of the birds. The male, female, and juvenile are all depicted perched on the common cocklebur (*Xanthium strumarium*), a food source for the birds. (© Natural History Museum, London; reproduced by courtesy.)

The approach to natural history in both the drawings and the writings by naturalists such as Bartram, Wilson, and Audubon was suggestive of the ecological movement that emerged in the late nineteenth century. The phenomenon of patterns in regional climates influencing the variety, number, and size of species in different geographical areas interested Catesby, Bartram, and Wilson. It was addressed more fully by Humboldt, who developed a theory of plant geography in which he demonstrated that the effects of altitude, temperature, climate, and geography determined where plants grew. This theory was beautifully portrayed by Humboldt in many of his drawings (figure 20-12).

In due course, it was Linnaeus's classification system that came to triumph and dominate in scientific illustration, and this re-asserted the traditional

N° 6

PLATE. 26.

Carolina Parrot. Males 1. F. 2. Young 3.

PSITACUS CAROLINENSIS,

Plant Vulgo. Cuckle Burr.

Drawn from Nature & Published by John J. Audubon F.R.S.E. M.W.S.

Engraved, Printed, & Coloured by R. Havell & Son, London.

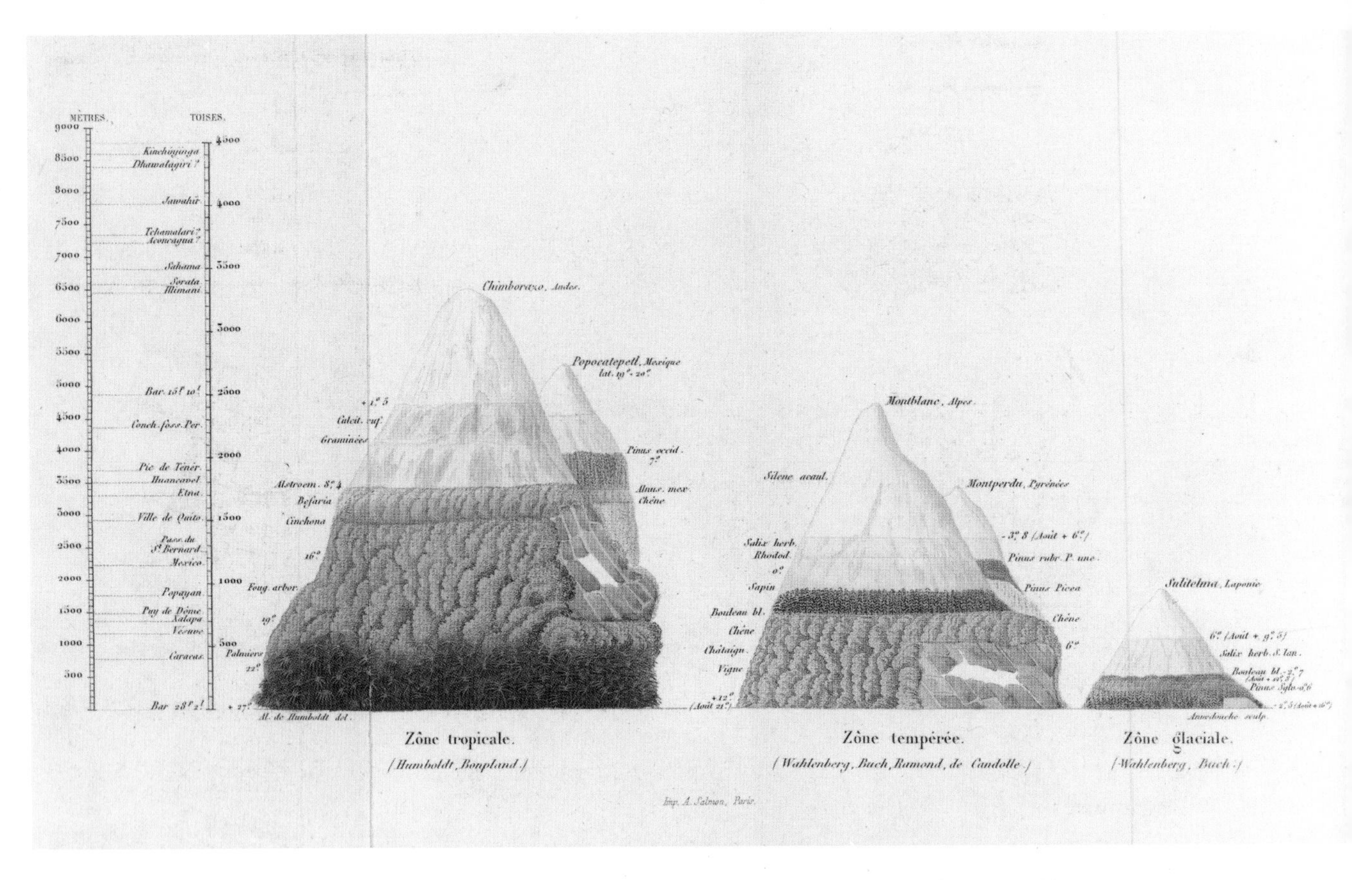

FIGURE 20-12. Alexander von Humboldt, 1865, "Tableaux de la nature." Humboldt was a master at portraying his ideas and conveying information graphically. Having established the influence of climate on plant geography and vegetation zonation, he then illustrated it simply and beautifully. Here he demonstrates how vegetation patterns correspond to changes in elevation and latitude. (© Natural History Museum, London; reproduced by courtesy.)

style of drawing, focusing on the essential features for identifying and classifying purposes. The relationship between the viewer and the subject, which had become an engaging experience with Catesby's, Bartram's, and Audubon's drawings, reverted to becoming distant, with the subject viewed only as a specimen, and any contextual understanding of the natural history of that specimen was lost.

Bartram, Wilson, and Audubon built upon the work done by Catesby. They went beyond what he did either in their drawings or in their text, and by so doing they kept alive what we today call the ecological view of nature. It was synonymous with Humboldt's view and was adopted by many of the scientists who recognized the struggle for existence taking place in nature and who developed an understanding of the processes that lead to such diversity of life on earth. Although scholars did not use the terms "ecology" and "environmental science" until the second half of the nineteenth century, the roots of the environmental approach can be found in the ideas and drawings of naturalists studying the North American flora and fauna from Catesby to Audubon.

21

Inspiration from *The natural history of Carolina . . .* by Mark Catesby

GHILLEAN T. PRANCE

I knew very little about Mark Catesby until I watched *The Curious Mister Catesby*. I have since read much about him. As I looked at Catesby's life I saw that I had a great deal in common with him, even though I have never painted what I have seen because of the ease of photography now. Catesby and I were both born in eastern England, near Sudbury in Suffolk. We both also spent extended periods in North America and a long time in the field observing natural history. We both became Fellows of the Royal Society in recognition of our work. One of Catesby's sponsors was William Sherard, who founded the Sherardian Chair of Botany at Oxford University, where I studied under the leadership of Cyril Darlington, the Sherardian Professor of Botany in the 1950s. Catesby's achievements are remarkable, as he worked in much more primitive conditions and with fewer resources and equipment than I.

Two hundred and fifty years after Mark Catesby traveled to North America, I also traveled westward and became equally entranced by the wonders of nature, mainly in South America. My sadness is that we are now losing so many of these wonders through the destruction of what was described and illustrated by the early naturalists such as Catesby. In the time since then, as I have browsed through Catesby's paintings, what has struck me the most is the mixture of plants and animals on each plate. In many cases, the animals and the plants are species that interact with each other in some way in their natural environment, and it is this aspect that I most want to explore.

The coco plum

Catesby's illustration of the coco plum (*Chrysobalanus icaco*, *NH* I: plate 25), a painting from the Bahamas, is my starting point (figure 21-1), because the Chrysobalanaceae to which it belongs was the plant family upon which I did my doctoral research at Oxford and that I have studied around the world.[1] The coco plum is an interesting species because it is a "transatlantic" plant, being

FIGURE 21-1. Coco plum (*Chrysobalanus icaco*)and white-crowned pigeon (*Patagioenas leucocephala*); plate 25, M. Catesby, 1730, *The natural history of Carolina* . . . , volume I. A painting from the Bahamas is my starting point. (Digital realization of original etchings by Lucie Hey and Nigel Frith, DRPG England; courtesy of the Royal Society ©.)

abundant around the islands of the Caribbean, the Atlantic coast of South America, and the coast of West Africa. The seeds are commonly dispersed by pigeons, so it was most interesting to see this species depicted by Catesby beside the white-crowned pigeon (*Patagioenas leucocephala*), a bird that is also widely distributed. The coco plum needs a widely distributed disperser for its seeds. This was certainly true of the white-crowned pigeon in Catesby's time, but the plant is now much rarer, as the pigeon has been hunted to near extinction in many places. Unlike the Carolina parakeet illustrated by Catesby,[2] the white-crowned pigeon is not extinct. This is a poignant example of the need to increase our efforts in species conservation. The endangerment of this species had already begun in Catesby's day, as he wrote: "They breed in great Numbers on all the *Bahama-Islands*, and are of great advantage to the Inhabitants, particularly when young. They are taken in great quantities from off the Rocks on which they breed."

Catesby recorded of the coco plum:

> They produce a Succession of Fruit most part of the Summer, which is of the Size and Shape of a large Damsin; most of them blue: Some Trees produce pale yellow, and some red. . . . The Fruit is esteemed wholesome, and hath a sweet luscious Taste. The *Spaniards* at *Cuba* make a Conserve of them, by preserving them in Sugar.

In most populations of the coco plum, I have observed trees with either yellow or dark red fruits. When Carl Linnaeus named this species from a painting by the French botanist and explorer of the West Indies Charles Plumier, he must also have noticed Sloane's reference to yellow fruits[3] and so chose to name it *chrysos* (= golden) *balanos* (= acorn, referring to the fruit). Icacos, as they are called, are still bottled in sugar and sold in Colombia and Venezuela.

There are many interesting relationships between species of the Chrysobalanaceae and animals. Catesby would have liked the genus *Couepia*, which is predominantly pollinated by hawk moths, except for two species that have flowers dangling beneath the crowns of the trees on long stalks, making them accessible to the bats that pollinate them.[4] Some species of the genus *Hirtella* have swollen cavities, called domatia, at the base of their leaves. Ants inhabit these cavities and protect the plant from other leaf predators, for example, by removing the eggs of other insects from the leaves. This mutualism, where ants and plants cooperate to the benefit of both organisms, has arisen in a number of different unrelated tropical plants of the Amazon rain forest.

I am glad that Catesby encountered at least one species of the pan-tropical family Chrysobalanaceae.

Brazil nuts and agoutis

The other plant family that I have studied in detail is the Brazil nut family, the Lecythidaceae. Catesby did not encounter this family because it does not reach as far north as the Bahamas or Jamaica, but his painting of the false mastic tree (*Sideroxylon foetidissimum*, *NH* II: plate 75) with an eastern chipmunk beneath it (figure 21-2) reminded me of the relationship between the Brazil nut (*Bertholletia excelsa*) and the agouti (*Dasyprocta leporina*). Later, I found out that Catesby did paint an agouti, calling it "The Java Hare" (*NH* II: Appendix, plate 18) (figure 21-3). The Brazil nuts are produced, arranged like the segments of an orange, in a round woody fruit that is slightly larger than a baseball or a cricket ball. The trees flower in October to November. After pollination by large bees that can lift the hood that protects the interior of the flowers to reach the nectar, it takes fourteen months for the fruits to mature and drop like bombs in January and February. Brazil nuts are hard to crack open, but the outer shell that encases the whole fruit is even harder. Once they have fallen from the tree and are lying on the ground, only one animal can gnaw them open: the

FIGURE 21-2. False mastic (*Sideroxylon foetidissimum*) and eastern chipmunk (*Tamias striatus*); plate 75, M. Catesby, 1739, *The natural history of Carolina* . . . , volume II. (Digital realization of original etchings by Lucie Hey and Nigel Frith, DRPG England; courtesy of the Royal Society ©.)

FIGURE 21-3. Brazilian agouti ("The Java Hare") (*Dasyprocta leporina*); plate 18, M. Catesby, 1747, *The natural history of Carolina* . . . , volume II: Appendix. (Digital realization of original etchings by Lucie Hey and Nigel Frith, DRPG England; courtesy of the Royal Society ©.)

agouti. The agouti gnaws at the top of the shell where there is a small hole created when a small plug at the apex (the operculum) falls inward. The rodent can then get a grip on the rim with its teeth. It enlarges this tiny hole until it is just large enough for the agouti to remove the nuts and take them away to bury them in caches some distance from the parent tree. This process of scatter hoarding by agoutis is how many species of Amazon rain forest trees get their seeds dispersed. The agouti forgets where it has hidden some of the nuts or it meets its end when it encounters a jaguar, and the seeds are free to germinate and produce new trees.

The ways in which rain forest trees disperse their seeds is a fascinating topic that would have enchanted Mark Catesby. Each genus in the Brazil nut family has a different method of seed dispersal. The sapucaia nut (*Lecythis pisonis*) also has a large woody fruit, but only a large lid or operculum falls off, and the rest of the fruit is left hanging on the tree rather like a bell. Inside the outer shell, the seeds are left dangling by their fleshy foul-smelling stalks (arils). As soon as the lid falls off, bats visit the fruit and remove the seeds. They eat the fleshy stalk and discard the seeds, often under their roosts. In this way, the seeds are dispersed around the forest. The cannonball tree (*Couroupita guianensis*) is so called because of its large round fruits, borne on the trunk and larger branches. When mature these fall to the ground and are often sought after by wild pigs, which eat the fleshy pulp surrounding the seeds and excrete the seeds around the forest. Two other genera of Lecythidaceae, *Cariniana* and *Couroupita*, have winged seeds that fall out when the operculum of the fruit detaches. The winged seeds are carried away by the wind and so are dispersed widely.[5]

Bignoniaceae, *Clusia*, and Margaret Mee

A beautiful Catesby painting is that of the orange-flowered trumpet vine (cross vine, *Bignonia capreolata*, *NH* II: plate 82) (figure 21-4). This reminds me of a painting by twentieth-century painter of Amazonian flowers Margaret Mee (1909–1988), who painted the orange-blossomed *Memora schomburgkii* in 1984; it belongs to the same family, Bignoniaceae (figure 21-5).

I met Margaret Mee several times in Amazonia as she set out on or returned from her expeditions to study and collect flowers to paint. *Memora* was painted without any background. In her later years, Mee became alarmed at the rapid loss of the Amazon rain forest and so began to paint the forest behind the plants she was illustrating. Two such paintings are of members of the Brazil nut family, *Gustavia pulchra* and *Gustavia augusta*. Unlike those of the Brazil nut, the flowers of *Gustavia* are more open and regular and are visited by pollen-gathering bees. These paintings were part of Mee's way of drawing attention to the wonders of the Amazon ecosystem, which she so courageously defended. When we were on the BBC Radio morning program *Start the week* together, she was asked what was her favorite plant to paint, and she immediately responded, "*Clusia*." She made many paintings of *Clusia*, including one of

FIGURE 21-4. Cross vine (*Bignonia capreolata*); plate 82, M. Catesby, 1743, *The natural history of Carolina . . .* , volume II. (Digital realization of original etchings by Lucie Hey and Nigel Frith, DRPG England; courtesy of the Royal Society ©.)

Clusia nemorosa as early as 1943 and one of the large-flowered *Clusia grandiflora* (figure 21-6) in 1982.

I was delighted to also find a Catesby illustration of a *Clusia*, the balsam fig (Scotch attorney) (*Clusia rosea*, *NH* II: plate 99) (figure 21-7), in the tenth part of *The natural history of Carolina . . .* , published in 1743. Catesby also liked *Clusia*, for he wrote of the "Balsam-Tree" as he called it: "The whole Plant is exceeding beautiful, and particularly the Structure of the Fruit in all it's Parts, is a most exquisite Piece of natural Mechanism." He was also aware that birds dispersed the seeds of this species: "These Trees grow on Rocks, and frequently on the Limbs and Trunks of Trees; occasioned by Birds scattering or voiding the Seeds, which being glutinous like those of Misletto, take Root and grow."

One time I encountered Margaret Mee as she returned from an expedition, and she proudly showed me specimens of the moonflower cactus (*Selenicereus wittii*). She was disappointed that she was there on the wrong night, as the flowers last for only a single night. She gave me a fruiting specimen to press. She vowed to return one day to find the moonflower open and achieved this

FIGURE 21-5. *Memora schomburgkii* (Bignoniaceae) by Margaret Mee, 1984.

FIGURE 21-6. *Clusia grandiflora* (Clusiaceae) by Margaret Mee, 1982. (© Copyright The Board of Trustees of the Royal Botanic Gardens, Kew.)

FIGURE 21-7. Scotch attorney (*Clusia rosea*); plate 99, M. Catesby, 1743, *The natural history of Carolina . . .* , volume II. (Digital realization of original etchings by Lucie Hey and Nigel Frith, DRPG England; courtesy of the Royal Society ©.)

on her very last expedition in 1988, shortly before she had an exhibition of her paintings at the Royal Botanic Gardens, Kew, and shortly before she died in a car accident in England. Her last painting of the moonflower is a spectacular one that also places it into its Amazon environment with the leaf-like stems pressed flat against the trunk of a tree of the igapó flooded forest, an unusual habit and habitat for a cactus.

Guaraná and bird-dispersed seeds

The paintings by Catesby of sweet-bay (*Magnolia virginiana*, *NH* I: plate 39) with a blue grosbeak reaching up toward a dangling seed (figure 21-8) and of coral bean (*Erythrina herbacea*) with a snake (*NH* II: plate 49) are of two species that expose their seeds to attract birds to disperse them. The published plate of the *Erythrina* issued by Mark Catesby in the spring of 1737 is the nomenclatural type upon which the name is based.[6] Many Amazonian plants exhibit this habit, such as *Ormosia*, the seeds of which are much used in native and tourist necklaces. These red seeds may be fleshy, with some nutrients for the birds, or they may have a hard seed-coat like the *Ormosia* and just pass through the bird

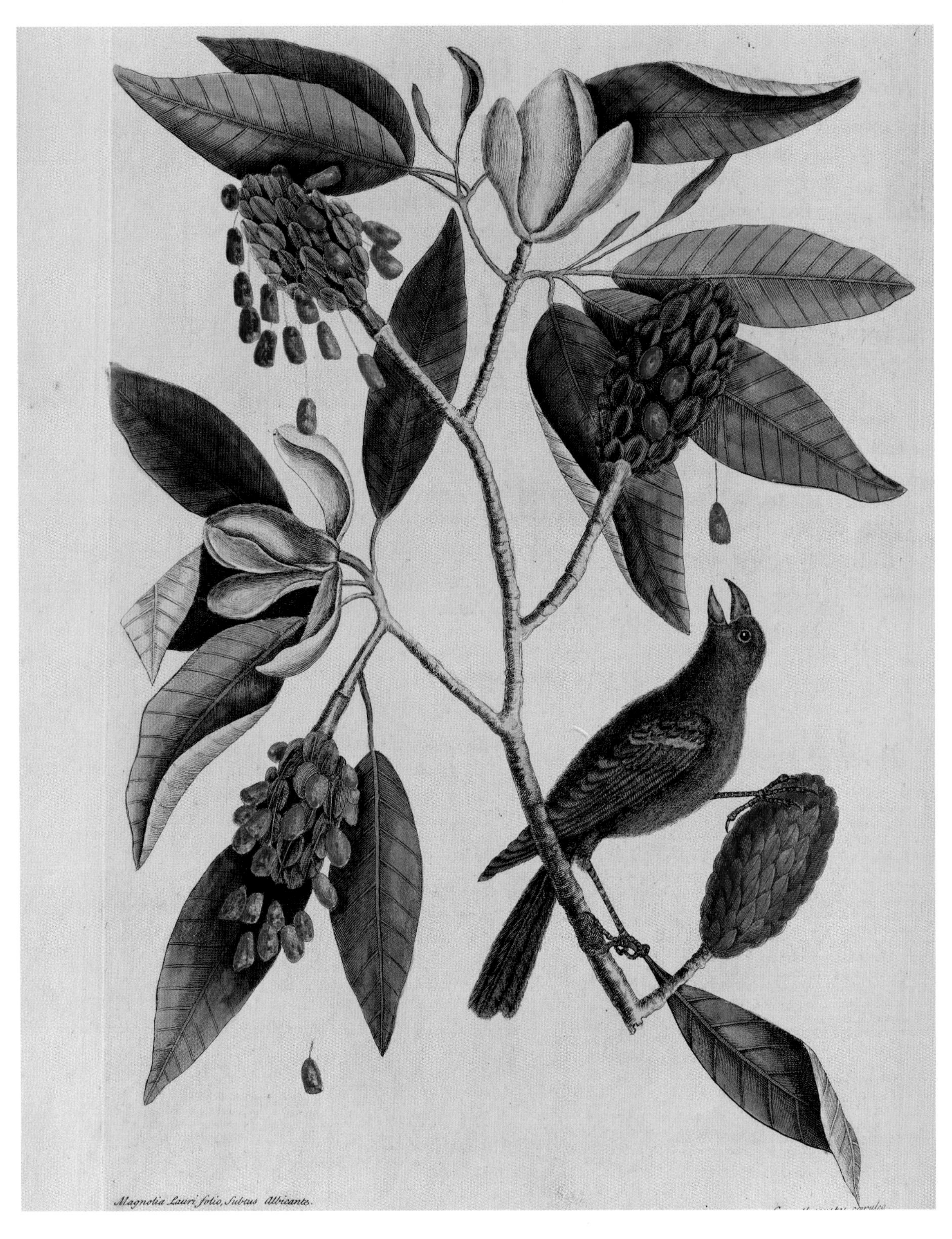

FIGURE 21-8. Sweet bay (*Magnolia virginiana*) and blue grosbeak (*Passerina caerulea*); plate 39, M. Catesby, 1730, *The natural history of Carolina . . .* , volume I. Note the bird reaching up towards a dangling seed. (Digital realization of original etchings by Lucie Hey and Nigel Frith, DRPG England; courtesy of the Royal Society ©.)

that has been coaxed into taking the seed. Catesby depicted the grosbeak about to take one of the seeds of the magnolia. He observed: "These red Seeds, when discharged from their cells, fall not to the ground, but are supported by small white threads of about two Inches long." The hanging seeds are easy for birds to gather.

These paintings remind me of another bird-dispersed seed, that of the guaraná (*Paullinia cupana* var. *sorbilis*). Guaraná has a caffeine-rich seed that was discovered by Amazonian natives many centuries ago. They used it as a stimulant, and it is now an important source of a caffeinated soda drink in Brazil. The fruit has a red outer shell that opens to expose the black seed with a white pulpy aril around its base. The birds, attracted by the red part, take the seeds, eat the pulp, and reject the seeds, thereby dispersing them. The Maués Indians have a legend about the origin of guaraná. At one time the Maués people were facing trouble. Their territory was shrinking because of invasion by other tribes, and they were constantly being attacked. The good god told the chief not to be concerned, because he would send someone to help them. Soon after, a very precocious child was born to the chief, and the boy soon became a leader who caused the misfortunes of the tribe to reverse. As the tribe began to prosper again, the evil god was jealous and resolved to harm the youthful leader. He turned himself into a poisonous snake and climbed into the favorite fruit tree of the boy. The boy climbed the tree to collect a fruit, and he was bitten by the snake and fell out of the tree dead. There was great mourning in the tribe, but the good god gave another message to the chief not to worry, as good would come out of this. He instructed the chief to plant the boy's eyes in his field. He obeyed, and out of the left eye sprouted the guaraná and out of the right eye the wild guaraná (figure 21-9). The resemblance between the seed of guaraná and an eye is quite obvious.

FIGURES 21-9. Guaraná. (© G. T. Prance.)

An abundance of tropical products

Catesby painted several tropical plants of economic importance, and these brought back many memories to me. These fruits, such as cacau, cashew, and passionfruit, are of great importance today as foods for residents of the tropics and as part of their efforts to achieve the sustainable use of tropical ecosystems.

Cacau (*Theobroma cacao*)

This painting on vellum by Claude Aubriet is among Catesby's paintings in the Royal Collection, Windsor Castle (figure 21-10). Catesby included the cacao tree in the Appendix (plate 6) (see figure 22-4, p. 321). He had seen it in Jamaica in 1714 and was fascinated by it:

> The fruit hangs pendant, and, when ripe, has a shell of a purple colour, in substance somewhat like that of a pomegranate, and furrowed from end to end, containing in the middle many kernels of the size of acorns, inclosed in a mucilaginous substance, and which are known amongst us by the name of *Cacao Nuts*, of which is made chocolate.

He then included a description of the cacao taken "from an author of great observation," William Dampier.[7]

Cacau is a native of the Amazonian rain forests and is widely distributed both naturally and by early human populations in the region. The pulp to

FIGURE 21-10. Cacao fruits. (© G. T. Prance.)

which Catesby referred was a favorite of my children when we lived in Brazil, and the seeds, after processing, are the source of chocolate. Cacau has a long and fascinating history of its uses as a food and as a currency in Mexico. No animal is depicted beside the cacau in the painting reproduced by Catesby, and he probably did not know about the tiny fly that is the main pollinator of the small flowers that are borne on the trunk and branches and that develop into the large round or oval fruits that hang from the trunk and branches.

Cashew (*Anacardium occidentale*)

The cashew that produces the popular nut has a most interesting fruit. It was painted by Catesby (*NH* II: Appendix, plate 9) in its full glory, with clusters of red and yellow cashew pears hanging from the branches and the nuts attached to the base (figure 21-11). The fruit-like cashew pear is actually a fleshy swollen stem, and the real fruit containing the nut is the curved structure attached beneath. This pear-shaped structure is juicy and tasty and is often used as a vitamin-rich drink. The nut is highly poisonous until it has been roasted. This is a clever adaptation for the dispersal of the seed. Animals and birds like to eat the pear and will often carry it away some distance from the tree to eat it. They do not touch the poisonous seeds, so they survive.[8]

The cashew is a native of the arid northeast of Brazil but has now been introduced all over the tropical world and must have reached some of the places visited by Catesby in the eighteenth century.

Vanilla (*Vanilla mexicana*)

Like the painting of cacau, this portrait on vellum (figure 21-12) is not by Catesby but was engraved by him and published in the Appendix to *The natural history of Carolina . . .* (II: Appendix, plate 7). This is also the lectotype for the species name.[9] Catesby wrote: "With this fruit the *Spaniards* perfume their chocolate, and employ *Indians* to cure the pods, which they do by laying them in the sun to dry; then dipping them in an oil drawn from the kernel of the *Acajou* nut." The extraction of vanilla from the pods is a lengthy and laborious process that requires patience to put them out in the sun by day and cover them at night. *Vanilla* is the only orchid with a use other than ornamental and is also unusual for that family in that it is a vine. It is to be found climbing up some of the trees in the forest. Catesby wrote that the seeds "are very small, and black, and are contained in a long pod, which when ripe splits open, and discharges them."

FIGURE 21-11. Cashew (*Anacardium occidentale*); plate 9, M. Catesby, 1747, *The natural history of Carolina . . .* , volume II: Appendix. Cashew in its full glory with clusters of red and yellow cashew pears hanging from the branches and the nuts attached to the base. (Digital realization of original etchings by Lucie Hey and Nigel Frith, DRPG England; courtesy of the Royal Society ©.)

FIGURE 21-12. (*Opposite*) Mexican vanilla (*Vanilla mexicana*); plate 7, M. Catesby, 1747, *The natural history of Carolina . . .* , volume II: Appendix. Note the tiny orchid seeds that are depicted at the bottom of the painting. (Digital realization of original etchings by Lucie Hey and Nigel Frith, DRPG England; courtesy of the Royal Society ©.)

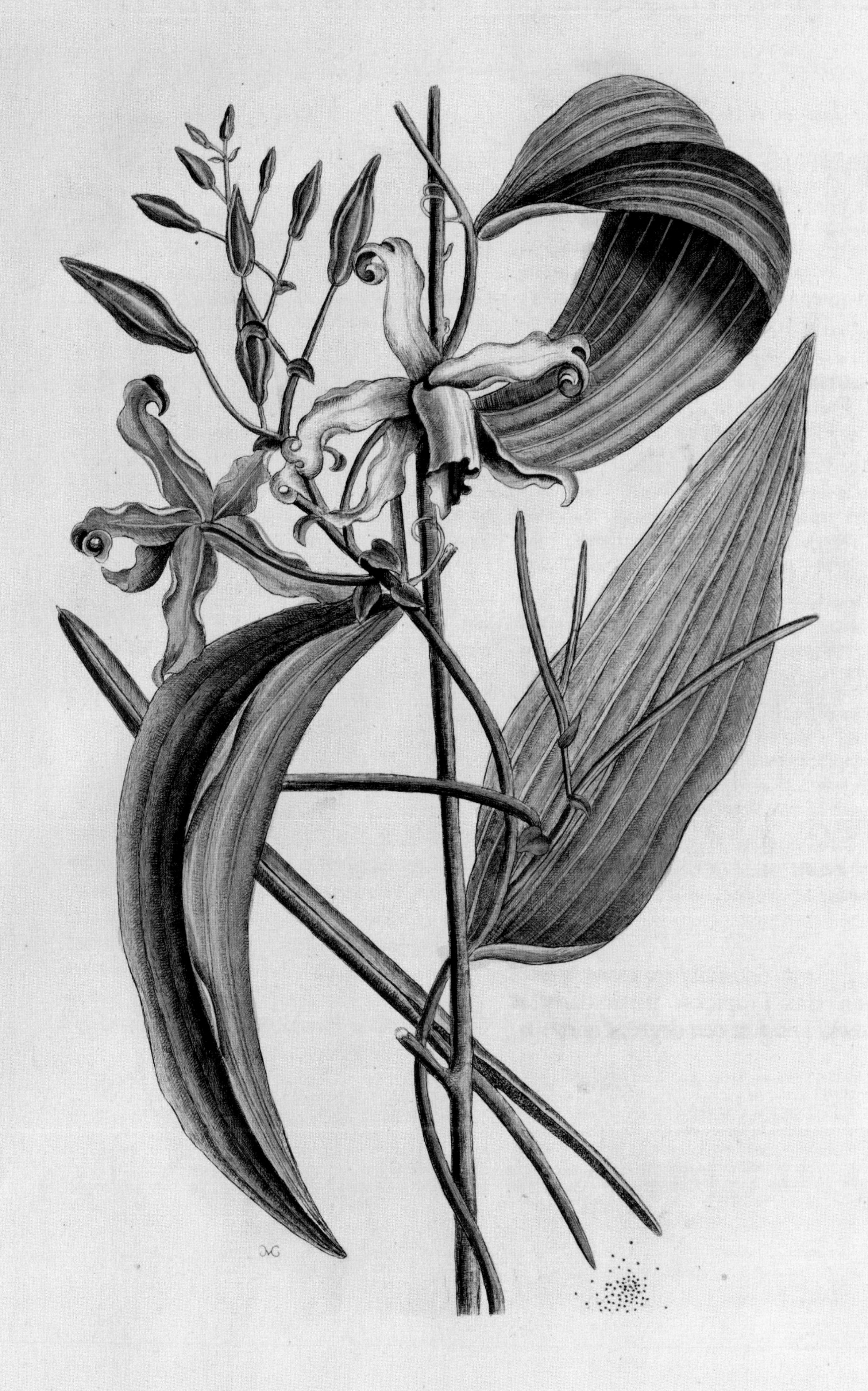

FIGURE 21-13. Frangipani (*Plumeria obtusa*) and devil's pumpkin (*Passiflora cuprea*); plate 93, M. Catesby, 1743, *The natural history of Carolina . . .* , volume II. (Digital realization of original etchings by Lucie Hey and Nigel Frith, DRPG England; courtesy of the Royal Society ©.)

Passionflowers and passionfruits

The passionflower and passionfruit genus is a large one, and I have come across many species in Amazonia, including some new ones. I have also enjoyed the fruits of several species of this now much cultivated plant. Catesby depicted *Passiflora cuprea* (devil's pumpkin) (*NH* II: plate 93) (figure 21-13),[10] referring to it as "Granadilla,"[11] and stated:

> The Leaves of this Kind of Passion-Flower are of an oblong oval Form, having three parallel Ribs, extending from the Stalk to the End.... The Flower is made up of ten narrow purple Petals, five of which are long, the other five about half as long: The Pointal arising from the Center of the Flower, is longer than in any other of this Tribe, that I have observed. The Embrio, at the End of it, swells to a Fruit, of the Size and Form of an Olive: These Plants, as likewise the Plant on which this is supported, grow plentifully on many of the *Bahama* Islands, where I painted them in the natural Appearance as is here represented.

This informative description, including both morphological and habitat details, must have been useful to Linnaeus when he described the species.

The royal water lily

The painting of the two pitcher plants (*Sarracenia*) with a southern toad (*NH* II: plate 69) reminded me of aquatic plants and my experiences studying the royal water lily. Catesby depicted *Sarracenia flava*, and this is the type specimen upon which the name is based.[12] The royal water lily (*Victoria amazonica*) (figure 21-14) was named after Queen Victoria. The white flowers open as

FIGURE 21-14. The royal water lily (*Victoria amazonica*). (© Royal Botanic Gardens, Kew.)

darkness falls. It is magical to sit in a canoe on a lake and watch these star-like flowers open around you. When the flowers open they are strongly scented, and the temperature inside the central cavity is about eleven degrees warmer than ambient temperature. Scarab beetles (*Cyclocephala* spp.) are attracted to the flowers and enter into the warm central cavity. They are quite contented there, because there is a circle of knob-like paracarpels full of starch and sugar for them to eat. By midnight the flowers close up again, and the beetles are trapped inside, warm and feeding on the paracarpels. During the next day, the flowers gradually change color from white to deep purple-red. On the second evening, the flowers are no longer scented and are at ambient temperature, and they reopen as darkness returns. At this time, pollen is released by the stamens, which are arranged in a circle around the exit. The beetles emerge, sticky from the plant juices below. Since the water lily produces flowers every second day, the beetles will fly on to a different plant, thereby effecting cross-pollination.[13]

One important aspect that I have in common with Mark Catesby is that he documented his work with herbarium specimens, many of which survive today. These are so important as a record of this sort of work. I hope that this review of a few aspects of the natural history of the Amazon, inspired by Catesby's plants of the Bahamas and Carolinas, will encourage us to continue our observations of the natural world around us. These observations are still vitally important to our understanding in this age, when excessive reductionism is common. Observations about the interactions between plants and animals are extremely important tools for developing the conservation of natural ecosystems.

22

Conclusions: Account, Appendix, *Hortus*, and other endings among Mark Catesby's work

DAVID J. ELLIOTT

In the middle decades of the eighteenth century, authors commonly concluded their books with "Finis" to signal their end, but Mark Catesby's "Finis" at the bottom of the text page accompanying the one-hundredth plate (portraying *Catesbaea*) in the second volume of *The natural history of Carolina, Florida and the Bahama islands* was, like some other aspects of his work, misleading. It was not the end.

Even though Mark Catesby developed an early interest in searching for plants "and other productions in nature," he was frustrated by living too far from London, "The Center of All Science."[1] While the distance from his childhood home to that city was less than seventy miles, he traveled well over twenty thousand miles by sailing ship, sloop, and periagua and on foot to reach his goal. His travels started when he was twenty-nine and ended when he was forty-three. These explorations, however, involved far more than merely drawing and describing the fauna and flora he encountered in Virginia, Bermuda, Jamaica, South Carolina, the northeastern corner of Georgia (then called Florida), and the Bahamas. He illustrated the area in *The natural history of Carolina*... with a map; by then the British colony of Georgia had been established (figure 22-1).

An ACCOUNT of *CAROLINA*, AND THE *BAHAMA* Islands

Usually bound at the front of volume I, "An Account..." was not published until after the tenth part of *The natural history of Carolina*... was issued in 1743. It demonstrated that Catesby's curiosity was pervasive and far from limited to simply illustrating and describing the new and strange plants and animals he encountered. He expressed his philosophy in the section of "An Account" dealing with sturgeons:

> Speculative Knowledge in Things meerly curious, may be kept secret without much Loss to Mankind. But the concealing Things of real Use is derogating them

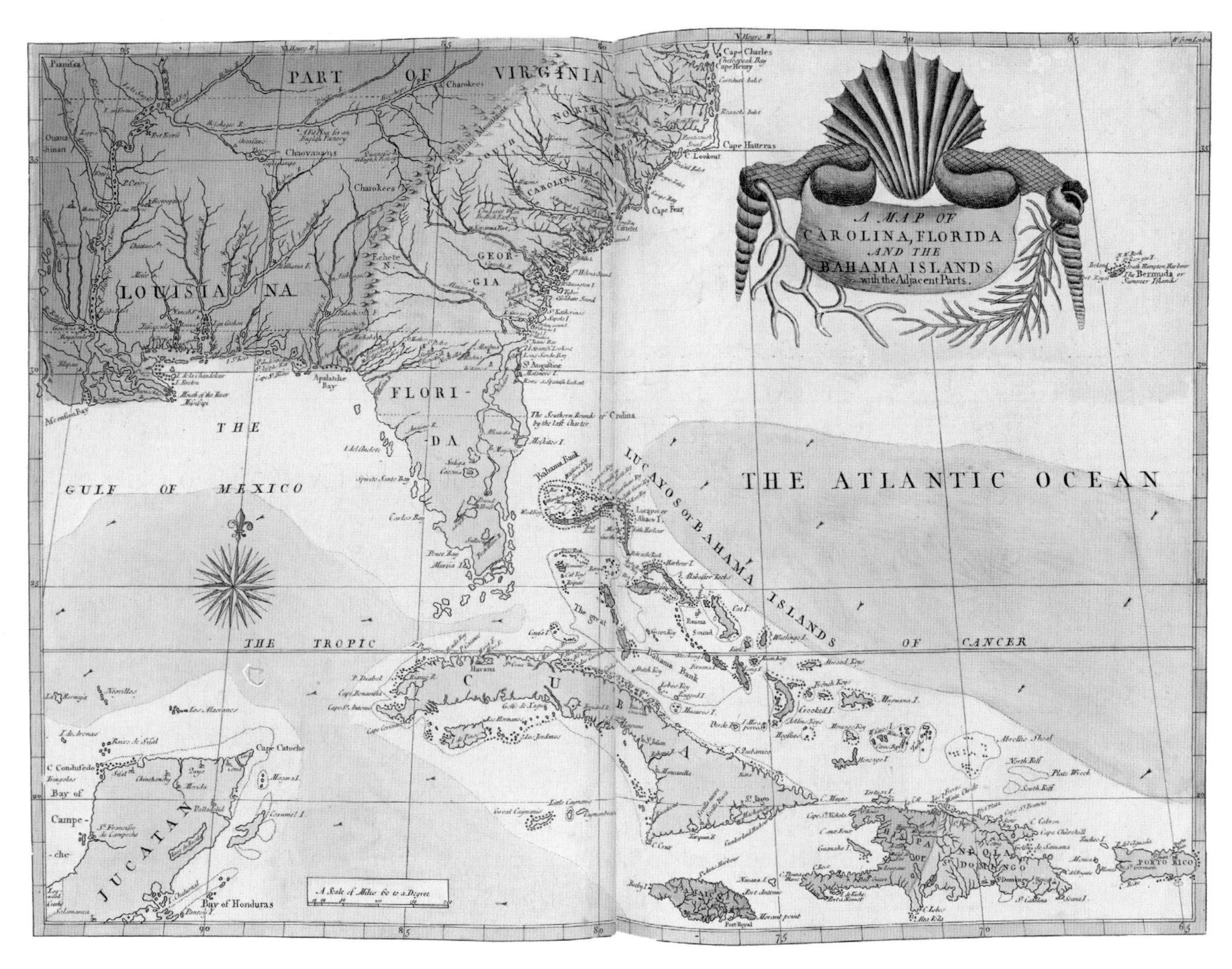

FIGURE 22-1. A map of Carolina, Florida, and the Bahama Islands with adjacent parts; M. Catesby, 1743, *The natural history of Carolina . . .* , volume II. (Digital realization of original etchings by Lucie Hey and Nigel Frith, DRPG England; courtesy of the Royal Society ©.)

from the Purposes we were created for, by depriving the Publick of a Benefit designed them by the Donor of all Things.[2]

Catesby's interest in anthropology was demonstrated in his general discussions of the American Indians, discussions that, he noted, were a mixture of his own and John Lawson's studies and his own observations, as well as specific comments about the economic, medicinal, and nutritional uses of various plants and animals. He deduced from Japanese maps belonging to Sir Hans Sloane that the Native peoples of the Americas derived from Asian Tartars and had come from Asia by island-hopping across the Aleutian Islands, which was an early view of their origins.[3] While the Bering Strait had been discovered in 1728, knowledge of it was not widespread, and the possibility of an ice- or land-bridge linking Asia and North America came later. In contrast, his views

FIGURE 22-2. Ornamental headpiece by Mark Catesby to "An ACCOUNT of *CAROLINA*, AND THE *BAHAMA* Islands"; M. Catesby, 1743, *The natural history of Carolina* . . . , volume II. (Courtesy of Smithsonian Institution Libraries and the Biodiversity Heritage Library.)

on the cultural homogeneity of American Indian culture and his disbelief in Spanish writers' descriptions of their attainments in statuary and architecture in Mexico were in error.[4]

While Catesby apologized for the digression of his "Account" in *The natural history of Carolina* . . . , he nevertheless headed it with his signed, but generally unnoticed, etching of the head of an American Indian accompanied by Native weapons, including stone-tipped arrows, a spear, and a tomahawk, as well as a quiver with a bison's tail, a long pipe (calumet and bowl), a ball-headed club, a clam-shell hoe, and a necklace probably of glass beads.[5] The surrounding decoration owes a little to corn (maize), their principal crop (figure 22-2).[6] Notably missing is what had become the American Indians' weapon of choice by Catesby's day, the musket. These were obtained from the colonists' traders in exchange for deerskins and captives from wars with other tribes.

"An Account . . ." provides a broad and deep view not only of the geography, topography, natural history, and manufactures of Carolina and the Bahama Islands with occasional diversions into Virginia, New England, and even the Arctic but also of what might usefully be produced there. The "Publick" for whom Catesby was writing was apparently British. For example, drawing on his 1714 visit to Jamaica, he recommended restoration of the Spanish-planted cocoa plantations that had been abandoned after the island's capture by the British. He suspected that cocoa produced there would be useful not only for home consumption but also for export markets.[7] He also suggested that conquest of Cuba by the British would be beneficial.[8] While this was indeed done, albeit temporarily, two decades later, nothing has been found to suggest it was instigated on Catesby's recommendation.

Catesby's thoughts on how birds of the same species occurred on both sides of the Atlantic Ocean suggest he would not be surprised by current understanding of continental drift and the movement of tectonic plates.[9] He was also intrigued by changing levels of the ocean, commenting, for example, on the discovery in Virginia of fossilized shark teeth about seventy feet down a well more than one hundred miles from the coast.[10] He also noted the excavation of what he believed to be elephant teeth at Stono near the Lowcountry coast, an opinion shared by the African slaves who had discovered them.[11] While the teeth of Siberian mammoths, thought to be elephants until 1796, had been studied by Sloane, this may well have been the first recognition of the mammoth in North America.[12] And Catesby commented on a dead "Grampus" (possibly a pygmy sperm whale (*Kogia breviceps*) since whales periodically beach themselves on the South Carolina coast) on the shore of the North Edisto River. The carcass was covered by sand in less than a month, reflecting the continuing movement of sand in the Lowcountry estuaries.[13]

While by no means an ardent conservationist, Catesby showed concern about wasting natural resources. He emphasized that, with a few exceptions, he only painted living birds.[14] And, after catching sixteen sturgeon above Fort Moore, he regretted leaving most of them behind to rot.[15] While then abundant, with Catesby seeing an opportunity for pickling and exporting them to the West Indies, the short-nose sturgeon (*Acipenser brevirostrum*) is now officially an endangered species.[16] It appears that he was the first, at least for North America, to express concern for the effects of habitat degradation on the survival of a species.[17] He wrote to William Sherard:

> I repent of my having Spake of and given you expectations of Bulbs perticularly of one very remarkable for its' Beauty I have gone in Search and have done what possibly I could to procure it but have as yet been unsuccesfull. in less than 20 years they were seen within the Inhabitants but have since that time dissappeared. Supposed to be Cattle or Hogs that root y^{m} up.[18]

The Appendix

In 1743 Catesby advised his subscribers that *The natural history of Carolina . . .* was complete and that they could proceed to have the ten parts bound into volumes. However, he also advised them that this should be done in a way that permitted the addition of an Appendix composed of an additional twenty plates with accompanying text.[19] Completing the Appendix took him another four years, which is not surprising, since he was by then an old man by the standards of the time. An "Advertisement" (see figure 12-6, p. 164) listed the insects and other animals and plants proposed for inclusion. However, while most were eventually included, some were not, such as the "Penguin," "a beautiful Duck from Newfoundland," and a species of "Auk." Several plants were either replaced or re-named.

In addition to unused watercolors of his own that Catesby had brought from America, the Appendix covered various plants and animals he had received since then. Mostly they were from John Clayton in Virginia and John Bartram in Pennsylvania.[20] However, the final plate of the rose locust (*Robinia viscosa*) was prepared using samples from a tree growing in the Exmouth garden of Sir John Colleton, formerly a Lord Proprietor who owned Fair Lawn Barony in South Carolina.[21] Some were grown in Collinson's garden in Peckham and Catesby's own in Fulham, but how they came to be there was not always specified (figure 22-3).[22] Two were accompanied by descriptions by John Mitchell (1711–1768), a Virginian doctor, botanist, and mapmaker who had contacts with Collinson. For reasons of health, Mitchell moved to London in 1746 (shortly before Catesby completed the Appendix). He was elected a Fellow of the Royal Society on 15 December 1748, and Mark Catesby was one of his sponsors for this honor.[23]

FIGURE 22-3. Lamb laurel or sheep laurel (*Kalmia angustifolia*) (*left*) and great laurel (*Rhododendron maximum*) (*right*); plate 17, M. Catesby, 1743, *The natural history of Carolina* . . . , volume II: Appendix. (Courtesy of Smithsonian Institution Libraries and the Biodiversity Heritage Library.)

Mark Catesby also covered insects, which, other than some butterflies and moths, hitherto he had largely ignored. One of these, the chigoe (*Tunga penetrans*, *NH* II: Appendix, plate 10), is a sand flea that he encountered in the Bahamas. It is one of the parasitic arthropods, which are commonly found in subtropical regions, and Catesby accurately described it as having particularly obnoxious parasitic characteristics. Breeding females burrow into exposed skin on the feet of mammals, and during the two weeks they remain while the eggs develop, they swell dramatically, sometimes causing intense irritation (a condition called tungiasis). If the flea is left within the skin, dangerous complications can occur, including secondary infections.[24] Governor George Phinney had been searching his feet for chigoes, which Catesby then examined. The minute size of these insects suggests that he drew them, along with another tiny insect that he did not identify (it remains unnamed), with the aid of a microscope while he was in the Bahamas.[25]

Catesby's statement in the "Advertisement" that "This additional Part of the Collection consists of such curious Subjects, that for the Reputation of the Work, I am loath to omit, and for no other Reason" (see figure 12-6, p. 164) appears somewhat disingenuous. In some cases Catesby expressed the reasons, in others they are apparent, and in the remainder they require speculation. Overall, the Appendix does not follow the style and approach of the main book and should be considered with these facts in mind. For example, the depiction of the "Cacao Tree" (*Theobroma cacao*) (figure 22-4), based on an original watercolor by Claude Aubriet, is essentially an illustrated reiteration of Catesby's argument in the Account that Britain should restore its cacao plantations in Jamaica, supported by a background quotation from William Dampier.[26] "Vanelloe" (*Vanilla mexicana*, *NH* II: Appendix, plate 7) (figure 21-12, p. 311), likewise by Aubriet and which came next, is another tropical plant and not native to the area covered by *The natural history of Carolina*. . . . It was, however, noted as being used by the Spanish for flavoring chocolate.[27] A further example is the "Cushew Tree" (*Anacardium occidentale*, NH II: Appendix, plate 9), which Catesby reported was native to Jamaica, Hispaniola, and other Caribbean islands. Commenting on its edibility, he also took the opportunity to comment that Maria Sybilla Merian had portrayed it incorrectly.[28]

The "American Swallow" (*Chaetura pelagica*, NH II: Appendix, plate 8), shown on a nest within a cross-section of a brick chimney, appears to have been included to support Catesby's theory of bird migration, then being actively debated in scientific circles.[29] There is no mention of the relevance of its pairing with the wood lily (*Lilium philadelphicum*) beyond the opportunity to note that it had flowered in Collinson's garden during 1743, one of a number of references to Collinson in the Appendix.

The "Viper-mouth" (Sloane's viperfish, *Chauliodus sloani*, *NH* II: Appendix, plate 19) (figure 22-5) may be the most unattractive species illustrated in the Appendix but it is also one of its several geographic anomalies. A specimen

FIGURE 22-4. Cacao (*Theobroma cacao*); plate 6, M. Catesby, 1743, *The natural history of Carolina . . .* , volume II: Appendix. (Digital realization of original etchings by Lucie Hey and Nigel Frith, DRPG England; courtesy of the Royal Society ©.)

FIGURE 22-5. Spiny catfish (*Acanthodoras cataphractus*) (*upper*) and viperfish (*Chauliodus sloani*) (*lower*); plate 19, M. Catesby, 1743, *The natural history of Carolina* . . . , volume II: Appendix. (Digital realization of original etchings by Lucie Hey and Nigel Frith, DRPG England; courtesy of the Royal Society ©.)

of this fish was sent from Gibraltar by Catesby's younger brother John, an officer in Grove's Regiment of Foot, which was stationed there.[30] He had obtained his commission, probably by purchase for around £500,[31] after obtaining his majority and Mark's return from Virginia. The regiment was transferred to Ireland in 1739, but whether John went with it or what happened to him has not been discovered. Re-named the 10th Regiment of Foot in 1751, this regiment engaged in the battles of Concord and Lexington, as well as other battles during the American Revolution.[32]

FIGURE 22-6. (*Opposite*) "Scolopendra" and Virginian witch-hazel (*Hamamelis virginiana*); plate 2, M. Catesby, 1743, *The natural history of Carolina* . . . , volume II: Appendix. (Digital realization of original etchings by Lucie Hey and Nigel Frith, DRPG England; courtesy of the Royal Society ©.)

Among other species from outside the geographic range of *The natural history of Carolina* . . . are the "Yellow and Black Pye" (*Icterus icterus*; troupial) of Jamaica, which was depicted apparently about to eat a wasp of indeterminate North American origin, together with a "Lilio-Narcissus," which Catesby had seen near the Savannah River in Georgia.[33] "Scolopendra," also from Jamaica, was a centipede with a bite as painful as that of a scorpion (figure 22-6).[34]

Closer to the geographical limits of the first two volumes of *The natural history of Carolina* . . . is the red-billed tropicbird (*Phaëthon aethereus*), which

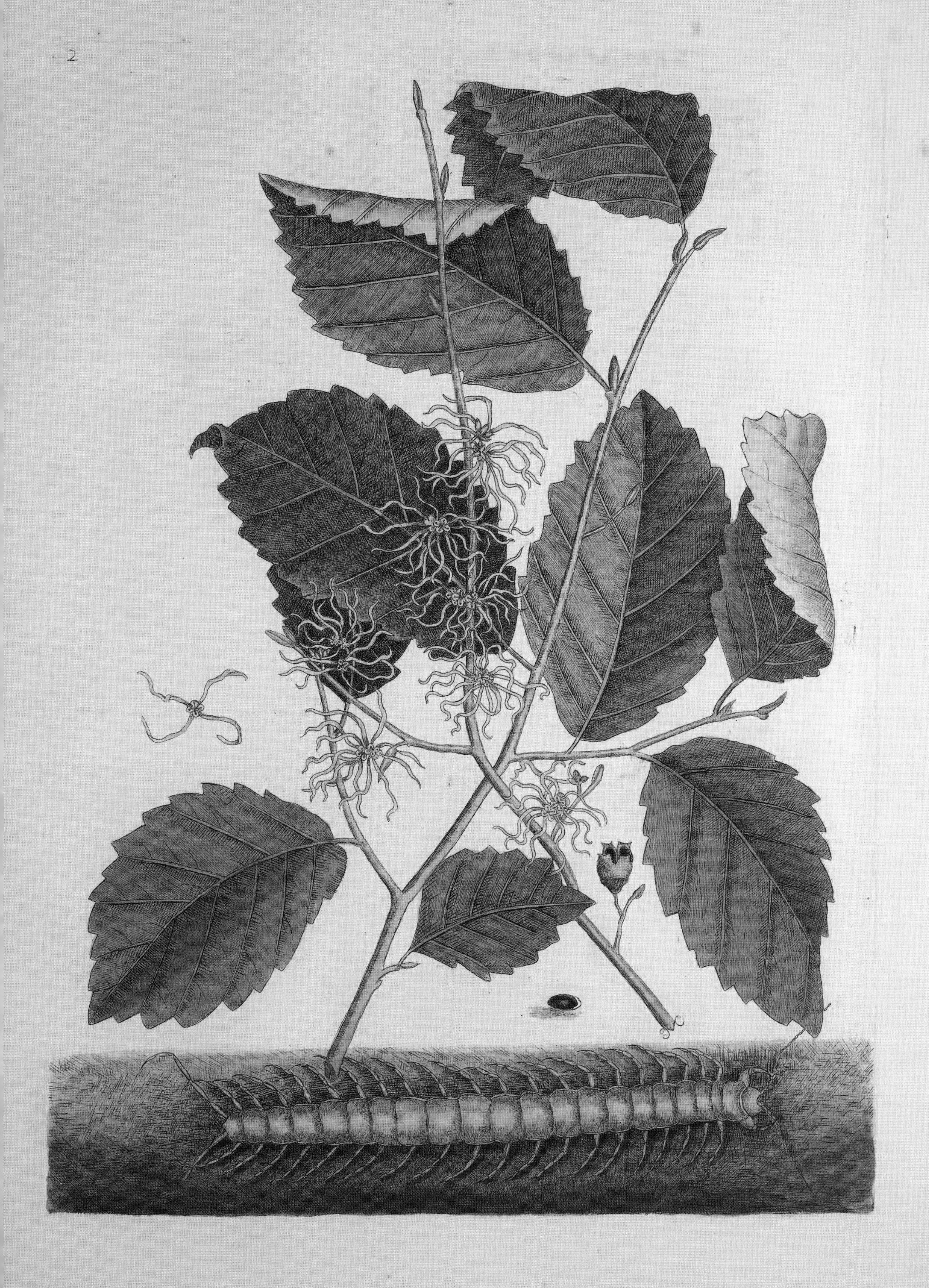
2

Catesby had seen in Bermuda and near Puerto Rico and which he believed was not usually found outside the tropics. He disagreed with Willughby's description, which had been derived from a dead specimen, stating that his description was of a living bird.[35]

Catesby also wrote a detailed report in the Appendix on the effects he felt after taking ginseng ("or Ninsin of the Chinese"); he reported that these effects were apparently little known in Europe. The illustration (*NH* II: Appendix, plate 16) is of the North American ginseng (*Panax quinquefolius*) that had been grown successfully in Collinson's Peckham garden since before 1746.[36] On the same plate Catesby illustrated the "Whip-poor-Will" (common nighthawk; *Chordeiles minor*), which concluded his depiction of land birds, making the total 113. He thought it possible that some more birds might be discovered in North America, but probably not many.[37]

Some of the other items in the Appendix in one way or another recognize a select few of Catesby's major supporters. Sir Hans Sloane, who had been a sponsor of Catesby's trip to South Carolina, purchased five copies of *The natural history of Carolina* . . . and provided Catesby with access to his collections. Sloane was mentioned several times and was honored in the Preface.[38] Notably, Sloane's collection was termed a "Musæum," an appropriate name for what was the basis of the British Museum in 1759.[39]

Dr Richard Mead, FRS, another patron of the expedition to South Carolina, a subscriber to *The natural history of Carolina* . . . , and a physician to King George II, had the genus *Meadia* named after him in recognition by Catesby of Mead's general support of the arts and sciences and "in particular for his generous assistance for carrying the general design of this work into execution."[40]

FIGURE 22-7. Richard Mead. (Reproduced by courtesy of the National Library of Medicine, U.S. National Institutes of Health.)

What Mead had done to earn this recognition by Catesby above that of Sloane and Collinson is not clear (figure 22-7).

Accompanying *Meadia* (now *Primula meadia*, pride-of-Ohio) was the greater prairie chicken (*Tympanuchus cupido*; *NH* II: Appendix, plate 1), which Catesby had seen in the Chiswick garden of another of his subscribers, the Earl of Wilmington (Spencer Compton, c. 1674–1743). Wilmington, who died before the Appendix was issued, told Catesby the bird had come from America, but he did not know the precise origin. From February 1742 until his death, Wilmington was the First Lord of the Treasury and Prime Minister of Great Britain, although merely a figurehead, because Lord Carteret, who had introduced Catesby at Court fourteen years before, was really in control.

The other person after whom Catesby named a plant was John Stuart, Earl of Bute (1713–1792), who was also a subscriber. The history of Catesby's "*Steuartia*" is very convoluted, for Carl Linnaeus must be given credit for first publishing the name in 1746, and his account was accompanied by an engraving based on a drawing by Georg Ehret (figure 22-8).[41] The plant illustrated by Catesby had been provided by John Clayton from Virginia and blossomed three months after it arrived in England in May 1742 in Catesby's Fulham garden. Given that

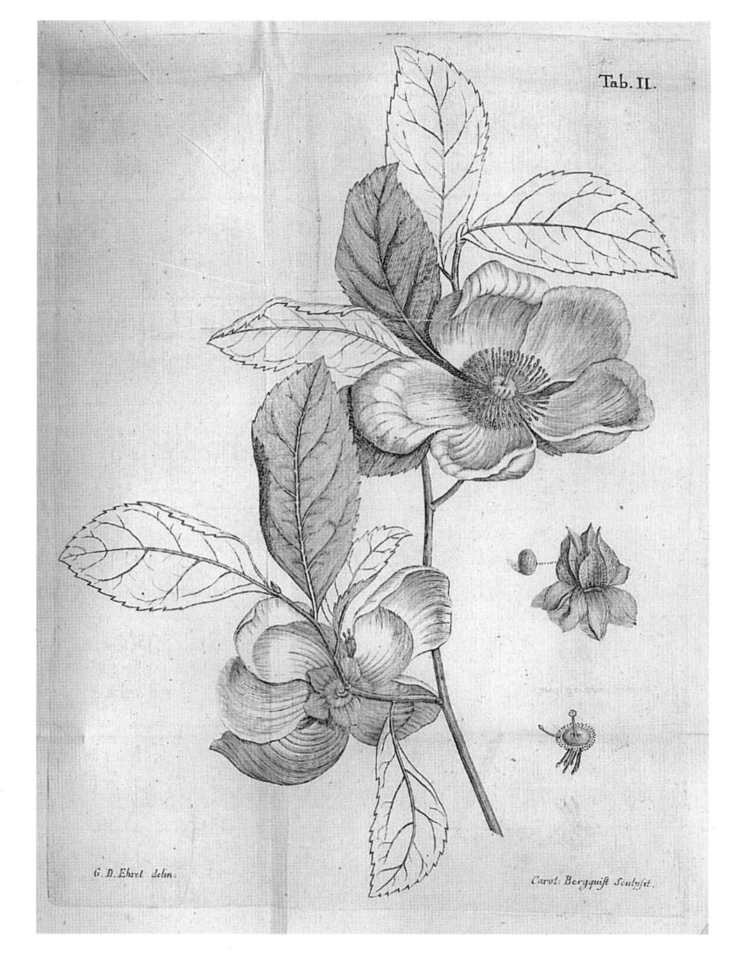

FIGURE 22-8. "*Steuartia*"; engraving after Georg Ehret from Carl Linnaeus, Decem plantarum genera, *Societas Regia Scientiarum Upsaliensis. Acta*, 1741 (published 1746). (By permission of the Linnean Society of London.)

Ehret's and Catesby's images are identical, it seems likely that Catesby took Ehret's drawing as the basis for the engraving he published (*NH* II: Appendix, plate 13). Immensely wealthy, an enthusiastic student of botany, and a benefactor of scholars, artists, and universities, Bute much later became Prime Minister.[42] However, at this time Bute was a politically obscure figure in Scotland, so why Catesby named a genus after him is a mystery.

While Catesby hoped that Bute would not object to his naming a plant for him, Charles Lennox, second Duke of Richmond (1701–1750), specifically suggested that Catesby include the "Java Hare," which from its decorative collar was very likely a pet.[43] There was reason to heed this request. Richmond, a grandson of King Charles II, was elected a Fellow of the Royal Society in 1724, had subscribed for two copies of *The natural history of Carolina . . .* , and from January 1735 was a member of the then powerful Privy Council. It has not been determined whether the Duke and Duchess of Richmond were involved in Catesby's presentation at Court in 1729, but both were important members of the royal household at the time, having been appointed a lord and a lady of the bedchamber (figure 22-9).[44]

FIGURE 22-9. Portrait of the Duke and Duchess of Richmond by Jonathan Richardson at the request of his grandmother Louise de Kérouaille, Duchess of Portsmouth. (By permission of the trustees of the Goodwood Collection.)

Catesby concluded the Appendix and *The natural history of Carolina . . .* by emphasizing the importance of field observation and the drawing of living plants and animals. He commented that the extended time the book took to produce was because "The whole work was done within my house, and by my own hands."[45] In a narrow sense, this is too strong a claim; for example, we know that a few of the illustrations were by Ehret, and some were developed from work by Kickius and others.[46] In addition, much of his text on American Indians came from Lawson (with credit, as Catesby generally gave to other writers).[47] In a broader sense, however, his claim appears to be appropriate.

Hortus britanno-americanus: or, a curious collection of trees and shrubs

Following his return to England, Catesby corresponded with John Clayton in Virginia and, at first through Peter Collinson and later directly, with John Bartram in Philadelphia, primarily to obtain plants that he could have grown in England and that would provide information for illustrating *The natural history of Carolina. . . .*[48] By the mid-1730s the focus of his horticultural interests had shifted westward from Fairchild's, later Bacon's, nursery in Hoxton to that of Christopher Gray in Fulham on the opposite side of London. Exactly why this change occurred is unclear, but it happened after Bacon died in 1734 and his wife sold the nursery to a John Sampson. There is no indication that Catesby was involved with Sampson or that Sampson had any interest in North American plants.

Gray had a nursery of nearly thirty acres on the King's Road in Fulham. His interest in North American plants apparently started in 1713, when he purchased some from the garden at Fulham Palace of the lately deceased Bishop of London, Henry Compton. Compton's garden included many North American plants. Catesby's connection with Gray was clearly established by 1737, and soon thereafter Gray issued a one-page catalog advertising American trees and shrubs that "will endure the Climate of England." This broadsheet, illustrated with Catesby's engraving of *Magnolia grandiflora*, is a masterpiece of etching, since every word had to be inscribed in reverse in order to be printed legibly (see figure 7-6, p. 94).[49]

Catesby was deeply committed to the introduction of hardy North American plants into England for both their utilitarian and their aesthetic values. The expression of his work in this field concluded in his posthumous book *Hortus britanno-americanus* (hereafter *Hortus*[50]), which, while completed after August 1748, was not published until thirteen years after he died. Why the delay and how it came to be published are among the enduring mysteries about Catesby. The book's first publisher, in 1763, was John Ryall of Hogarth's Head, Fleet Street, London, who also published, inter alia, the proceedings of the Old Bailey Criminal Court and diverse engravings and caricatures. As far as is known, he

FIGURE 22-10. In most copies of *Hortus britanno-americanus*, the etching of *Magnolia grandiflora* has a black background. Unusually, in the copy at the University of South Carolina the background was not painted black, demonstrating the variability in hand-colored eighteenth-century books. (© David J. Elliott.)

was not involved in publishing any parts or edition of *The natural history of Carolina. . . .* Ryall dedicated Catesby's *Hortus* to Henry Seymer of Hanford, Dorset, the dedication being dated 2 May 1763. Henry Seymer (1714–1785) and his family have no identifiable connection to Catesby; for example, they were not among the "Encouragers" of *The natural history of Carolina. . . .* They were, nevertheless, enthusiastic gardeners and amateur naturalists. A beautifully illustrated but unpublished essay on plants and butterflies, compiled by Henry Seymer between 1755 and 1783, included paintings copied from Catesby's *The natural history of Carolina. . . .*[51]

Catesby's *Hortus* contained discussions of eighty-five "Curious Trees and Shrubs" suitable, as the title-page stated, for cultivation in England. Catesby's proudest horticultural achievement, "Magnolia altissima . . . The Laurel-tree of Carolina" (*Magnolia grandiflora*, southern magnolia), was given a whole page (figure 22-10), while sixty-two more plants were illustrated, generally four to a page.[52] In two instances there are five plants on a page: in one the fifth is simply a nut, and in the other the leaves of two species of oak are shown on opposite sides of a single common mid-vein (figure 22-11).[53] The etchings were usually revised extracts from *The natural history of Carolina . . .* , with ten bearing Catesby's monogram. Catesby noted:

FIGURE 22-11.
Five species of oak; facing p. 6, M. Catesby, *Hortus europae-americanus*, 1767. (© David Elliott.)

> Few people have opportunities of procuring these things from America; wherefore, lest I should seem to treat of what cannot be got at all, or with very great difficulty, it seems proper to mention, that Mr. Gray at Fulham has for many years made it his business to raise and cultivate the plants of America (from whence he has annually fresh supplies) in order to furnish the curious with what they want; and that through his industry and skill a greater variety of American forest-trees and shrubs may be seen in his gardens, than in any other place in England.[54]

Four years later, in 1767, *Hortus* was re-issued by J. Millan, with a re-set title-page and amended title: *Hortus europae americanus: or, a collection of 85 curious trees and shrubs, the produce of North America; adapted to the climates and soils of Great Britain, Ireland, and most parts of Europe, &c. Together with their blossoms, fruits and seeds; observations on their culture, growth, constitution and virtues. With directions how to collect, pack up and secure them in their passage.* Otherwise it was identical—even with Ryall's dedication, although he was no longer the publisher, and the heading of the section discussing the plants retained the previous title.

Conclusion

During the course of a century, Mark Catesby had both developed new knowledge and applied it to functional purposes. The next century saw further technological development, such as the railroad and the steamship and, of course, Charles Darwin's *On the origin of species*, published in 1859. Darwin had read "Catesby Nat Hist of Carolina" and made notes from it about the

"Tumble-Turds" (*NH* II: Appendix, plate 11; see figure 4-13, p. 54).[55] The following century brought us, for example, the theory of relativity, the jet aircraft, antibiotics, and space rocketry that enabled man to land on the moon. It also brought George Frick's and Raymond Stearns's landmark biography *Mark Catesby: the colonial Audubon* in 1961. Desk-top computers and the Internet, which have made the present volume possible, were still in the future.

What will be developed over the next century and, specifically, how it may relate to Catesby are unknown, but while thinking about possibilities is speculative (something we have generally avoided), there is one that excites the imagination. Advances in forensic techniques for analyzing minute amounts of inorganic and organic material, linked with the developing understanding of DNA in the cells of plants and animals, have made it possible to use antique herbarium samples that are preserved in such places as the Royal Botanic Gardens, Kew, to identify the specific strain of the blight fungus that caused the Irish potato famine.[56] What, for example, are the possibilities with the plant specimens that Catesby collected? Might those with identified medicinal properties be studied in pharmaceutical research? Could we find out exactly where they were gathered by analyzing the dust particles trapped on the leaves? Might fingerprints left on the hand-colored plates of his books tell us if his wife was his colorist too?

APPENDIX

Identification of the plants and animals illustrated by Mark Catesby for *The natural history of Carolina, Florida and the Bahama islands*

JAMES L. REVEAL

This two-part appendix provides up-to-date identifications of the plants and animals illustrated by Mark Catesby in *The natural history of Carolina, Florida and the Bahama islands* and also those depicted on the associated watercolors now preserved in the Royal Library, Windsor Castle.[1]

Attempts to associate Catesby's images with modern scientific names began with his friend and editor George Edwards who published a catalog of names in each of the two volumes of the third edition of Catesby's *Natural history . . .* ,[2] and such efforts have continued up to the present day. The most recent efforts, since the early 1970s, by Joseph Andorfer Ewan,[3] Richard A. Howard and George W. Staples,[4] Robert L. Wilbur,[5] and the present author,[6] have resolved the identity of nearly all Catesby's organisms. Still, some remain questionable and, as may be seen in this listing, a few are still not assigned any modern name. Taxonomic disagreements among experts, especially regarding the identity of some of Catesby's fishes and snakes, exist and are bound to continue. Likewise, historical confusion among some of the birds, one of the best studied group,[7] failed to note that for two and a half centuries the blue-winged warbler was nomenclaturally confused with Catesby's pine warbler, meaning that the former, known to science since 1763, was actually not given a modern name until 2009.[8] Thus, what is done here is just another report, a report that will no doubt be subject to corrections in the future.

PART I. Plants and animals depicted in the published etchings arranged numerically by plate number and volume

Names of plants are in bold.

Plate Number	*Scientific Name*	*Common Name*	*RL inventory no.*
Volume 1. Part 1 (May 1729)			
1	*Haliaeetus leucocephalus*	bald eagle	24814
	Mugil cephalus	gray mullet	24814
	Pandion haliaetus [background]	osprey	24814
2	*Pandion haliaetus*	osprey	24815
	Mugil cephalus	gray mullet	24815
3	*Falco columbarius*	merlin	24816
4	*Elanoides forficatus*	swallow-tail kite	24817
5	*Falco sparverius*	American kestrel	24818
6	*Cathartes aura*	turkey vulture	24819
7	*Megascops asio*	eastern screech-owl	24820
8	composite of three birds		24821
	(*Chordeiles minor*	(common nighthawk	
	+ *Caprimulgus carolinensis*	+ chuck's-will-widow	
	+ *Caprimulgus vociferus*)	+ whip-poor-will)	
	Neocurtilla hexadactyla	northern mole cricket	24821
9	*Coccyzus americanus*	yellowbilled cuckoo	24822
	Castanea pumila	**Allegheny chinkapin, chinquapin**	24822
10	*Amazona leucocephala*	Cuban parrot	24823
	Colubrina elliptica	**smooth snake-bark, soldierwood**	24823
11	*Conuropsis carolinensis*	Carolina parakeet	24824
	Taxodium distichum	**bald cypress, swamp cypress**	24824
12	*Quiscalus quiscula*	common grackle	24825
13	*Agelaius phoeniceus*	red-winged blackbird	24826
	Morella caroliniensis	**southern bayberry**	24826
14	*Dolichonyx oryzivorus*	bobolink	24827
	Oryza sativa	**rice**	24827
15	*Cyanocitta cristata*	bluejay	24828
	Smilax laurifolia	**laurel greenbrier**	24828
16	*Campephilus principalis*	ivory-billed woodpecker	24829
	Quercus phellos	**willow oak**	24829
17	*Dryocopus pileatus*	pileated woodpecker	24830
	Quercus virginiana	**live oak**	24830
18	*Colaptes auratus*	northern flicker	24831
	Quercus michauxii	**swamp chestnut oak**	24831
19	*Melanerpes carolinus* [left]	red-bellied woodpecker	24832
	Picoides villosus [right]	hairy woodpecker	24832
	Quercus marilandica	**blackjack oak**	24832
20	*Melanerpes erythrocephalus*	red-headed woodpecker	24833
	Quercus nigra	**water oak**	24833
	Mitchella repens [lower]	**partridgeberry**	24833

Plate Number	*Scientific Name*	*Common Name*	*RL inventory no.*
Volume 1. Part 2 (January 1730)			
21	*Sphyrapicus varius* [left]	yellow-bellied sapsucker	24834
	Picoides pubescens [right]	downy woodpecker	24834
	Quercus alba [left + acorn on right]	**white oak**	24834
	Quercus falcata [right]	**southern red oak**	24834
22	*Sitta pusilla* [upper]	brown-headed nuthatch	24835
	Sitta carolinensis [lower]	white-breasted nuthatch	24835
	Quercus incana	**bluejack oak**	24835
23	*Ectopistes migratorius*	passenger pigeon	24836
	Quercus laevis	**(American) turkey oak**	24836
24	*Zenaida macroura*	mourning dove	24837
	Podophyllum peltatum	**mayapple**	--
25	*Patagioenas leucocephala*	white-crowned pigeon	24838
	Chrysobalanus icaco	**coco plum**	24838
26	*Columbina passerina*	common ground-dove	24839a & 24839b
	Zanthoxylum clava-herculis	**Hercules' club**	24839a
27	*Mimus polyglottos*	northern mockingbird	24840
	Cornus florida	**(eastern) flowering dogwood**	24840
28	*Toxostoma rufum*	brown thrasher	24841
	Prunus virginiana	**choke cherry**	24841
	Prunus serotina	**black cherry**	--
29	*Turdus migratorius*	American robin	24842
	Aristolochia serpentaria	**Virginia dutchmanspipe, Virginia snakeroot**	24842
	(probably ***Carya ovata***)	**(shagbark hickory)**	24842
30	*Turdus plumbeus*	red-legged thrush	24843
	Bursera simaruba	**gumbo limbo, gum-elemi**	24843
31	(perhaps *Catharus guttatus*)	(hermit thrush)	24844
	Ilex cassine	**dahoon**	24844
32	*Eremophila alpestris*	horned lark	24845
	Uniola paniculata	**(North American) sea oats**	--
33	*Sturnella magna*	eastern meadowlark	24846
	Hypoxis hirsuta	**common goldstar**	25892
34	*Pipilo erythrophthalmus* [upper]	eastern towhee	24847
	Molothrus ater [lower]	brown-headed cowbird	24847
	Populus heterophylla	**swamp cottonwood**	24847
35	*Spizella passerina*	chipping sparrow	24848
	Hypercompe scribonia	giant leopard moth	26077
	Ipomoea sagittata	**saltmarsh morning-glory**	24849
36	***Monotropa uniflora*** [left]	**one-flower Indian-pipe**	24850
	Junco hyemalis [centre]	dark-eyed junco	24855
	Geoglossum glabrum [right]	**black adder tongue fungus**	24850
37	*Tiaris bicolor*	black-faced grassquit	24851
	Tabebuia bahamensis	**white dwarf tabebuia, beef-bush**	24851
38	*Cardinalis cardinalis*	northern cardinal	24852
	Carya tomentosa [branch & large nut]	**mockernut hickory**	24852
	Carya glabra [small nut]	**pignut, pignuthickory**	24852
39	*Passerina caerulea*	blue grosbeak	24853
	Magnolia virginiana	**sweetbay**	24853
40	*Loxigilla violacea*	greater Antillean bullfinch	24854
	Metopium toxiferum	**Florida poisontree, poisonwood**	24854

Plate Number	*Scientific Name*	*Common Name*	*RL inventory no.*
Volume 1. Part 3 (November 1730)			
41	*Carpodacus purpureus*	purple finch	24855
	Nyssa sylvatica	**black gum, black tupelo**	24855
42	*Spindalis zena*	stripe-headed tanager	24856
	Jacaranda caerulea	**boxwood, cancer tree**	24856
43	*Carduelis tristis*	American goldfinch	24857
	Gleditsia aquatica	**water honeylocust, water locust**	24857
44	*Passerina ciris*	painted bunting	25875
	Gordonia lasianthus	**loblolly bay**	--
45	*Passerina cyanea*	indigo bunting	25877
	Trillium catesbaei	**bashful wakerobin**	25876
46	*Bombycilla cedrorum*	cedar waxwing	25878
	Calycanthus floridus	**Carolina allspice, eastern sweetshrub**	25877
47	*Sialia sialis*	eastern bluebird	25879
	Smilax pumila	**sarsaparilla vine**	25879
48	*Icterus galbula*	Baltimore oriole	25880
	Liriodendron tulipifera	**tulip poplar, tulip tree**	25880
49	*Icterus spurius*	orchard oriole	25881
	Catalpa bignonioides	**common catalpa, southern catalpa**	25881
50	*Icteria virens*	yellow-breasted chat	25882
	Trillium maculatum	**spotted wakerobin**	25883
51	*Progne subis*	purple martin	25884
	Cocculus carolinus	**Carolina coralbead, Carolina moonseed**	--
52	*Myiarchus crinitus*	great crested flycatcher	25886
	Smilax tamnoides	**bristly greenbrier**	25886
53	*Sayornis phoebe*	eastern phoebe	25886
	Gelsemium sempervirens	**Carolina jessamine, false jasmine**	25887
54	*Contopus virens* [upper]	eastern wood-pewee	25888
	Vireo olivaceus [lower]	red-eyed vireo	25888
	Symplocos tinctoria	**common sweetleaf**	25888
55	*Tyrannus tyrannus*	eastern kingbird	25890
	--	unidentified insect	
	Sassafras albidum	**sassafras**	25890
56	*Piranga rubra*	summer tanager	25891
	Platanus occidentalis	**American sycamore**	25891
57	*Baeolophus bicolor*	tufted titmouse	25892
	Rhododendron canescens	**mountain azalea**	25892
58	*Setophaga coronata*	yellowrumped warbler	25893
	Cleistes divaricata [upper]	**rosebud orchid, spreading pogonia**	25893
	Echites umbellatus [lower]	**devil's potato, devil's potato-root**	--
59	*Coereba flaveola bahamensis*	Bahama bananaquit	25894
	Casasia clusiifolia	**sevenyear-apple**	25894
60	*Setophaga citrina*	hooded warbler	25895
	Nyssa aquatica	**water tupelo**	25895
Volume 1. Part 4 (November 1731)			
61	*Setophaga pinus*	pine warbler	25897
	Cartrema americana	**American devilwood**	25896
62	*Setophaga dominica*	yellow-throated warbler	25897
	Acer rubrum	**red maple**	25897
63	*Setophaga petechia*	yellow warbler	25898
	Persea borbonia* var. *pubescens [9]	**swamp redbay**	25898

Plate Number	Scientific Name	Common Name	RL inventory no.
64	*Setophaga americana*	northern parula	25899
	Halesia tetraptera	**mountain silverbell**	25899
65	*Archilochus colubris*	ruby-throated hummingbird	25900
	Campsis radicans	**trumpetcreeper**	25900
66	*Dumetella carolinensis*	gray catbird	25901
	Clethra alnifolia	**(coastal) sweet pepper bush**	--
67	*Setophaga ruticilla*	American redstart	25902
	Juglans nigra	**black walnut**	25902
68	*Melopyrrha nigra*	Cuban bullfinch	25903
	Chionanthus virginicus	**fringetree**	25903
69	*Megaceryle alcyon*	belted kingfisher with unidentified fish	25904
	Morella cerifera	**bay-berry, wax myrtle**	--
70	*Porzana carolina*	sora	25905b
	Gentiana catesbaei	**Elliott's gentian**	25905a
71	*Charadrius vociferus*	killdeer	25906
	Oxydendrum arboreum	**sorrel tree, sourwood**	25906
72	*Arenaria interpres*	ruddy turnstone	25907
	Salmea petrochioides	shanks	25907
73	*Phoenicopterus ruber*	American flamingo	25908
	Plexaurella dichotoma [background]	double-forked plexaurella	25908
74	*Phoenicopterus ruber*	American flamingo	25909
	Plexaura flexuosa [background]	bent sea-rod	25909
75	*Grus americana*	whooping crane	25910
	Reynosia septentrionalis	**darlingplum**	--
76	*Egretta caerulea*	little blue heron	25911
77	*Egretta caerulea*	little blue heron	
	Phymosia abutiloides	**Bahamian phymosia**	--
78	*Botaurus lentiginosus*	American bittern	25912
79	*Nyctanassa violacea*	yellow-crowned night-heron	25913
	Scaevola plumieri	**gullfeed**	25913
80	*Butorides virescens*	green heron	25914
	Fraxinus caroliniana	**Carolina ash, swamp ash**	25914
Volume 1. Part 5 (November 1732)			
81	*Mycteria americana*	wood stork	25915
82	*Eudocimus albus*	white ibis	25916
	Orontium aquaticum	**goldenclub**	25917
83	*Eudocimus albus*	white ibis	25918
	Peltandra virginica	**green arrow arum**	--
84	*Eudocimus ruber*	scarlet ibis	25919
85	*Haematopus palliatus*	(American) oystercatcher	25920
	Avicennia germinans	**black mangrove**	25920
86	*Morus bassanus*	northern gannet	25921
	Laguncularia racemosa	**white mangrove**	25921
87	*Sula leucogaster*	(brown) booby	25922
88	*Anous stolidus*	(brown) noddy	25923
89	*Larus atricilla*	laughing gull	25924
90	*Rynchops niger*	black skimmer	25925
91	*Podilymbus podiceps*	pied-billed grebe	25926
92	*Branta canadensis*	Canada goose	25927
	Wedelia bahamensis	**rong bush**	25927
93	*Anas bahamensis*	white-cheeked pintail	25928
	Borrichia arborescens	**sea ox-eye, tree seaside tansy**	25928

Plate Number	*Scientific Name*	*Common Name*	*RL inventory no.*
94	*Lophodytes cucullatus*	hooded merganser	25929
95	*Bucephala albeola*	bufflehead	25930
96	*Anas clypeata*	northern shoveler	25931
97	*Aix sponsa*	wood duck	25932
98	*Bucephala albeola*	bufflehead	25933
	Jacquinia keyensis	**joewood**	--
99	*Anas discors*	blue-winged teal	25934
100	*Anas discors*	blue-winged teal	25935
Volume 2. Part 6 (April 1734)			
1	*Sphyraena barracuda* [upper]	(great) barracuda	25936
	Albula vulpes [lower]	bonefish	25936
2	*Haemulon album* [upper]	margate (fish)	25937
	Synodus foetens [lower]	inshore lizardfish	25938
3	*Micropogonias undulatus* [upper]	Atlantic croaker	25939
	Holocentrus rufus [lower]	longspine squirrelfish, squirrel	25940
4	*Anisotremus virginicus* [upper]	porkfish	25941
	Lutjanus apodus [lower]	schoolmaster	25942
5	*Mycteroperca venenosa*	(princess) rockfish, yellowfin grouper	25943
6	*Haemulon plumieri* [upper]	(white) grunt	25944
	(*Mugil cephalus*) [lower]	(probably (gray) mullet)	25945
7	*Cephalopholis fulva* [upper]	coney	25946
	Haemulon melanurum [lower]	cottonwick	25946
8	Exocoetidae (unidentified) [upper]	(flying fish)	25947
	Kyphosus saltatrix [center]	Bahama rudderfish	25948
	Lepomis gibbosus [lower]	pumpkinseed	25949
9	*Lutjanus griseus*	gray snapper	25950
10	*Acanthurus coeruleus* [upper]	(blue) tang	25951
	Cephalopholis fulva [lower]	coney	25952 [lower]
11	*Bodianus rufus* [upper]	(Spanish) hogfish	25952
	Gerres cinereus [lower]	yellowfin mojarra	25953
12	*Halichoeres radiatus* [upper]	puddingwife	25954
	Menticirrhus americanus [lower]	southern kingfish	25955
13	(probably *Ulaema lefroyi*)	(longfinned silverbiddy)	25956
	Gorgoniidae (unidentified)		--
14	*Epinephelus guttatus* [upper]	(red) hind	25957
	Pomatomus saltatrix [lower]	bluefish	25958
15	*Lachnolaimus maximus*	(great) hogfish	25959
16	*Calamus calamus*	(saucereye) porgy	25960
17	*Lutjanus synagris* [upper]	lane snapper	25961
	Fistularia tabacaria [lower]	tobacco trumpetfish	25962
18	*Scarus coeruleus*	blue parrotfish	25963
19	*Aluterus scriptus*	scrawled filefish	25964
20	*Gymnothorax funebris*	green moray	25965
Volume 2. Part 7 (January 1736)			
21	*Gymnothorax moringa*	(spotted) muray	--
	Gorgoniidae (unidentified)		--
22	*Balistes vetula*	queen triggerfish, turbot	25966
23	*Ameiurus catus*	white catfish	25966
24	(probably *Elops saurus*)	(ladyfish)	--
	Dalbergia ecastaphyllum	**coinvine**	--

Plate Number	*Scientific Name*	*Common Name*	*RL inventory no.*
25	*Lutjanus analis*	mutton fish, mutton snapper	25968
26	*Remora remora*	remora	25970
	Phyllanthus epiphyllanthus	**Abraham-bush, swordbush**	25969
27	*Bothus lunatus*	peacock flounder	25971
28	*Sphoeroides testudineus*	checkered puffer	25973
	Nectandra coriacea [upper]	**bastard torch, lancewood**	25972
	Galactia rudolphioides [lower]	**red milk-pea**	25972
29	*Sparisoma viride*	(stoplight) parrotfish	25974
30	*Lepisosteus platyrhincus*	Florida gar	25975
	--	unidentified	--
	Leucothoe axillaris	**coastal doghobble** [fruiting branch]	--
31	*Holacanthus ciliaris*	(queen) angelfish	25976
32	*Gecarcinus ruricola*	purple land crab	25977
	Picrodendron baccatum	**blackwood**	25977
33	*Coenobita clypeatus*	land hermit crab	25978
	Cittarium pica	West Indian topsnail, whelk	25978
	Conocarpus erectus [upper]	**button mangrove, buttonwood**	25979
	Amyris elemifera [lower]	**(sea) torchwood**	25979
34	*Petrochirus diogenes*	(giant or sea) hermit crab	25980
	Pterogorgia cf anceps	angular sea-whip	--
35	*Ocypode quadrata* [upper]	Atlantic ghost crab	25981
	Pseudopterogorgia cf acerosa	purple sea-plume	--
36	*Grapsus grapsus* [upper]	sally lightfoot crab	25982
	Calappa flammea [lower]	flame box crab	25982
37	*Dromia erythropus* [fore]	redeye sponge crab	25978
	Muricea muricata	spiny sea-fan	25983
38	*Chelonia mydas*	green (sea) turtle	25984
	Thalassia testudinum	**turtlegrass**	--
39	*Eretmochelys imbricata*	hawksbill (sea) turtle	25985
40	*Caretta caretta*	loggerhead sea turtle	25986
Volume 2. Part 8 (April 1737)			
41	*Crotalus horridus*	timber rattlesnake	25987
42	*Sistrurus miliarius*	pygmy rattlesnake	25989
	Lysiloma sabicu [left]	**horseflesh mahogany, sabicu, wild tamarind**	--
	Banara minutiflora [right]	**banara**	--
43	*Agkistrodon piscivorus*	eastern cottonmouth	25990
	Leucothoe racemosa [left]	**swamp doghobble**	25885
44	*Heterodon platirhinos*	eastern hog-nosed snake	25991
	Vachellia tortuosa	**poponax**	--
45	*Heterodon platirhinos*	eastern hog-nosed snake	25992
	(probably *Ambystoma talpoideum*)	(mole salamander)	--
	Xanthosoma sagittifolium	**arrowleaf elephant's-ear, tannia**	26000v
46	*Nerodia erythrogaster*	plain-bellied watersnake	25993
	Croton eluteria	**cascarilla, cascarilla bark**	--
47	*Opheodrys aestivus*	rough greensnake	25994
	Callicarpa americana	**American beautyberry**	25994
48	*Pantherophis alleghaniensis*	eastern ratsnake	25995
	Cissus obovata	**spoonleaf treebine, warty cissus**	25996
49	*Storeria dekayi*	Dekay's brownsnake	25997
	Erythrina herbacea	**cardinal-spear, coral bean**	25997
50	*Thamnophis sauritus*	eastern ribbonsnake	25998
	Canella winterana	**wild cinnamon**	25998

Plate Number	*Scientific Name*	*Common Name*	*RL inventory no.*
51	*Thamnophis sirtalis*	common gartersnake	25999
	Caesalpinia bahamensis	**Bahama caesalpinia**	25999
	Passiflora suberosa	**corkystem passionflower**	25999
52	*Lampropeltis getula*	eastern kingsnake	26000
	Leucothoe axillaris	**coastal doghobble**	25885
53	*Thamnophis sirtalis*	common gartersnake	26001
	Pentalinon luteum	**hammock viper's-tail**	--
54	*Coluber flagellum*	eastern coachwhip	26002
	Silene virginica	**fire pink**	26002
55	*Pantherophis guttata*	red cornsnake	26003
	Polystachya concreta	**greater yellowspike orchid**	26003
56	*Heterodon platirhinos*	eastern hog-nosed snake	26004
	Lilium superbum [habit, buds + right flower]	**turk's-cap lily**	26005
	Lilium michauxii [whorl of leaves + left flower]	**Carolina lily**	26006
57	*Opheodrys vernalis*	smooth greensnake	26007
	Ilex vomitoria	**yaupon**	26007
58	*Farancia abacura*	eastern mudsnake	26008
	Lilium catesbaei	**leopard lily, pine lily**	26008, 26009
59	*Ophisaurus ventralis*	eastern grass lizard	26010
	Echinacea laevigata	**smooth purple coneflower**	--
60	*Cemophora coccinea*	scarletsnake	26011
	Ipomoea batatas	**sweet potato**	26011
Volume 2. Part 9 (June 1739)			
61	***Magnolia grandiflora***	**large-flowered magnolia, southern magnolia**	
62	*Spilogale putorius*	eastern spotted skunk	26013
	Commelina erecta [lower]	**erect dayflower**	25994
63	*Alligator mississippiensis*	American alligator	--
	Rhizophora mangle	**red mangrove(-tree)**	26014
64	*Cyclura cychlura*	Andros Island ground iguana, guana	26015
	Annona glabra	**pond-apple**	26015
65	*Anolis carolinensis*	green anole	26016
	Liquidambar styraciflua	**sweetgum**	26016
66	*Anolis grahami*	Graham's anole	26018
	Haematoxylum campechianum	**bloodwood tree, logwood**	26017
67	*Plestiodon fasciatus*	common five-lined skink	26018
	Annona glabra	**pond-apple**	--
68	*Leiocephalus carinatus*	northern curly-tailed lizard	--
	Epidendrum nocturnum	**night scented orchid**	26019
69	*Anaxyrus terrestris*	southern toad	26020
	Sarracenia flava [left]	**yellow pitcherplant, yellow trumpet**	26020
	Sarracenia minor [center & right]	**nodding pitcherplant, hooded pitcherplant**	26020
70	*Lithobates sphenocephalus*	southern leopard frog	26021
	Sarracenia purpurea	purple pitcherplant, common pitcher plant	26021
71	*Hyla cinerea*	green tree frog	26023
	Peucetia viridans	green lynx spider	26024
	Symplocarpus foetidus	**skunk cabbage**	26024
72	*Lithobates catesbeianus*	American bullfrog	26025
	Cypripedium acaule	**pink lady's-slipper**	26026
73	*Sciurus niger*	eastern fox squirrel	26028
	Cypripedium pubescens	**greater yellow lady's-slipper**	26027

Plate Number	Scientific Name	Common Name	RL inventory no.
74	*Sciurus carolinensis*	eastern gray squirrel	26029
	Prosthechea boothiana	**dollar orchid**	26030
75	*Tamias striatus*	eastern chipmunk	26032
	Sideroxylon foetidissimum	**false mastic, mastic, mastic ironwood**	26031
	Calycanthus floridus (fruit: right)	**Carolina allspice, eastern sweetshrub**	[25877v]
76	*Glaucomys volans*	southern flying squirrel	26034
	Diospyros virginiana	**American persimmon, common persimmon**	26033
77	*Glaucomys volans*	southern flying squirrel	26036
	Catopsis berteroniana	**mealy wild pine, powdery strap airplant**	26035
78	*Urocyon cinereoargenteus*	common gray fox	26038
	Spigelia marilandica	**Indianpink**	26037
79	*Geocapromys ingrahami*	Bahamian hutia	26039
	Bourreria succulenta	**bodywood**	25996
80	***Magnolia tripetala***	**umbrella magnolia, umbrella tree**	26041, 26012 [fruit only]
Volume 2. Part 10 (December 1743)			
81	***Swietenia mahagoni*** [upper right]	**mahogany tree, West Indian mahogany**	26042
	Phoradendron rubrum [lower left]	**mahogany mistletoe, narrow-leaved mistletoe**	26058
82	***Bignonia capreolata***	**cross vine**	26043
83	*Papilio glaucus*	tiger swallowtail	26045
	Ptelea trifoliata	**(common) hop-tree**	26045
84	*Actias luna*	luna moth + cocoon [upper right]	26046
	Philadelphus inodorus [upper]	**scentless mock-orange**	26046
	Smilax smallii [lower]	**lanceleaf greenbrier**	26046
85	***Asimina triloba***	**pawpaw**	26047
86	*Hyalophora cecropia*	cecropia moth + cocoon [lower right]	26048
	Annona reticulata	**custard-apple, netted pawpaw**	26048
87	***Ipomoea microdactyla*** [left, twining]	**calcareous morning-glory, wild potato**	26049
	Manilkara jaimiqui subsp. ***emarginata*** [right]	**wild dilly**	26049
88	*Danaus plexippus*	monarch butterfly	26050
	Prosthechea cochleata [fore]	**clamshell orchid**	26050
	Encyclia plicata [background]	**pleated encyclia**	26050
89	*Dissosteira carolinus*	Carolina grasshopper	26051
	Tillandsia balbisiana	**northern needleleaf**	26051
90	*Antheraea polyphemus*	polyphemus moth	26053
	Talipariti tiliaceum	**tree hibiscus**	26052
91	*Antheraea polyphemus*	polyphemus moth	26053
	Cordia sebestena [upper]	**geiger-tree**	26054
	Ipomoea carolina [lower, twining]	**tievine**	26054
92	***Plumeria rubra***	**Spanish jasmine, templetree**	26055
93	***Plumeria obtusa*** [upper]	**frangipani, Singapore graveyard flower**	26056
	Passiflora cuprea [lower, twining]	**devil's pumpkin**	26056
94	*Citheronia regalis*	horned devils caterpillar	25988
	Coccoloba diversifolia	**pigeon-plum**	25957
95	*Zerynthia rumina*	Spanish festoon [butterfly]	
	Hippomane mancinella [lower]	**manchineel (tree)**	25958
	Dendropemon purpureus	**smooth leechbush, smooth mistletoe**	25958
96	*Utetheisa bella*	ornate moth	26059
	Coccoloba uvifera	**sea-grape**	26059
97	*Papilio glaucus*	tiger swallowtail	26044
	Pithecellobium × bahamense	**Bahamian (Bahama) cat's-claw**	26060

Plate Number	*Scientific Name*	*Common Name*	*RL inventory no.*
98	***Kalmia latifolia***	**mountain laurel**	26061
99	***Clusia rosea***	**pitch apple, Scotch attorney**	26063
100	*Protographium marcellus*	zebra swallowtail	26064
	Catesbaea spinosa	**lily thorn, prickly apple**	26064
Appendix. [Part 11] (July 1747)			
1	*Tympanuchus cupido*	greater prairie chicken	--
	Primula meadia	**pride-of-Ohio**	26065
2	(probably *Scolopendra alternans*)	(Florida Keys centipede)	26067
	Hamamelis virginiana	**(Virginian) witch-hazel**	26066
3	*Crotophaga ani*	smooth-billed ani	26068
	Cypripedium acaule	**pink lady's-slipper**	26069
4	*Megarhyssa atrata*	giant ichneumon wasp	26069
	Rhus glabra	**smooth sumac**	26071
5	*Icterus icterus*	troupial	26070
	Chalybion californicum	blue mud wasp	26070
	Pancratium maritimum	**sea-daffodil**	26070
6	***Theobroma cacao***	**cacao**	26072
7	***Vanilla mexicana***	**Mexican vanilla**	26073
8	*Chaetura pelagica*	chimney swift	26075
	Lilium philadelphicum	**wood lily**	26074
9	***Anacardium occidentale***	**cashew**	26076
10: figure 1	*Ardea herodias*	great blue heron (with spotted salamander bill)	26078
figure 2	*Ambystoma maculatum*	spotted salamander	26018
figure 3	*Tunga penetrans*	chigoe flea + egg	26077
figure 4	(possibly Carabidae)	(beetle)	26077
figure 5	*Periplaneta americana*	American cockroach	26077
figure 6	*Blaptica dubia*	dubia cockroach	26077
figure 7	*Necrophila americana*	American carrion beetle	26077
11	*Canthon pilularius* [lower left]	dung beetle	26077
	Phanaeus vindex [lower right]	rainbow scarab beetle	26077
	Lilium canadense	**Canada lily, meadow lily**	26079
12	*Colinus virginianus*	northern bobwhite	26080
	Zephyranthes atamasca	**Atamasco lily**	26081
13	*Regulus calendula*	ruby-crown kinglet	26081
	Sceliphron caementarium	black and yellow mud-dauber	26077
	Stewartia malacodendron	**silky camellia**	26082
14	*Phaëthon aethereus* [upper]	red-billed tropicbird	26083
	Hydrobates pelagicus [lower]	European storm-petrel	26083
15	*Dasymutilla occidentalis*	velvet ant	26081
	Magnolia virginiana [flower]	**sweet-bay**	26084
	Magnolia acuminata [foliage]	**cucumber-tree**	26084
16	*Chordeiles minor*	common nighthawk	26085
	Panax quinquefolius	**American ginseng**	26085
17	***Kalmia angustifolia*** [left]	**lamb laurel, sheep laurel**	26086
	Rhododendron maximum [right]	**great laurel**	26086
18	*Dasyprocta leporina*	Brazilian agouti	26087
	Ficus citrifolia	**shortleaf fig, Jamaica cherry-fig**	--
19	*Acanthodoras cataphractus* [upper]	spiny catfish	26088
	Chauliodus sloani [lower]	viperfish	26089
20	*Bison bison*	American buffalo	26090
	Robinia hispida	**rose locust**	26090

PART 2 Plants and animals depicted in the watercolor paintings in Royal Library, Windsor, arranged by Royal Collection Inventory Number

× = unpublished

App = Appendix to volume 2

Names of flowering plants and fungi are in bold; the common name is in parentheses.

RL	*Vol: Plate*	*Scientific Name*
24814	1 : 1	*Haliaeetus leucocephalus* (bald eagle) + (probably *Mugil cephalus*) (gray mullet) + *Pandion haliaetus* [background] (osprey)
24815	1 : 2	*Pandion haliaetus* (osprey) + (probably *Mugil cephalus*) (gray mullet)
24816	1 : 3	*Falco columbarius* (merlin)
24817	1 : 4	*Elanoides forficatus* (swallow-tail kite)
24818	1 : 5	*Falco sparverius* (American kestrel)
24819	1 : 6	*Cathartes aura* (turkey vulture)
24820	1 : 7	*Megascops asio* (eastern screech-owl)
24821	1 : 8	*Neocurtilla hexadactyla* (northern mole cricket) with composite of three birds *Caprimulgus carolinensis* (chuck's-will-widow) + *Caprimulgus vociferus* (whip-poor-will) + *Chordeiles minor* (common nighthawk)
24822	1 : 9	***Castanea pumila*** (Allegheny chinkapin, chinquapin) + *Coccyzus americanus* (yellowbilled cuckoo)
24823	1 : 10	*Amazona leucocephala* (Cuban parrot) + ***Colubrina elliptica*** (smooth snake-bark, soldierwood)
24824	1 : 11	*Conuropsis carolinensis* (Carolina parakeet) +***Taxodium distichum*** (bald cypress, swamp cypress)
24825	1 : 12	*Quiscalus quiscula* (common grackle)
24826	1 : 13	*Agelaius phoeniceus* (red-winged blackbird) + ***Morella caroliniensis*** (southern bayberry)
24827	1 : 14	*Dolichonyx oryzivorus* (bobolink) + ***Oryza sativa*** (rice)
24828	1 : 15	*Cyanocitta cristata* (bluejay) + ***Smilax laurifolia*** (laurel greenbrier)
24829	1 : 16	*Campephilus principalis* (ivory-billed woodpecker) + ***Quercus phellos*** (willow oak)
24830	1 : 17	*Dryocopus pileatus* (pileated woodpecker) + ***Quercus virginiana*** (live oak)
24831	1 : 18	*Colaptes auratus* (northern flicker) + ***Quercus michauxii*** (swamp chestnut oak)
24832	1 : 19	*Picoides villosus* [left] (hairy woodpecker) + *Melanerpes carolinus* [right] (red-bellied woodpecker) + ***Quercus marilandica*** (blackjack oak)
24833	1 : 20	*Melanerpes erythrocephalus* (red-headed woodpecker) + ***Quercus nigra*** (water oak) + ***Mitchella repens*** [lower] (partridgeberry)
24834	1 : 21	*Picoides pubescens* [left] (downy woodpecker) + *Sphyrapicus varius* [right] (yellow-bellied sapsucker) + ***Quercus alba*** [right + acorn on right] (white oak) + ***Quercus falcata*** [left] (southern red oak)
24835	1 : 22	*Sitta pusilla* [upper] (brown-headed nuthatch) + *Sitta carolinensis* (white-breasted nuthatch) + ***Quercus incana*** (bluejack oak)
24836	1 : 23	*Ectopistes migratorius* (passenger pigeon) + ***Quercus laevis*** ((American) turkey oak)
24837	1 : 24	*Zenaida macroura* (mourning dove)
24838	1 : 25	*Patagioenas leucocephala* (white-crowned pigeon) + ***Chrysobalanus icaco*** (coco plum)
24839a	1 : 26	*Columbina passerina* (common ground-dove) + ***Zanthoxylum clava-herculis*** (Hercules' club)
24839b	1 : 26	*Columbina passerina* (common ground-dove) {+ unpublished ***Anacardium occidentale*** (cashew)}
24840	1 : 27	*Mimus polyglottos* (northern mockingbird) + ***Cornus florida*** ((eastern) flowering dogwood)
24841	1 : 28	*Toxostoma rufum* (brown thrasher) + ***Prunus virginiana*** (choke cherry)
24842	1 : 29	*Turdus migratorius* (American robin) + ***Aristolochia serpentaria*** (Virginia dutchmanspipe, Virginia snakeroot) + (probably ***Carya ovata*** (shagbark hickory))

RL	*Vol: Plate*	*Scientific Name*
24843	1 : 30	*Turdus plumbeus* (red-legged thrush) + ***Bursera simaruba*** (gumbo limbo, gum elemi)
24844	1 : 31	perhaps *Catharus guttatus* (hermit thrush) + ***Ilex cassine*** (dahoon)
24845	1 : 32	*Eremophila alpestris* [upper] (horned lark) {+ unpublished *Anthus rubescens* [lower] (American pipit)}
24846	1 : 33	*Sturnella magna* (eastern meadowlark)
24847	1 : 34	*Pipilo erythrophthalmus* [upper] (eastern towhee) + *Molothrus ater* [lower] (brown-headed cowbird) + ***Populus heterophylla*** (swamp cottonwood)
24848	1 : 35	*Spizella passerina* [upper] (chipping sparrow) {+ unpublished *Setophaga dominica* (yellowrumped warbler) with an unidentified two-winged insect}
24849	1 : 35	***Ipomoea sagittata*** (saltmarsh morning-glory)
24850	1 : 36	***Monotropa uniflora*** [left] (one-flower Indian-pipe) + ***Geoglossum glabrum*** [right] (black adder tongue fungus)
24851	1 : 37	*Tiaris bicolor* (black-faced grassquit) + ***Tabebuia bahamensis*** (beef-bush, white dwarf tabebuia)
24852	1 : 38	*Cardinalis cardinalis* (northern cardinal) + ***Carya glabra*** [small nut] (pignut (hickory)) + ***Carya tomentosa*** [branch and large nut] (mockernut (hickory))
24853	1 : 39	*Passerina caerulea* (blue grosbeak) + ***Magnolia virginiana*** (sweet-bay)
24854	1 : 40	*Loxigilla violacea* (greater Antillean bullfinch) + ***Metopium toxiferum*** (Florida poisontree, poisonwood)
24855	1 : 36	*Carpodacus purpureus* [upper] (purple finch) + *Junco hyemalis* [lower] (dark-eyed junco) + ***Nyssa sylvatica*** (black gum, black tupelo)
24856	1 : 42	*Spindalis zena* (stripe-headed tanager) + ***Jacaranda caerulea*** (cancer tree, boxwood)
24857	1 : 43	*Carduelis tristis* (American goldfinch) + ***Gleditsia aquatica*** (water honeylocust, water locust)
25875	1 : 44	*Passerina ciris* (painted bunting) {+ unpublished *Passerina cyanea* (indigo bunting) + ***Magnolia virginiana*** (sweet-bay)}
25876	1 : 45	***Trillium catesbaei*** (bashful wakerobin)
25877	1 : 45	*Passerina cyanea* (indigo bunting)
25877	1 : 46	***Calycanthus floridus*** (Carolina allspice, eastern sweetshrub)
25878	1 : 46	*Bombycilla cedrorum* (cedar waxwing) {+ unpublished ***Magnolia virginiana*** (sweet-bay)}
25879	1 : 47	*Sialia sialis* (eastern bluebird) + ***Smilax pumila*** (sarsaparilla vine)
25880	1 : 48	*Icterus galbula* (Baltimore oriole) + ***Liriodendron tulipifera*** (tulip poplar, tulip tree)
25881	1 : 49	*Icterus spurius* (orchard oriole) + ***Catalpa bignonioides*** (common catalpa, southern catalpa)
25882	1 : 50	*Icteria virens* (yellow-breasted chat) {+ unpublished ***Acer rubrum* var. *trilobum*** (Carolina red maple)}
25883	1 : 50	***Trillium maculatum*** (spotted wakerobin)
25884	1 : 51	*Progne subis* (purple martin)
25885	2 : 43	***Leucothoe racemosa*** [left] (swamp doghobble) +
	2 : 52	***Leucothoe axillaris*** [right] (coastal doghobble)
25886	1 : 52	*Myiarchus crinitus* [lower] (great crested flycatcher) + ***Smilax tamnoides*** (bristly greenbrier) +
	1 : 53	*Sayornis phoebe* [upper] (eastern phoebe)
25887	1 : 53	***Gelsemium sempervirens*** (Carolina jessamine, false jasmine)
25888	1 : 54	*Contopus virens* [upper] (eastern wood-pewee) + *Vireo olivaceus* [lower] (red-eyed vireo) + ***Symplocos tinctoria*** (common sweetleaf)
25889		{unpublished *Sassafras albidum* (sassafras)}
25890	1 : 55	*Tyrannus tyrannus* (eastern kingbird) + ***Sassafras albidum*** (sassafras)

RL	*Vol: Plate*	*Scientific Name*
25891	1 : 56	*Piranga rubra* (summer tanager) + ***Platanus occidentalis*** (American sycamore)
25892	1 : 33	***Hypoxis hirsuta*** (common goldstar) +
	1 : 57	*Baeolophus bicolor* (tufted titmouse) + ***Rhododendron canescens*** (mountain azalea)
25893	1 : 58	*Setophaga coronata* (yellowrumped warbler) + ***Cleistes divaricata*** (rosebud orchid, spreading pogonia)
25894	1 : 59	*Coereba flaveola bahamensis* (Bahama bananaquit) + ***Casasia clusiifolia*** (sevenyear-apple)
25895	1 : 60	*Setophaga citrina* (hooded warbler) + ***Nyssa aquatica*** (water tupelo)
25896	1 : 61	***Cartrema americanus*** (American devilwood)
25897	1 : 61	*Setophaga pinus* [lower] (pine warbler) +
	1 : 62	*Setophaga dominica* [upper] (yellow-throated warbler) + ***Acer rubrum*** (red maple)
25898	1 : 63	*Setophaga petechia* (yellow warbler) + ***Persea borbonia* var. *pubescens*** (swamp redbay)
25899	1 : 64	*Setophaga americana* (northern parula) + ***Halesia tetraptera*** (mountain silverbell)
25900	1 : 65	*Archilochus colubris* (ruby-throated hummingbird) + ***Campsis radicans*** (trumpetcreeper)
25901	1 : 66	*Dumetella carolinensis* (gray catbird) {+ unpublished four-winged insect}
25902	1 : 67	*Setophaga ruticilla* (American redstart) + ***Juglans nigra*** (black walnut)
25903	1 : 68	*Melopyrrha nigra* (Cuban bullfinch) + ***Chionanthus virginicus*** (fringetree)
25904	1 : 69	*Megaceryle alcyon* (belted kingfisher) + unidentified fish
25905a	1 : 70	***Gentiana catesbaei*** (Elliott's gentian) {+ unpublished ***Nyssa sylvatica*** (black tupelo)}
25905b	1 : 70	*Porzana carolina* (sora)
25906	1 : 71	*Charadrius vociferus* (killdeer) + ***Oxydendrum arboreum*** (sorrel tree, sourwood)
25907	1 : 72	*Arenaria interpres* (ruddy turnstone) + ***Salmea petrochioides*** (shanks)
25908	1 : 73	*Phoenicopterus ruber* [fore] ((American) flamingo) + *Plexaurella dichotoma* [back] (double-forked plexaurella)
25909	1 : 74	*Phoenicopterus ruber* [fore] ((American) flamingo) + *Plexaura flexuosa* [back] (bent sea-rod)
25910	1 : 75	*Grus americana* (whooping crane)
25911	1 : 76	*Egretta caerulea* (little blue heron)
25911v	2 : 77	*Egretta caerulea* (little blue heron)
25912	1 : 78	*Botaurus lentiginosus* (American bittern)
25913	1 : 79	*Nyctanassa violacea* (yellow-crowned night-heron) + ***Scaevola plumieri*** (gullfeed)
25914	1 : 80	*Butorides virescens* (green heron) + ***Fraxinus caroliniana*** (Carolina ash, swamp ash)
25915	1 : 81	*Mycteria americana* (wood stork)
25916	1 : 82	*Eudocimus albus* (white ibis)
25917	1 : 82	***Orontium aquaticum*** (goldenclub)
25918	1 : 83	*Eudocimus albus* (white ibis)
25919	1 : 84	*Eudocimus ruber* (scarlet ibis)
25920	1 : 85	*Haematopus palliatus* ((American) oystercatcher) + ***Avicennia germinans*** (black mangrove)
25921	1 : 86	*Morus bassanus* (northern gannet) + ***Laguncularia racemosa*** (white mangrove)
25922	1 : 87	*Sula leucogaster* ((brown) booby)
25923	1 : 88	*Anous stolidus* ((brown) noddy)
25924	1 : 89	*Larus atricilla* (laughing gull)
25925	1 : 90	*Rynchops niger* (black skimmer)
25926	1 : 91	*Podilymbus podiceps* (pied-billed grebe)
25927	1 : 92	*Branta canadensis* (Canada goose) + ***Wedelia bahamensis*** (rong bush)
25928	1 : 93	*Anas bahamensis* (white-cheeked pintail) + ***Borrichia arborescens*** (sea ox-eye, tree seaside tansy)

RL	Vol: Plate	Scientific Name
25929	1 : 94	*Lophodytes cucullatus* (hooded merganser)
25930	1 : 95	*Bucephala albeola* (bufflehead)
25931	1 : 96	*Anas clypeata* (northern shoveler)
25932	1 : 97	*Aix sponsa* (wood duck)
25933	1 : 98	*Bucephala albeola* (bufflehead)
25934	1 : 99	*Anas discors* (blue-winged teal)
25935	1 : 100	*Anas discors* (blue-winged teal)
25936	2 : 1	*Sphyraena barracuda* [upper] ((great) barracuda) + *Albula vulpes* [lower] (bonefish)
25937	2 : 2	*Synodus foetens* (inshore lizardfish)
25938	2 : 2	*Haemulon album* (grunt, margate fish)
25939	2 : 3	*Micropogonias undulatus* (Atlantic croaker)
25940	2 : 3	*Holocentrus rufus* (longspine squirrelfish, squirrel)
25941	2 : 4	*Anisotremus virginicus* (porkfish)
25942	2 : 4	*Lutjanus apodus* [upper] (schoolmaster) + {unpublished *Albula vulpes* (bonefish)}
25943	2 : 5	*Mycteroperca venenosa* ((princess) rockfish, yellowfin grouper)
25944	2 : 6	*Haemulon plumieri* ((white) grunt, margat)
25945	2 : 6	probably *Mugil cephalus* ((gray) mullet)
25946	2 : 7	*Cephalopholis fulva* [upper] (coney) + *Haemulon melanurum* [lower] (cottonwick)
25947	2 : 8	unidentified Exocoetidae (flying fish)
25948	2 : 8	*Kyphosus saltatrix* (Bahama rudderfish)
25949	2 : 8	*Lepomis gibbosus* (pumpkinseed)
25950	2 : 9	*Lutjanus griseus* (gray snapper)
25951	2 : 10	*Acanthurus coeruleus* ((blue) tang)
25952	2 : 10	*Cephalopholis fulva* [lower] (coney) +
	2 : 11	*Bodianus rufus* [upper] ((Spanish) hogfish)
25953	2 : 11	*Gerres cinereus* (yellowfin mojarra)
25954	2 : 12	*Halichoeres radiatus* (puddingwife)
25955	2 : 12	*Menticirrhus americanus* (southern kingfish)
25956	2 : 13	probably *Ulaema lefroyi* (longfinned silverbiddy)
25957	2 : 14	*Epinephelus guttatus* ((red) hind)
25958	2 : 14	*Pomatomus saltatrix* (bluefish)
25959	2 : 15	*Lachnolaimus maximus* ((great) hogfish)
25960	2 : 15	*Calamus calamus* ((saucereye) porgy)
25961	2 : 17	*Lutjanus synagris* (lane snapper)
25962	2 : 17	*Fistularia tabacaria* (tobacco trumpetfish)
25963	2 : 18	*Scarus coeruleus* (blue parrotfish)
25964	2 : 19	*Aluterus scriptus* (scrawled filefish)
25965	2 : 20	*Gymnothorax funebris* (green muray)
25966	2 : 22	*Balistes vetula* (queen triggerfish, turbot)
25967	2 : 23	*Ameiurus catus* (white catfish)
25968	2 : 25	*Lutjanus analis* (mutton fish, mutton snapper)
25969	2: 26	***Phyllanthus epiphyllanthus*** (Abraham-bush, swordbush)
25970	2 : 26	*Remora remora* (remora)
25971	2 : 27	*Bothus lunatus* (peacock flounder)
25972	2 : 28	***Nectandra coriacea*** [upper] (bastard torch, lancewood) + ***Galactia rudolphioides*** [lower] (red milk-pea)
25973	2 : 28	*Sphoeroides testudineus* (checkered puffer)
25974	2 : 29	*Sparisoma viride* ((stoplight) parrotfish)
25975	2 : 30	*Lepisosteus platyrhincus* (Florida gar)
25976	2 : 31	*Holacanthus ciliaris* ((queen) angel fish)
25977	2 : 32	*Gecarcinus ruricola* (purple land crab) + ***Picrodendron baccatum*** (blackwood)
25978	2 : 33	*Cittarium pica* [center] (West Indian topsnail, whelk) + *Coenobita clypeatus* [lower & in shell] (land hermit crab)

RL	Vol: Plate	Scientific Name
	2 : 37	*Dromia erythropus* [upper] (redeye sponge crab)
25979	2 : 33	***Conocarpus erectus*** [upper] (button mangrove, buttonwood) + ***Amyris elemifera*** [lower] ((sea) torchwood)
25980	2 : 34	*Petrochirus diogenes* ((giant or sea) hermit crab)
25981	2 : 35	*Ocypode quadrata* [upper](Atlantic ghost crab) {+ unpublished *Aphonopelma* sp. (tarantula)}
25982	2 : 36	*Grapsus grapsus* [upper] (sally lightfoot crab) + *Calappa flammea* [lower] (flame box crab)
25983	2 : 37	*Muricea muricata* (spiny sea-fan)
25983v	1 : 10	***Colubrina elliptica*** (smooth snake-bar, soldierwood) +
	1 : 85	***Avicennia germinans*** (black mangrove)
25984	2 : 38	*Chelonia mydas* (green (sea) turtle)
25985 & 25985v	2 : 39	*Eretmochelys imbricata* (hawksbill (sea) turtle)
25986	2 : 40	*Caretta caretta* (loggerhead sea turtle) {+ unpublished ***Vachellia choriophylla*** (cinnecord))
25987	2 : 41	*Crotalus horridus* (timber rattlesnake)
25988	2 : 42	*Banara minutiflora* [right] (banara) {+ [left] unpublished ***Capparis cynophallophora*** (black willow)} +
	2 : 94	*Citheronia regalis* (horned devils caterpillar)
25989	2 : 42	*Sistrurus miliarius* (pygmy rattlesnake) {+ unpublished two-winged insect}
25990	2 : 43	*Agkistrodon piscivorus* (eastern cottonmouth)
25991	2 : 44	*Heterodon platirhinos* (eastern hog-nosed snake)
25992	2 : 45	*Heterodon platirhinos* (eastern hog-nosed snake) {+ unpublished *Bos taurus* (domestic cattle)}
25993	2 : 46	*Nerodia erythrogaster* (plain-bellied watersnake)
25994	2 : 47	*Opheodrys aestivus* (rough greensnake) + ***Callicarpa americana*** [upper] (American beautyberry) +
	2 : 62	***Commelina erecta*** [lower] (erect dayflower)
25995	2 : 48	*Pantherophis alleghaniensis* (eastern ratsnake)
25996	2 : 48	***Cissus obovata*** [lower, twinning] (spoonleaf treebine; warty cissus) +
	2 : 79	***Bourreria succulenta*** [upper] (bodywood)
25997	2 : 49	*Storeria dekayi* (Dekay's brownsnake) + ***Erythrina herbacea*** (cardinal-spear coral bean)
25998	2 : 50	*Thamnophis sauritus* (eastern ribbonsnake) + ***Canella winterana*** (wild cinnamon)
25999	2 : 51	*Thamnophis sirtalis* (common gartersnake) + ***Caesalpinia bahamensis*** (Bahama caesalpinia) + ***Passiflora suberosa*** [lower right] (corkystem passionflower, juniper-berry)
26000	2 : 52	*Lampropeltis getula* (eastern kingsnake)
26000v	2 : 45	***Xanthosoma sagittifolium (***arrowleaf elephant's-ear, tannia)
26001	2 : 53	*Thamnophis sirtalis* (common gartersnake)
26002	2 : 54	*Coluber flagellum* (eastern coachwhip) + ***Silene virginica*** (fire pink)
26003	2 : 55	*Pantherophis guttata* (red cornsnake) + ***Polystachya concreta*** (greater yellowspike orchid)
26004	2 : 56	*Heterodon platirhinos* (eastern hog-nosed snake)
26005	2 : 56	***Lilium superbum*** (turk's-cap lily)
26006	2 : 56	***Lilium michauxii*** (Carolina lily)
26007	2 : 57	*Opheodrys vernalis* (smooth greensnake) + ***Ilex vomitoria*** (yaupon)
26008	2 : 58	*Farancia abacura* (eastern mudsnake) + ***Lilium catesbaei*** (leopard lily, pine lily)
26009	2 : 58	***Lilium catesbaei*** (pine lily) {+ unpublished ***Lilium superbum*** (turk's-cap lily)}
26010	2 : 59	*Ophisaurus ventralis* (eastern grass lizard)
26011	2 : 60	*Cemophora coccinea* (scarletsnake) + ***Ipomoea batatas*** (sweet potato)
26012	2 : 80	***Magnolia tripetala*** (umbrella magnolia, umbrella tree)

RL	*Vol: Plate*	*Scientific Name*
26013	2 : 62	*Spilogale putorius* (eastern spotted skunk)
26014	2 : 63	***Rhizophora mangle*** (red mangrove)
26015	2 : 64	*Cyclura cornuta* (guana, horned ground iguana) + ***Annona glabra*** (pond-apple)
26016	2 : 65	*Anolis carolinensis* (green anole) + ***Liquidambar styraciflua*** (sweetgum)
26017	2 : 66	***Haematoxylum campechianum*** (bloodwood tree, logwood)
26018	2 : 66	*Anolis grahami* (Graham's anole) +
	2 : 67	*Plestiodon fasciatus* [lower] (common five-lined skink) +
	App: 10	*Ambystoma maculatum* (spotted salamander)
26019	2 : 68	***Epidendrum nocturnum*** (night scented orchid)
26020	2 : 69	*Anaxyrus terrestris* (southern toad) {+ unpublished *Pyrophorus noctilucus* (fire beetle)} + ***Sarracenia minor*** [left] (nodding pitcherplant, hooding pitcher plant) + ***Sarracenia flava*** [right] (yellow pitcherplant, yellow trumpet)
26021	2 : 70	*Lithobates sphenocephalus* (southern leopard frog) + ***Sarracenia purpurea*** (purple pitcherplant, common pitcher plant, including leaf on left from 26022)
26022	×	{unpublished ***Sarracenia purpurea*** (purple pitcherplant), left and non-flowering stalk on right + ***Sarracenia flava*** (flowering stalk on right)}
26023	2 : 71	*Hyla cinerea* [lower] (green tree frog) + {unpublished *Anaxyrus terrestris* (southern toad) + *Lithobates sphenocephalus* [centre] (southern leopard frog)}
26024	2 : 71	*Peucetia viridans* (green lynx spider) + ***Symplocarpus foetidus*** (skunk cabbage)
26025	2 : 72	*Lithobates catesbeianus* (American bullfrog)
26026	2 : 72	***Cypripedium acaule*** (pink lady's-slipper)
26027	2 : 73	***Cypripedium pubescens*** (greater yellow lady's-slipper)
26028	2 : 73	*Sciurus niger* (eastern fox squirrel)
26029	2 : 74	*Sciurus carolinensis* (eastern gray squirrel)
26030	2 : 55	***Polystachya concreta*** (greater yellowspike orchid) +
	2 : 74	***Prosthechea boothiana*** [right] (dollar orchid)
26030v	2 : 65	***Liquidambar styraciflua*** (sweetgum) +
	2 : 66	*Anolis grahami* (Graham's anole)
26031	2 : 75	***Sideroxylon foetidissimum*** (false mastic, mastic, mastic ironwood)
26032	2 : 75	*Tamias striatus* (eastern chipmunk) {+ unpublished ***Carya*** sp. (hickory)}
26033	2 : 76	***Diospyros virginiana*** (American or common persimmon)
26034	2 : 76	*Glaucomys volans* (southern flying squirrel) + ***Diospyros virginiana*** (American or common persimmon)
26035	2 : 77	***Catopsis berteroniana*** (mealy wild pine, powdery strap airplant)
26036	2 : 77	*Glaucomys volans* (southern flying squirrel)
26037	2 : 78	***Spigelia marilandica*** (Indianpink)
26038	2 : 78	*Urocyon cinereoargenteus* (common gray fox)
26039	2 : 79	*Geocapromys ingrahami* (Bahamian hutia)
26040	×	{unpublished ***Magnolia virginiana*** (sweet-bay)}
26041	2 : 80	***Magnolia tripetala*** (umbrella magnolia, umbrella tree)
26042	2 : 81	***Swietenia mahagoni*** ((West Indian) mahogany)
26043	2 : 82	***Bignonia capreolata*** (cross vine)
26044	2 : 97	*Papilio glaucus* (tiger swallowtail)
26045	2 : 83	*Papilio glaucus* (tiger swallowtail) + ***Ptelea trifoliata*** ((common) hoptree)
26046	2 : 84	***Actias luna*** [upper right] (luna moth) + ***Philadelphus inodorus*** (scentless mock-orange) + ***Smilax smallii*** [lower] (lanceleaf greenbrier) +
	2 : 98	***Kalmia latifolia*** (mountain laurel)
26047	2 : 85	***Asimina triloba*** (pawpaw)
26048	2 : 86	*Hyalophora cecropia* (cecropia moth + cocoon [lower right]) + ***Annona reticulata*** (custard-apple, netted pawpaw)
26049	2 : 87	***Manilkara jaimiqui subsp. emarginata*** [left] (wild dilly) + ***Ipomoea microdactyla*** [right] (calcareous morning-glory, wild potato)

RL	*Vol: Plate*	*Scientific Name*
26050	2 : 88	*Danaus plexippus* (monarch butterfly) + ***Encyclia plicata*** (pleated encyclia) {+ unpublished ***Prosthechea boothiana*** [upper left] (dollar orchid) + ***Prosthechea cochleata*** [foreground] (clamshell orchid)}
26051	2 : 89	*Dissosteira carolinus* (Carolina grasshopper) + ***Tillandsia balbisiana*** (northern needleleaf)
26052	2 : 90	***Talipariti tiliaceum*** (tree hibiscus)
26053	2 : 90 &	*Antheraea polyphemus* [upper] (polyphemus moth)
	2 : 91	*Antheraea polyphemus* [lower]
26054	2 : 91	***Cordia sebestena*** [upper] (geiger-tree) + ***Ipomoea carolina*** [lower] (tievine)
26055	2 : 92	***Plumeria rubra*** (Spanish jasmine, templetree)
26056	2 : 93	***Passiflora cuprea*** [lower, twining] (devil's pumpkin) + ***Plumeria obtusa*** [upper] (frangipani, Singapore graveyard flower)
26057	2 : 94	***Coccoloba diversifolia*** (pigeon-plum)
26058	2 : 81	***Phoradendron rubrum*** [upper left] (mahogany mistletoe, narrow-leaved mistletoe)
	2 : 95	***Dendropemon purpureus*** [lower] (smooth leechbush, smooth mistletoe) + ***Hippomane mancinella*** [upper] (manchineel)
26059	2 : 96	*Utetheisa bella* (ornate moth) + ***Coccoloba uvifera*** (sea-grape)
26060	2 : 97	***Pithecellobium × bahamense*** (Bahamian cat's claw)
26061	2 : 98	***Kalmia latifolia*** (mountain laurel)
26062	2 : 98	***Kalmia latifolia*** (mountain laurel)
26063	2 : 99	***Clusia rosea*** (pitch apple, Scotch attorney)
26064	2 : 100	*Protographium marcellus* (zebra swallowtail) + ***Catesbaea spinosa*** (lily thorn, prickly apple)
26065	App: 1	*Primula meadia* (pride-of-Ohio)
26066	App: 2	*Hamamelis virginiana* ((Virginian) witch-hazel)
26067	App: 2	(probably *Scolopendra alternans* (Florida Keys centipede))
26068 & 26068v	App: 3	*Crotophaga ani* (smooth-billed ani)
26069	App: 3	***Cypripedium acaule*** (pink lady's-slipper) +
26069	App: 4	*Megarhyssa atrata* [right] (giant ichneumon wasp) + *Chalybion californicum* [upper left] (blue mud wasp)
26070	App: 5	*Chalybion californicum* (blue mud wasp) + *Icterus icterus* (troupial) + ***Pancratium maritimum*** (sea-daffodil)
26071	App: 4	***Rhus glabra*** (smooth sumac)
26072	App: 6	***Theobroma cacao*** (cacao)
26073	App: 7	***Vanilla mexicana*** (Mexican vanilla)
26074	App: 8	***Lilium philadelphicum*** (wood lily)
26075	App: 8	*Chaetura pelagica* (chimney swift)
26076	App: 9	***Anacardium occidentale*** (cashew)
26077	1 : 35	*Hypercompe scribonia* (giant leopard moth) +
	App: 10	*Hypercompe scribonia* [bottom center; cut-out] (giant leopard moth) + unidentified Carabidae [lower right; cut-out] + *Blaptica dubia* [upper, left, two views] (dubia cockroach) + *Tunga penetrans* [middle, right; cut-out] (chigoe flea & egg) + *Necrophila americana* [second row; dorsal & ventral views] (American carrion beetle) +
	App: 11	*Canthon pilularius* [lower left] (dung beetle) + *Phanaeus vindex* [center] (rainbow scarab beetle) +
	App: 13	*Sceliphron caementarium* [center right, 2 images] (black and yellow mud-dauber) +
	App: 15	*Dasymutilla occidentalis* [second row, right] (velvet ant)
26078	App: 10	*Ardea herodias* (great blue heron)
26079	App: 11	***Lilium canadense*** (Canada or meadow lily)
26080	App: 12	*Colinus virginianus* (northern bobwhite)

RL	*Vol: Plate*	*Scientific Name*
26081	App: 12	***Zephyranthes atamasca*** (Atamasco lily, naked ladies) +
	App: 13	*Regulus calendula* (ruby-crown kinglet) +
	App: 15	*Dasymutilla occidentalis* (velvet ant)
26082	App: 13	***Stewartia malacodendron*** (silky camellia)
26083	App: 14	*Phaëthon aethereus* [upper] (red-billed tropicbird) + *Hydrobates pelagicus* [lower] (European storm-petrel)
26084	App: 15	***Magnolia virginiana*** [flower] (sweet-bay) + ***Magnolia acuminata*** [foliage] (cucumber-tree)
26085	App: 16	*Chordeiles minor* (common nighthawk) + ***Panax quinquefolius*** (American ginseng)
26086	App: 17	***Rhododendron maximum*** [left] (great laurel) + ***Kalmia angustifolia*** [right] (lamb or sheep laurel)
26087	App: 18	*Dasyprocta leporina* (Brazilian agouti)
26088	App: 19	*Acanthodoras cataphractus* (spiny catfish)
26089	App: 19	*Chauliodus sloani* (viperfish)
26090	App: 20	*Bison bison* (American bison) + ***Robinia hispida*** (rose locust)
26091	×	{unpublished ***Robinia hispida*** (rose locust)}
26092	×	{unpublished *Bison bison* (American bison) + ***Robinia hispida*** [sketch] (rose locust)}
26093	×	{unpublished *Pristis pristis* (common sawfish) + ***Ficus citrifolia*** (Jamaica cherry-fig, shortleaf fig)}

NOTES TO CHAPTERS AND AUTHORS' ACKNOWLEDGMENTS

Abbreviations

MC	Mark Catesby
NH	M. Catesby, *The natural history of Carolina, Florida and the Bahama islands* (all references are to the original edition, unless otherwise stated). Text and/or image are referred to as "plate."

INDIVIDUAL CORRESPONDENTS

CL	Carl Linnaeus
HS	Hans Sloane
PC	Peter Collinson
WS	William Sherard

ARCHIVES

BL	British Library, London
BM	British Museum, London
ERO	Essex Record Office, Chelmsford
LMA	London Metropolitan Archive, London
LS	Linnean Society of London
NAUK	National Archives, Kew
NHM	Natural History Museum, London
NHM-HS	Sloane Herbarium, Natural History Museum, London
OBL-S	Sherardian Library of Plant Taxonomy, one of the Bodleian Libraries of the University of Oxford, Oxford
OUH	Oxford University Herbaria
RLW	Royal Library, Windsor Castle
RS	Royal Society of London
SRO	Sussex Record Office, Bury St. Edmunds

CHAPTER 1. "The truly honest, ingenious, and modest Mr Mark Catesby, F.R.S.": documenting his life (1682/83–1749)

NOTES

1. Anonymous 1750. Although it was printed without an author being named, this was undoubtedly written by Peter Collinson (see Turner 1835: 401–402).

2. Costa 1812: 206.

3. Anonymous 1750 (see note 1 above).

4. ERO: Register of St. Nicholas, Castle Hedingham, (available online, D/P48/1/1/, image 71).

5. The use of the Julian calendar in England at this period continues to cause confusion among those writing about Mark Catesby. Allen (1951: 469) was so mistaken by the entry in the parish register that she stated that "He was thus a year old at baptism" when he was in fact only six days old.

6. Vicar General Marriage Licences, Boyd's first miscellaneous series 1538–1775.

7. SRO: Borough charter (Letters Patent of James II), EE 501/1/11, 26 March 1685 (extract naming Catesby on http://www.nationalarchives.gov.uk/A2A/, accessed 1 April 2013).

8. NAUK: PROB/11/480.

9. Nelson 2013.

10. NAUK: PROB/11/480. Of note also was the land "lying behind my House in Mr Gainsborowes occupation in Sudbury," for Sudbury was to be the birthplace, two decades later, of Thomas Gainsborough, who became one of the most renowned artists of the late eighteenth century. Whether the house mentioned in John Catesby's will was occupied by Thomas's father, John Gainsborough (c. 1683–1748), publican, clothier, and postmaster, or another Gainsborough is not known. (Among the aldermen named in the Letters Patent in 1685 [see note 7 above] was one Richard Gainsborough.)

11. SRO: deed ref. 1674/14.

12. The original manuscript of Byrd's diary is in the Huntington Library, San Marino, California; for the decoded transcript, see Wright & Tinling 1941.

13. Frick & Stearns 1961: 12.

14. Wright & Tinling 1941: 518.

15. The name of the ship is sometimes given as *Hanover*, but this appears to be a recent error; for example, see Mary A. Stephenson, 1961 (Cocke-Jones Lots), Block 31, Colonial Williamsburg Foundation Library research report series, 1614 (http://research.history.org/DigitalLibrary/View/index.cfm?doc=ResearchReports%5CRR1614.xml).

By 1710 William and Elizabeth Cocke had had at least seven children born at roughly yearly intervals after August 1701: Elizabeth, Catesby, William, Anne, Rachel, Lucy, and Susan. For some unknown reason only two accompanied their mother on the *Harrison* for the long voyage to Virginia.

16. NAUK: PROB/11/480:

> I give to my disobedient Daughter Elizabeth one Anuity or yearly rent charge of twenty pounds to be issuing out of my said Messuage and Lands called Holgate to be paid Quarterly by equal portions upon the four most usuall ffast days or time of payment in the year (that is to say) Lady-day Midsomer Michmas and Christmas The ffirst payment to begin at such of the said ffeasts as shall next happen after the death of William Cock the Younger her supposed Husband But in case the said William shall be imprisoned for debt or other cause or shall get beyond Sea or otherwise absent or abscound himself from her so that she have no maintenance from him Then for such time of his absence I do give her ffive shillings per week to be paid her out of the said ffarent during the time of the said Cock imprisonment or absence and no longer And my will is that for want of paymt of the said yearly or weekly sums it shall be lawfull for the said Elizabeth to enter and distraint and to detain and keep such distress until payment . . .

(This is an exact transcription; the double *f* was used.)

17. ERO: N. Jekyll to W. Holman, 20 September 1712, D/Y1/1/111/11.

18. NHM: Botany Department, Dale/Catesby collection (no. 30: ". . . it's fruit are black berries ye bignes of Rouncival peas of which I have sent . . ."). The sending of the seed is confirmed by the manuscript list (see note 19 below) of "Seeds from Virginia sent by Mr Catesby to Mr Dale" (item 2, "Passiflora lutea Hedera fol. fr. parvo nigro").

19. BL: Sloane MS 3339: fols. 73b–75a (see Frick & Stearns 1961: 14 note 20).

20. Boulger 1883: 225; NHM: Botany Department, Dale collection (*Lathyrus* sp. "Lathyrus latifolius C.B. 344. . . . in horto D^ni^ Jekyll apud Hedingham ad castrum collegi Anno 1711.").

21. Alexander Spotswood to Henry Compton, 16 November 1713 (Brock 1882: II: 44–45; Frick & Stearns 1961: 14 note 18).

22. *NH* I: Account, v.

23. *NH* II: Appendix, plate 6.

24. Boulger 1883: 225; Frick & Stearns 1961: 15. NHM: Botany Department, Dale/Catesby collection contains several specimens from Jamaica, for example, *Cleome spinosa* (no. 4).

25. NHM: Botany Department, Dale/Catesby collection. The specimen is labeled "Bermudiana Iridis folio fibrosa radice. . . . In Insula Jamaicae invenit M. Catesby Anno 1715," but the species, being endemic to Bermuda, does not occur in Jamaica.

26. *NH* II: Appendix, plate 14.

27. *NH* I: Account, xli.

28. OBL-S: MC to J. J. Dillenius, not dated (c. 1737), Sherard MS 202, fol. 15, transcribed by Serena Marner (Herbarium Manager, OUH); on juniper, see also O'Malley 1998: 167–168, 170.

29. Petiver 1714: 357 (no. 106), 358 (no. 109), 359 (no. 112), 362 (no. 119).

30. NHM: Botany Department, Dale/Catesby collection (no. 44).

31. NHM: Botany Department Dale/Catesby collection (no. 40).

32. RS CCLIII 211: S. Dale to WS, 15 October 1719.

33. Frick & Stearns 1961: 18.

34. Frick & Stearns 1961: 19.

35. This copy was sold at auction (sale no. 303, lot 1) on 24 February 2005 by PBA Galleries, San Francisco.

36. Frick & Stearns 1961: 20.

37. Jones 1891: 218.

38. ERO: N. Jekyll to W. Holman, 14 December 1722 (D/Y 1/1/111/82) (for drawing this to my attention I gratefully acknowledge information from Mrs. Valerie Herbert).

39. RS CCLIII 163: MC to WS, 5 May 1722.

40. RS CCLIII 163: MC to WS, 5 May 1722.

41. RS CCLIII 163: MC to WS, 5 May 1722.

42. *NH* I: Preface, v.

43. Jones 1891: opposite p. 118.

44. RS CCLIII 178: MC to WS, 16 August 1724; for the earlier letter dated 16 January 1723/24, see RS CCLIII 174.

45. BL: Sloane MS 4046, fols. 307–308: MC to HS, 5 January 1724/25.

46. *NH* I: Account, xxxix.

47. *NH* I: Preface, xi. On Goupy, see O'Connell 2008. He was listed in "A list of the Encouragers" in *NH* 1 but his surname was printed as "Gopy."

48. *NH* I: Preface, xii.

49. See Frick & Stearns 1961: 39; Meyers & Pritchard 1998: 14 note 31; see Chaplin 1998: 54–55 on Catesby's capacity to read and comprehend R. A. Réaumur's *Mémoires pour servir à l'histoire des insectes*, which he reviewed for the Royal Society. He must also have had a thorough knowledge of Latin, for he agreed to review Linnaeus's *Systema naturae* (1735); see Chaplin 1998: 46.

50. Jones 1891: 217; Jones erred in making Rutherford the brother-in-law of Mark Catesby.

51. *Derby mercury*, 29 May 1729: 4; *Newcastle courant*, 31 May 1729: 3.

52. *Newcastle courant*, 31 May 1729: 3.

53. LS: PC to CL, 15 April 1747.

54. Overstreet p. 158 (this volume).

55. Meyers 1997: 17, figure 1.

56. NAUK: PROB/11/632, fols. 268r–269v; see Leapman 2000.

57. *NH* I: title-page. Bacon was acknowledged six times in the first edition of *NH* I: plates 49, 53, 55, 57, 62, and 66 (see note 64 below); Fairchild was acknowledged three times in *NH* I: plates 22, 27, and 39.

58. "Without" signifies that the church, when originally built, was outside the walls of the city, near the gate called Cripplegate. Saint Giles is the patron saint of cripples and beggars.

59. LMA: P69/GIS/A/002/MS06419, item 016: St. Giles Cripplegate, Composite register, 1726/27–1733.

60. LMA: P76/LUK, item 001: St. Luke, Old Street, Composite register: baptisms, marriages, burials, October 1733–April 1742.

61. Boyd's marriage index, http://www.findmypast.co.uk/content/sog/misc-series, accessed 1 March 2013.

62. Henrey 1986: 135. Knowlton had been gardener to James Sherard (1666–1738) at Eltham in Kent during the period when Catesby was in the Carolinas and Bahamas, and later he helped get subscribers for *The natural history of Carolina* . . . (see Henrey 1986).

63. ERO: N. Jekyll to W. Holman, 14 August 1730, D/Y1/1/111/187.

64. Raphael 2004. Catesby acknowledged Gray in *NH* II: plate 78 and Appendix: plate 4, as well as in the revised text (in which Gray's name replaced Bacon's) for *NH* I: plates 49, 53, 55, 57, and 66.

65. Henrey 1975: 2:opposite p. 348.

66. *NH* II: plate 98 and Appendix: plate 13, respectively.

67. Allen (1937: 470) was skeptical, pointing out that "No letters that have yet come to light were written in Fulham."

68. Willson (1982: 18–19) did not give any archival source for this and probably based her statement on earlier sources, of which Lysons (1811: 829) is perhaps the earliest: "Catesby, the naturalist, had a botanic garden within the site of the nursery grounds, lately occupied by Mr. Birchall [*sic* Burchell], (now by Whitley and Brames,) where some of the American trees planted by him are still remaining."

69. Allen 1937. There was no ceremonial induction, as supposed by Dr. Elsa Allen.

70. Catesby 1747; see Krech pp. 219–250 (this volume).

71. Overstreet p. 158 (this volume).

72. *NH* II: Appendix, plate 20.

73. Lucas 1892: 119.

74. Henrey 1986: 209.

75. *Caledonian mercury*, no. 4,561 (Tuesday, 2 January 1750): 1.

76. Transcribed in Frick 1960.

77. As shown by the baptismal register of St. Giles-without-Cripplegate (see note 59 above), the "Boy" was in fact more than eighteen years old when his father died.

78. LMA: P76/LUK/002/002.

79. Frick 1960: 173–174.

80. See note 1 above.

81. Henrey 1986: 292.

82. NAUK: PROB/11/803. Probate was granted on 29 August 1753.

83. LMA: P76/LUK, item 003/1: Saint Luke, Old Street, Composite register: baptisms January 1753– November 1759; marriages January 1753–March 1754; burials January 1753–November 1759.

84. Boyle et al. 2005.

CHAPTER 3. Mark Catesby's botanical forerunners in Virginia

ACKNOWLEDGMENTS

I am grateful to Anne-Marie Catterall (Sherardian Librarian, Sherardian Library of Plant Taxonomy); Michael Eck (Library Director, Mount Holly Library and Burlington County Lyceum of History and Natural Sciences); Roy Goodman (Assistant Librarian and Curator of Printed Materials, American Philosophical Society); Chris Lippa (Photographer and Imaging Supervisor, Rare Book and Manuscript Library, University of Pennsylvania); Kerry Magruder (Curator, University of Oklahoma Libraries); Leslie Overstreet (Curator, Smithsonian

Institution); Stacey Peeples (Curator, Pennsylvania Hospital Historic Collections); and John Pollack (Library Specialist for Public Services, Rare Book and Manuscript Library, University of Pennsylvania).

NOTES

1. Harriot 1588, 1590; Sloan 2007.

2. Sloan 2007; Harkness 2007: 37; 2009.

3. Gerard 1598: 752; Reeds 2009; Rickman 2011. On the identification of White's and Gerard's figure, see Ewan & Ewan 1970: 40, 170–172, 397–399; Larry Allain, 2010, *Grasse silke*, in "Roanoke colonies illuminated" (http://digital.lib.ecu.edu/hakluyt/view.aspx?p=Grasse%20silke&id=847, accessed 4 June 2013, Joyner Library, East Carolina University).

4. Gerard 1598: 752; Jackson 1876.

5. Gerard 1633; Parkinson 2007; Raven 1947: 248–273; Gunther 1922; Henrey 1975: 1:45–54; Potter 2006; Smith 1624: 26–32.

6. Gerard 1633: 898–900.

7. Sloan 2007; McBurney 1997: 33, 39.

8. *NH* II: plate 97.

9. Potter 2006; Leith-Ross 2006.

10. Potter 2006; Parkinson 1629: 151–153.

11. Tradescant 1634, 1656: 73–178; Leith-Ross 2006: appendixes 2, 3.

12. Parkinson 1629: dedication, fol. i c recto; Parkinson 2007; Henrey 1975: 1: 161–167; Boulger 1918; Raven 1947: 248–273.

13. Potter 2006; Gerard 1633: 260, 848–849.

14. These are based on my tallies of plants identified as coming from Virginia in the indexes of Parkinson (1629, 1640); see also Gunther (1922: 370–371).

15. For four plants, Catesby gave Latin names attributed to Banister (*NH* I: 9, 17, 26, 57, plate 17). Banister is also invoked through Plukenet's name "Frutex Virginianus trifolius, Ulmi samaris; Banisteri" (*NH* II: plate 83). The manuscript inscriptions on Catesby's original watercolors for *NH* I: plates 9 and 57, also used Banister's names (see McBurney 1997: 42–43, 58–59).

16. Tjaden 1977; Ewan & Ewan 1970.

17. Byrd II was in England from 1681 to 1705. After returning to Virginia, he became both friend and employer to Banister's son (also named John Banister) (Ewan & Ewan 1970: 106–107).

18. Vines & Druce 1914; Clokie 1964: 1–3, 10–13.

19. Switzer 1718: 70, 356; Ewan 1970: 73–75, 77, 84–86; Winnington-Ingram 1908.

20. Ewan & Ewan 1970: 38–46, 166.

21. Ewan & Ewan 1970: 80–84.

22. Ewan & Ewan 1970: 94; Ray 1686: 1: 674; 1688: 2: praefatio, 1928–[1929].

23. Harriot 1588: fol. F 4 recto; Quinn 1955: 1: 260, 288, 293, 387 note 3.

24. Harriot 1588: fol. B 3 verso; Quinn 1955: 1: 334 notes 1, 3.

25. Quinn 1955: 1: 260, 293; Batho 2000: 31; Trevor-Roper 2006: 207.

26. Quinn 1955: 2: 515–543, 615; Sloan 2007; Reeds 2009.

27. Potter 2006: 113 note 2; Leith-Ross 2006.

28. Tradescant 1656; Boulger 1918; Leith-Ross 2006.

29. Ray 1688 2: Praefatio; Ewan & Ewan 1970: 94.

30. Ewan & Ewan 1970: 61, 89–96. An inquest found death by misadventure and acquitted the woodsman.

31. Ewan & Ewan 1970: 98–108, 160–165; Plukenet 1691–1694, 1696, 1696–1725, 1700, 1705; Plukenett 1942; Riley 2006; Vines & Druce 1914.

32. BL: Sloane MS 4002, J. Banister, "Catalogus Stirpium Rariorum in Virginiâ sponte nascentium Inchoatus Dmi. M.DC.LXX.IIX. Quibus praefigitur Asteriscus, sunt plantae a Nostratibus, vel Indigenis Cultae." Previously, Banister had sent a longer version ("Inchoatus

Dmi. M.DC.LXXIX") to England, which survives with Bobart's annotations (OBL-S: Sherard MS 37, fols. 22–45). The edition of Banister's "Plant Catalogue" published by Ewan & Ewan (1970: 161–163, 167–253, 402–403) was based on that longer manuscript; they regarded Sherard MS 37 and Sloane MS 4002 as "essentially" the same and did not record every variation between the two.

33. Ewan 1970: 119–126; Beverley 1705: ix.

34. Ewan 1970: 45.

35. Monardes 1577: 1: 100; Harriot 1588: fol. B 2 recto–verso; Quinn 1955: 2: 329 and notes 2–5; Reeds 2009; Gerard 1598: 750–751.

36. Reeds 2009; Quinn 1955: 2: 615–616.

37. Sloane 1696: 129; *NH* II: plate 95.

38. Monardes 1593, 1605; Tradescant 1656; Leith-Ross 2006: 266–306.

39. The Byrd library had two copies of each of the first and second volumes of Ray's *Historia plantarum* (1686, 1688) but none of the third (Ray 1704) (see Hayes 1997: 3–4; 268, no. 786; 470, no. 1917; Plukenett 1942; Ewan & Ewan 1970: 137).

40. Hayes 1997: 37–38; Byrd 1941: 203.

41. In 2006, I had the good luck to recognize the Parkinson herbal now in the Burlington County Lyceum of History and Natural Science collection of the Mount Holly Public Library, Mount Holly, New Jersey, as the long-missing Banister and Byrd library copy (see Ewan & Ewan 1970: 158; Hayes 1997: 294, no. 916).

42. NHM: Sloane Herbarium, HS 168 (see Ewan & Ewan 1970). See also Banister's companion notebook to his herbarium (OBL-S: Sherard MS 30, "Catalogus Plantarum in Horto et circa Oxoniam crescentium" [not dated; before 1678]).

43. BL: Sloane MS 4002; Ewan & Ewan 1970: 167–253.

44. Banister's sole reference to "Ray in Edit. nov." (for a moss) shows he was still working on his draft catalog in 1686 and almost certainly in 1688, when Byrd I delivered the copies of Ray's *Historia plantarum* (BL: Sloane MS 4002, fol. 22r; OBL-S: Sherard MS 37; Ray 1686: 113; Ewan & Ewan 1970: 218).

45. In 1696, at the age of twenty-two, Byrd II was elected a Fellow of the Royal Society (Marambaud 1971).

46. Ewan & Ewan 1970: 114–117; Tinling 1977: 1: 267.

47. Wright & Tinling 1941: 523, 534, 544; Hayes 1997: 37–47.

48. Using Hayes (1997), my tally of books on these subjects published by 1712 comes to more than 150 titles.

49. For Byrd II's references to Catesby, see Wright & Tinling 1941. The journals for 1713 to late 1717 are not extant. Byrd left Virginia early in 1715 for England. He set sail for Virginia on 10 December 1719, arriving in February 1720. Catesby had returned to England by mid-October 1719. Byrd's London diary for 1719 does not mention Catesby (Hayes 1997: 48–57; Wright & Tinling 1958). The Byrd family correspondence has no mention of Catesby until 1736 (Tinling 1977: 3: 498). Byrd's references to his new-built library from 1709 to 1715 imply that he ordinarily used it by himself in the afternoon and occasionally for socializing in the evening; otherwise, he kept it locked.

50. *NH* I: v–vi.

51. *NH* I: v–vi.

52. Morris 1974; Taylor 1913: iv–v.

53. Nelson note 20, p. 352 (this volume); Sperling 1895a: 260.

54. Frick & Stearns 1961: 5–10, 14; Raven 1947.

55. Ballard 1708; Raven 1947: 481. Ray's library held key sources for Virginian plants: Ray's own works, Gerard (1598, 1633), Parkinson (1629, 1640), Tradescant (1656), *Philosophical transactions*,

and books by Petiver and Plukenet that had incorporated John Banister's material. It did not, however, include a copy of Harriot (1588, 1590); and Catesby did not see the Sloane album of copies of White's watercolors until he settled in London. He and Sloane believed that the drawings originated in Raleigh's expedition to Virginia (see McBurney 1997: 33, 39).

56. Boulger 1883. The *Pharmacologia* references to "Banis. MSS. Cat." for "Asarum" and "Aristolochia" (Dale 1693: 146, 289) probably rely on the pre-1691 Banister material incorporated into Plukenet 1691–1694: tables 15 and 148 (see also table 78). Ewan & Ewan 1970: 76, 100, 174, 231–233, figs. 16 and 17. However, it is also just possible that Banister's personal draft plant catalog, sent posthumously from Virginia in late June 1692, reached London in time for Dale (or his London colleagues) to work information from it into the *Pharmacologia* (BL: Sloane MS 4002, fols. 10r, 26r).

57. Ray 1698; Petiver 1707; Ewan & Ewan 1970: 265–271.

58. *NH* I: x.

59. *NH* I: xix.

60. *NH* I: xix.

61. If Catesby saw Banister's annotations in his copy of Parkinson, there is no sign of it in *The natural history of Carolina . . .* references. *NH* I: 34, 53; Parkinson 1640: 1410, 1465. See also the heading in *NH* I: 21. See note 15 above.

62. Chaplin 1998: 35.

63. Catesby [1729?]. See also this volume, p. 157 (figure 12-1).

64. *NH* I: Dedication.

65. *NH* I: xi.

CHAPTER 4. Maria Sibylla Merian (1647–1717): pioneering naturalist, artist, and inspiration for Catesby

ACKNOWLEDGMENTS

We thank Jo Francis and John Fuegi for their comments and suggestions. For help in various ways, many thanks are also due to Leslie Overstreet (Smithsonian Libraries); Diny Winthagen, Hans Mulder, and Jip Binsbergen (Artis Library, University of Amsterdam); Godard Tweehuysen (Library of the Dutch Entomological Society, Amsterdam); Sandrine Ulenberg (now at NCB Naturalis, Leiden); and Niels de Boer (Art Gallery P. de Boer, Amsterdam). Furthermore, we wish to express our gratitude to the owners of the artwork reproduced here for giving us permission to use it.

NOTES

1. Merian [1705].

2. Named on their respective marble plaques as "ARISTOTELES," "PLINIUS," and "C. v. LINNÉ" (Pieters 1988).

3. Since 1939 the Artis Library has been part of the University of Amsterdam, whereas the Zoological Museum was merged in 2011 into the Netherlands Centre for Biodiversity *Naturalis* at Leiden.

4. Bürger & Heilmeyer 1999: 8.

5. Segal 2012: 19–28.

6. Marrell witnessed "tulipomania," the economic phenomenon of 1637, when many Dutch people speculating on the price of tulip bulbs went bankrupt. At the time the price of a tulip bulb was sometimes higher than that of a painting in oils by Marrell and could be ten times the annual salary of a skilled laborer (Bott 2001: 127–149; Pavord 1999).

7. Bott 2001: 131.

8. For instance, in Nuremberg (Wettengl 1998: 19).

9. Jonston 1653, 1660; caterpillars were treated as separate organisms, toward the end of the book.

10. The collective title of *Neues Blumenbuch* was issued with the third installment in 1680 (see Merian 1680); first and second installments were published in 1675 and 1677, respectively (see Bürger & Heilmeyer 1999).

11. Merian 1679.

12. Merian 1683.

13. See, for instance, the title of John Ray's book *The wisdom of God manifested in the works of creation* (1691), a book also known to Catesby. Likewise, Jan Swammerdam considered his research in natural history as the search after God in his *Biblia Naturae* (*Bible of Nature*). Thus natural theology was a way of studying nature that was seen as equaling the study of the Bible (Deppermann 2002; Trepp 2009; Jorink 2010).

14. After its rediscovery in the Russian Academy of Sciences in St. Petersburg, Merian's research journal was published in facsimile (Merian 1976). For Merian's description of the metamorphosis of frogs, see Merian 1976: 1: plate 77; 2: 293/76; see also Etheridge 2010.

15. Deppermann 2002: 144–150; Trepp 2009: 210–305.

16. Merian [1705]: preface: "However, in Holland I marveled to see what beautiful creatures were brought in from the East and West Indies, particularly when I had the honor of seeing the fine collection of the Most Honorable Heer Meester Nicolaas Witsen, mayor of the city of Amsterdam and director of the East India Company, &c., as well as that of the Honorable Heer Jonas Witsen, secretary of that city. In addition, I saw the collection of Heer Fredericus Ruisch, M.D., *Anatomes et Botanices Professor*, that of Heer Livinus Vincent, and of many others."

17. Merian [1705]: preface (translated by Michael Ritterson and Florence Pieters).

18. Merian 1717. The edition in Latin was published the next year (Merian 1718).

19. Pieters & Winthagen 1999: 6; Lebedeva 1976.

20. Both daughters remained active as artists during the rest of their lives; however, they were apparently not interested in the study of metamorphoses. The eldest daughter, Johanna Helena, died in Surinam in 1728, while the younger one died in St. Petersburg in 1743 (Reitsma & Ulenberg 2008: 237–238).

21. Merian 1679.

22. Merian even conducted simple experiments on larval food choice, offering a variety of plants to caterpillars and recording their preferences. These types of observations form a cornerstone of modern plant-animal ecology (Etheridge 2011: 41).

23. Merian 1679: preface. Merian's study notes (reproduced in facsimile in Beer 1976: 2) establish that her early work on insect reproduction was contemporaneous with that of Francesco Redi (1626–1697), Marcello Malpighi (1628–1694), and Jan Swammerdam (1637–1680). Jan Goedaert (1662–1669) depicted European moths and butterflies with their pupal and larval forms but usually did not portray food plants or eggs.

24. Jonston 1653, 1660.

25. Sloane 1707, 1725.

26. Merian [1705]: 12.

27. Merian [1705]: 36.

28. Merian [1705]: 48.

29. Merian [1705]: 4, included a description of the tegu (*Tupinambis merianae*), its breeding, and its food habits. Merian commented: "Another whole book could follow about such creatures, if I see that this work is appreciated by amateur naturalists and sells well."

30. Etheridge 2010: 24.

31. Merian [1705]: plate 4.

32. Merian [1705]: plate 18.

33. Etheridge 2010: 17, 21.

34. Examples include James Petiver (1663–1718), Moses Harris (1731–1785), René-Antoine Ferchault de Réaumur (1683–1757), John Abbot (1751–1840), and August Johann Rösel von Rosenhof (1705–1759) (Etheridge 2011: 43–47).

35. Albin 1720. Albin was a protégé of Sloane, but he also may have seen Merian's work in his native Germany.

36. Merian [1705]: plate 56; *NH* II: Appendix, plate 11.

37. Meyers 1997: 21–22.

38. *NH* II: Appendix, plate 9.

39. *NH* I: Account, xxix.

40. A version with seventy-two plates instead of Merian's original sixty was reprinted a number of times in various languages. The additional plates were added by a publisher in 1719 after Merian's death, and some of these were not her work.

41. Merian 1726.

42. As pointed out by Chaplin (1998: 87), "Merian's work reached its widest audience from 1719 to 1730, just before Catesby began to publish his volumes."

43. Merian [1705].

44. Merian 1719.

45. Etheridge 2010: 25. The "reverse metamorphosis" of frogs changing into fish is figured on the lower part of plate 71 in editions of *Metamorphosis* from 1719; Merian undoubtedly would have been appalled at its inclusion in her carefully researched work.

46. *NH* II: Appendix, plate 11.

47. *NH* II: plates 84 and 86 depict, respectively, a luna moth and a cecropia moth and their cocoons, and *NH* II: plate 94 includes a horned devils caterpillar.

48. In cases where Merian showed insects on plants other than a larval host, she usually gave a reason. In some instances this was because she wished to showcase a specific plant, such as the pineapple, and in other cases she inserted an unrelated plant rather than repeat an illustration of a species that had multiple herbivores.

CHAPTER 5. William Dampier (1651–1715): the pirate of exquisite mind

NOTES

1. Sprat 1667: 1–2, 126.

2. G. Edwards to T. Pennant, 5 December 1761 (quoted in Frick 1960: 173; McBurney 1997: 12; Frick & Stearns 1961: 9).

3. Dampier 1697: 61–62.

4. Dampier 1697: 49.

5. *NH* I: plate 87.

6. *NH* II: plate 96.

7. These birds were the greater flamingo (*Phoenicopterus roseus*), which does not occur in the Americas. The American flamingo (*Phoenicopterus ruber*) was the bird observed and depicted by Catesby (*NH* I: plates 73 and 74).

8. Dampier 1697: 70–71.

9. Wilkinson 1929: 4.

10. Preston & Preston 2004: 11. Dampier does now have a memorial in the village church near that of the poet T. S. Eliot (1888–1965), whose ancestor Andrew Eliot immigrated to

America from East Coker in 1660 when Dampier was nine years old. They must have known each other.

11. Among his companions was Lionel Wafer (1640–1705), who also wrote an account of their exploits and some of whose loot was later seized by the authorities in Virginia and used to found the renowned College of William & Mary at Williamsburg.

12. Dampier 1697: 463.

13. 6 August 1698 (Dobson 1906: 3: 336).

14. Beck 1987: 90.

15. Barrett et al. 1987: 22.

16. 4 June 1836 (Keynes 1988: 425).

17. Dampier 1699.

18. Dampier 1699: 103.

19. Dampier 1699: 93–94.

20. Dampier 1697: 414–415.

21. See Rediker 1987: 182.

22. Dampier 1697: 415.

23. Dampier 1697: 533.

24. Dampier 1699: 126.

25. Dampier 1697: 296–267.

26. Bligh 1790.

27. Dampier 1697: 202–203.

28. Dampier 1697: 202–203.

29. Dampier 1697: 228–229.

30. Dampier 1703: 75.

31. The "red notebook" is one of a series of notebooks kept by Charles Darwin during and immediately following his service as naturalist on the 1831–1836 voyage of HMS *Beagle*.

32. Dampier 1703: 122.

33. Dampier 1709: 137.

34. Dampier 1697: 408.

35. Dampier 1699: 112–113.

36. *History of the works of the learned* 1 (February 1699): 94–98.

37. Coleridge 1917: 168.

38. Coleridge 1917: 280.

39. Somerset Record Office, Taunton: DD/WHh 1090, pt. 2: W. Whaley to W. Helyar, 27 January 1675.

40. Ray 1704.

41. *NH* II: plate 85: "Anona fructu lutescente, laevi, scrotum Arietis referente."

42. Morton 1703; see also Krech pp. 219–250 (this volume).

CHAPTER 6. John Lawson's *A new voyage to Carolina* and his "Compleat History": the Mark Catesby connection

ACKNOWLEDGMENTS

Alan S. Weakley (Director, University of North Carolina Herbarium, Chapel Hill, North Carolina) provided important assistance in identification of Lawson's plants in the Sloane Herbarium from photos by the author, 1993–2009.

NOTES

1. Simpson & Simpson 2008: 224–225. In 1712, the Lords Proprietors started appointing separate governors for North Carolina due to its distance from Charleston (Weir 1997).

2. *NH* I: Account, viii.

3. BL: J. Lawson to J. Petiver, 30 "XBER." 1710, Sloane MS 4064, fols. 249–250, 264 (transcribed in Lefler 1967a: 272).

4. Briceland 2000; Lefler 1967a: xi–liv; Ransome 2004; Wright & Tinling 1941: 523.

5. Lawson 1709: 1–2 (pagination herein is to the 1709 edition, which differs from that in the recent edition by Lefler 1967a).

6. Stearns 1952: 241–365; Armytage 1954.

7. Lawson 1709: 6–60.

8. Stearns 1952: 363–365.

9. BL: J. Lawson to J. Petiver, 12 April 1701, Sloane MS 4063, fol. 79 (transcribed in Lefler 1967a: 267–268).

10. Stearns 1952: 334–354.

11. Lawson 1709: Preface.

12. Dickinson 1967; Lefler 1967a: xiv–xx, xxiv–xxviii; Snapp 2000. Thousands of refugees from the Palatinate of the Rhine valley fled to England in 1708–1709, and Lawson became a central participant with the Swiss land speculator von Graffenried in efforts, supported by Queen Anne, to transport some 650 of the "poor Palatines" to a new life in the colony of Carolina.

13. The "Glorious Revolution" of 1688–1689 brought the deposing of King James II and the accession of King William III and Queen Mary II to the English throne (Miller 1997; Simpson & Simpson 2008: 223–234).

14. Simpson & Simpson 2008: 226–227.

15. See Overstreet pp. 155–172 (this volume).

16. Simpson & Simpson 2008: 225–229.

17. Simpson & Simpson 2008: 229–233.

18. Lawson 1709: 6–60, 169–238.

19. Lawson 1709: 89–114.

20. Lawson 1709: 115–163.

21. Lawson 1709: 135–151.

22. McAtee 1955, 1956.

23. *NH* I: plate 16.

24. Simpson & Simpson 2008: 225.

25. The most notable example was John Brickell's *Natural history of North-Carolina*, published in 1737 (for others, see Lefler 1967a: lii–liv).

26. BL: J. Petiver to G. London, 7 September 1709, Sloane MS 3337, fol. 56 (transcribed by Lefler 1967a: xli). George London served as gardener to William III and Mary II and later to Henry Compton (1632–1713), Bishop of London. Coleby 2004; Dandy 1958: 157–159; Ewan & Ewan 1970: 24; Harris 2004; Royle 1995; Switzer 1718: 79–84.

27. Simpson et al. 2010: 338.

28. Coues 1878: 577–578.

29. Wright & Tinling 1941: 441. Catesby cited the 1714 edition of Lawson's *History* rather than *A new voyage* (Lawson 1709), indicating that his working copy was the later release.

30. Ray 1686.

31. *NH* I: Account, viii.

32. *NH* I: plate 11; Lawson 1709: 96.

33. Lawson 1709: 132; *NH* II: plate 57.

34. Lawson 1709: 141; *NH* II: plate 23.

35. Lawson 1709: 117; *NH* I: Account, xxvi.

36. Lawson 1709: 123; *NH* I: Account, xxviii.

37. Lawson 1709, 158; *NH* I: Account, xxxiii.

38. Lawson 1709: 257.

39. Simpson et al. 2010: 336.

40. Lawyer, statesman, scientist, and philosopher, Francis Bacon (1561–1626) argued for an empirical method of scientific investigation based on direct observation and experimentation. His principal work, *Novum organum* (1620), strongly influenced the rationalist philosophy advocated by members of the Royal Society (Peltonen 2004).

41. Lawson to Petiver, 30 "XBER." 1710.

42. Simpson et al. 2010: 335.

43. Lawson to Petiver, 30 "XBER." 1710.

44. Lawson to Petiver, 30 "XBER." 1710.

45. Lawson to Petiver, 30 "XBER." 1710.

46. Simpson et al. 2010: 335–336.

47. Dandy 1958: 154. NHM: specimens from Sloane's *hortus siccus* HS 145, fols. 47, 58, 60; HS 242, fols. 123, 126, 129, 132, 133, 134, 135 (identifications by M. B. Simpson and A. S. Weakley).

48. Simpson et al. 2010: 345.

49. Lefler 1967a: xxxvii.

50. Simpson et al. 2010: 341.

CHAPTER 7. Mark Catesby's world: England

NOTES

1. Cook 1993, 2007; Allen 1976.

2. Smith & Findlen 2002.

3. See, for example, Secord 1994b; Gasgoigne 1994; Findlen 2008.

4. Harris 2008.

5. Minter 2000: 11–28; Le Rougetel 1971.

6. O'Neill & McLean 2008.

7. Gunther 1922: 279; James 2004.

8. Johns 2008.

9. Allen 1976: 10; Secord 1994a; Riley 2006.

10. Christakis 2009.

11. Spary 2000. Cook (2012) discussed eighteenth-century botanical correspondence networks and described these networks as a "commonwealth of learning."

12. These examples are imaginary. For an analysis of gift-giving practices up and down the social scale, see Secord 1994b.

13. The entry for Catesby in the *Oxford dictionary of national biography* confuses Catesby's uncle Nicholas Jekyll with his grandfather of the same name and mistakes Catesby's sister Elizabeth for his aunt. For Catesby's early contacts, see Meyers & Pritchard 1998; Brigham 1998; Frick & Stearns 1961: 3–8; French 2000: 56. A comparable trajectory has been mapped out by Cook (2012) relating to Samuel Dale's nephew Thomas (1700–1750).

From Catesby's letters we know he sent material from Carolina to Isaac Rand at the Chelsea Physic Garden and to the Earl of Oxford. He was in correspondence with Sir George Markham too. There is no full listing of Catesby's extant correspondence, but letters are known to have been sent in his lifetime to Linnaeus, J. F. Gronovius, Collinson, John Bartram, Sloane, William Byrd II, William Sherard, and Dillenius. His letter-writing was not on the grand scale of Collinson's (but see Frick & Stearns 1961: 86–98). Laird (1998) detailed the garden plants introduced by Catesby.

14. Leapman 2000: 155–184.

15. Chaplin 1998.

16. On this significant period in Catesby's career, see Meyers & Pritchard 1998: 5–7; Frick & Stearns 1961: 17–21.

17. Frick & Stearns 1961: 19.

18. Brigham (1998) discussed in detail the patronage that Catesby received.

19. Calmann 1977.

20. See Nelson 2014.

21. Leapman 2000: 143.

22. Overstreet p. 156 (this volume).

23. O'Malley 1998; Meyers & Pritchard 1998: 10 note 21.

24. Catesby 1767: 2. This work was evidently taken through publication by John Ryall, who signed the dedication to Henry Seymer of Hanford, Dorset.

CHAPTER 8. Mark Catesby's world: Virginia

ACKNOWLEDGMENTS

I am grateful to Marianne Martin (Visual Resources Librarian, Colonial Williamsburg Foundation) for helping me to locate images of early Williamsburg.

NOTES

1. *NH* I: Preface, v.

2. *NH* I: Preface, v.

3. Jones 1724: 32; Lefler 1967b.

4. Kornwolf 1993.

5. Lounsbury 2000.

6. Thomas Jones's will is transcribed in Stephenson 1961. On purchasing habits, see Shammas 1980.

7. Plows were not necessary in part because colonists grew maize and tobacco in "hills" or turned-up soil (Greene 1965: 442).

8. Wells 1993: 7–9: "Early Virginia countryside was indeed dominated by houses built of wood . . . [and] . . . the majority of advertised houses had wooden chimneys and footings as well as wooden structures and siding."

9. For more on how and why colonists changed their diets, see Meacham 2009; for Byrd, see Meacham 2009: 27–28.

10. RS CCLIII 165: MC to WS, 9 December 1722.

11. Russo & Russo 2012: 93.

12. Wenger 1981.

13. See *NH* I: plate 65.

14. Wright & Tinling 1941: entries for 25 May 1712, 26 May 1712; Tinling 1977: 1: 266 (W. Byrd to HS, 10 September 1708).

15. Wright & Tinling 1941: entry for 5 June 1712; Pennsylvania Historical Society: Bartram Papers, box F. 21, 18 July 1740.

16. Wright & Tinling 1941: entry for 14 June 1712.

17. Rountree 1990: 159, 169; Wright & Tinling 1941: entry for 23 September 1712.

18. Wright & Tinling 1941: entry for 29 September 1712.

19. *NH* II: plate 76, copied from a sketch of a flying squirrel by Everard Kick; see also McBurney p. 151 (this volume).

20. See Beatty & Mulloy 1940: 56–58, 64.

21. Stiverson & Butler 1977: 18–44, 39.

22. Kalm 1770: 165.

23. *NH* I: Preface, v.

24. Tinling 1977: 2: 498 (Byrd to Posford, c. August 1736), 518 (Byrd to MC, 27 June 1737), 523 (Byrd to PC, 5 July 1737).

25. Oberg 2002: 29: 298 (Jefferson to the American Philosophical Society, 10 February 1797); Onuf 2010.

NOTES

1. For an excellent first-person account by Janet Schaw, "a lady of quality," of a voyage to Carolina, see Andrews & Andrews 1939: 48–49.

2. Linder 2000: 29–30; Frick & Stearns 1961: 23.

3. Smallpox was certainly recognizable, but many other diseases were not differentiated and were called fevers or fluxes. RS CCLIII 166: MC to WS, 10 December 1722.

4. For correspondence with Ellis, Bohun, and Francklin, see Armytage 1954.

5. Hannah English, Abstract of Land Grant, 500 acres in Berkeley County, 1695, series S213019, vol. 0038, p. 00285, item 004, South Carolina Archives, Columbia, South Carolina.

6. Petiver 1704: 1: decade 4, plate XXXIII, figure 11 (Smith 1986: 83, 85); Smith 1918: 51.

7. Stearns 1970: 296–302.

8. *NH* I: Preface, i.

9. Meyers 1997: 14; Edgar 1998: 178.

10. Weir 1997: 54–56.

11. Edgar 1998: 102–105.

12. Crane 1956: 162–186; Edgar 1998: 100.

13. Kovacik & Winberry 1987: 77; Crane 1956: 131.

14. Hall 1991: 165.

15. McDowell 1992: 103.

16. Edgar 1998: 73–74.

17. Margaret Kennett to Mrs. Thomas Brett (Spring Grove, Wye, Ashford, Kent), 20 January 1724/25 (transcribed in Enright 1960: 15–18).

18. Linder 2006: 712.

19. Frick & Stearns 1961: 29–31.

20. RS CCLIII 171: MC to WS, 10 May 1723.

21. Crane 1956: 198–199.

22. Edgar & Bailey 1977: 468–469.

23. Frick & Stearns 1961: 23.

24. Heitzler 2010: 36–39; Swanton 1922: plate 3.

25. Feduccia 1985: 125–126.

26. Edgar & Bailey 1977: 468–469, 700–701.

27. *NH* I: plate 43.

28. *NH* I: plate 31.

29. Feduccia 1985: 7–8, 134, 101; Meyers 1997: 14.

30. Cooper 1838: 2: 396–399; Edgar & Bailey 1977: 618.

31. Smith 1905: 80; Cooper 1838: 3: 214–216.

32. Smith 1905: 82.

33. RS CCLIII 174: MC to WS, 16 January 1723; *NH* II: plate 41; Frick & Stearns 1961: 24.

34. RS CCLIII 174: MC to WS, 16 January 1724; Frick & Stearns 1961: 26.

35. Edgar 2006: 476.

36. *NH* I: Preface, viii–ix.

37. Ivers 1970: 28–29, 62–63.

38. RS CCLIII 171: MC to WS, 10 May 1723.

39. *NH* I: Account, xxxiv; Feduccia 1985: 161.

40. P. D. McMillan, pers. comm., 18 January 2012; McMillan et al. 2013.

41. Crane 1956: 44.

42. *NH* I: Account, iii.

43. Zierden et al. 1999: 42; Ivers 1972: 126.

44. Feduccia 1985: 161; McMillan et al. 2013.

45. Frick & Stearns 1961: 29.

46. Linder 2000: 73–77.

47. *NH* I: Account, xiii–xiv.

48. Merrens 1969: 546–549. In 1815, the legislature of South Carolina passed a tax bill that categorized land in a similar way. Prime tidal rice fields were taxed at $26 per acre, and the valuations declined down to $1 per acre for pine barrens or salt marsh (McCord 1839: 6: 7).

49. *NH* I: plate 23.

50. *NH* I: plate 11.

51. Feduccia 1985: 55–56.

52. *NH* I: plate 16.

53. Feduccia 1985: 88.

54. *NH* I: plate 14.

55. Robertson pp. 127–140 (this volume); Krech pp. 219–250 (this volume).

56. *NH* I: Preface, xi.

CHAPTER 10. Mark Catesby's Bahamian natural history (observed in 1725–1726)

NOTES

1. *NH* I: Account, xxxviii–xliv, contains Catesby's comments on the Bahamas.

2. *NH* I: Account, xli.

3. *NH* I: Account, xl.

4. *NH* I: Account, xxxix.

5. *NH* I: Account, xxxix.

6. *NH* I: Account, xxxix.

7. Morison 1942.

8. *NH* I: plate 40.

9. *NH* I: plate 59.

10. OUH have specimens of about a dozen marine algae, one gorgonian, and ferns and flowering plants from the Bahamas collected by Catesby (see http://herbaria.plants.ox.ac.uk/bol/catesby). Catesby did not mention algae or ferns in his book.

11. *NH* I: Account, xxxviii.

12. Frick & Stearns 1961.

13. *NH* II: plate 95.

14. *NH* I: plate 73.

15. Britton & Millspaugh 1920; Cates 1998; Reveal 2013.

16. *NH* II: plate 81.

17. *NH* II: plate 95.

18. *NH* I: plate 30.

19. *NH* I: plate 25.

20. *NH* I: plate 41.

21. *NH* I: plate 25.

22. *NH* I: plate 25.

23. *NH* I: plate 79.

24. *NH* II: plate 32.

25. *NH* II: plate 51.

26. *NH* II: plate 46.

27. *NH* II: plate 50.

28. *NH* I: plate 98.

29. *NH* II: plate 33.
30. *NH* II: plate 38.
31. *NH* II: plate 1.
32. *NH* II: plate 5.
33. *NH* I: Account, xli–xlii.
34. *NH* II: plate 33.
35. Lister 1685–1692.
36. *NH* I: Account, xlii.
37. Robertson 2005, 2011b.
38. *NH* II: plate 63.
39. *NH* II: plate 79.
40. Day 1989: 218–219.
41. Day 1989: 218–219.
42. *NH* II: plate 100.
43. Robertson 2011a.

CHAPTER 11. Mark Catesby's preparatory drawings for *The natural history of Carolina, Florida and the Bahama islands*

ACKNOWLEDGMENTS
I would like to thank Stephen Harris, Charlie Jarvis, and Leslie Overstreet for their assistance. My text is based on a spoken presentation made at the Catesby Commemorative Trust's conference on Mark Catesby held at the Smithsonian Institution, Washington, D.C., in November 2012.

NOTES

1. Frick & Stearns 1961: 19.

2. McBurney 1997. Most of Catesby's drawings in the Royal Collection can be viewed online (www.royalcollection.org.uk, accessed 11 May 2013).

3. Original drawings are in the British Museum, London; British Library, London; Oxford University Herbaria (see note 13 below); Sloane Herbarium, Natural History Museum, London; Pierpont Morgan Library, New York; and the collection of the Earl of Derby, Knowsley Hall, Knowsley, Lancashire.

4. *NH* I: Preface, iv.

5. Petiver, not dated (c. 1695). This broadsheet is bound into some copies of *Musei Petiveriani. Centuria prima*, issued in London in 1695.

6. Harris pp. 175–182 (this volume).

7. Petiver not dated (c. 1695) (see note 5 above).

8. *NH* I: Preface, xx.

9. Edwards 1750: 217. Much of this information was taken from earlier manuals, such as Anonymous 1729.

10. Edwards 1750: 217.

11. Anonymous 1729: 14–15.

12. Anonymous 1729: 14–15. The text continues: "Some of this Water is put in the Shell with the Colour you would temper, and diluted with the Finger till it be very fine. If it be too hard, you must let it soften in the Shell with the said Water, before you dilute it. Afterwards let it dry. . . ."

13. The watercolors were lifted during the 1990s from the Royal Library's unique set of Catesby's *The natural history of Carolina . . .*; they are now mounted and kept in boxes (see McBurney 1997: 29–32).

14. Another example of a highly worked up preliminary drawing in pen and ink is that of the American lotus (*Nelumbo lutea*) found amongst Catesby's plant specimens in OUH (see Harris p. 182, figure 13-2, this volume).

15. *NH* I: Preface, iv.

16. *NH* I: Preface, xi.

17. Tinling 1977; Wright & Tinling 1941.

18. Edwards 1750: 215.

19. *NH* I: Preface, vii. When referring to plants or animals such as birds, "specious" is defined as "having brilliant, gaudy, or showy colouring" (*Oxford English dictionary online*, www.oed.com).

20. *NH* I: Preface, vii.

21. Stevenson 1961: 2: 486–487; Overstreet pp. 155–172 (this volume).

22. Fairchild 1722: 70: "The many experiments I am now making in my Gardens, for the Improvement of all sorts of Fruits, Flowers, and Trees, at the Request of several Gentlemen in the Country, who are my Customers." See also Bradley 1726.

23. BM: Department of Prints and Drawings, Sloane MS 5261, no. 121.

24. Sloan 2012.

25. BM: Department of Prints and Drawings, Sloane MS 5261.

26. Etheridge & Pieters pp. 39–56 (this volume).

27. *NH* I: Preface, iv.

28. Sloan 2007: 170–223.

29. Although we do not know what drawings Catesby made during his first visit to America, we do know from a letter Samuel Dale wrote on 23 December 1719 to William Sherard that they included "drawings of Birds and Plants" (RS CCIII 212).

30. McBurney 1997: figures 23 and 23.1.

31. Unlike models that Catesby studied in Sloane's collections, this drawing was acquired by Catesby and was kept with another drawing by Kickius amongst his own drawings for *The natural history of Carolina. . . .*

32. *NH* II: plates 61 and 91.

33. McBurney 1997: 142–145 (numbers 46 and 47).

34. McBurney 1997: 29–33.

35. Overstreet pp. 166–171 (this volume).

36. McBurney 1997: 68–69 (number 16).

37. Edwards (1750: xix) wrote that his aim in his backgrounds was to "decorate the Birds with airy Grounds, having some little Invention in them. . . ."

CHAPTER 12. The publication of Mark Catesby's *The natural history of Carolina, Florida and the Bahama islands*

ACKNOWLEDGMENTS

I am most grateful to the Smithsonian Libraries for encouragement and for granting both leave to pursue this research and extra time for work trips to examine copies at host institutions. The dozens of copies of Catesby's book that I have seen over the past eighteen years involved too many libraries and staff to name individually, but my thanks go to every one. I also thank the private owners and their librarians who graciously allowed me to examine their copies. Those wishing anonymity will, I hope, know who they are; and in England, Mark Purcell (National Trust), John Gandy (Blickling Hall), Dr. Suzanne Reynolds (Holkham Hall), James Peill (Goodwood), Bridget Wright (Windsor Castle), and Dr. Clemency Fisher (for an introduction to Knowsley Hall). For specific and greatly appreciated help, hospitality, enthusiasm, and advice, I am happy to acknowledge my debts to colleagues and friends Gina Douglas and John Parmenter, Henrietta McBurney, David Elliott, Amy Meyers, and Andrew Arnold.

NOTES

1. BL: Sloane MS 4047, fol. 213: MC to HS, 15 August 1724: "My Sending Collections of plants and especially Drawings to every of My Subscribers is what I did not think would be expected from me My design was S^{r} (ti'l you'l please to give me your advice) to keep my Drawing intire that I may get them Graved, in order to give a gen^{ll} History of the Birds and other Animals, which to distribute Seperately would wholly ffrustrate that designe, and be of little value to those who would have so small fragments of the whole" (see Frick & Stearns 1961: 28). Catesby's plan was likely inspired to some degree by Maria Sybilla Merian's *Metamorphosis insectorum Surinamensium* (1705); see Etheridge and Pieters pp. 39–56 (this volume).

2. See Gaskell 1972 on book production in the handpress period (1450–1800).

3. *NH* I: Preface, xi. Engraving and etching are both intaglio processes that produce incised lines, like grooves, on a copper plate. In engraving, the lines of the image are physically carved out of the surface of the copper with a burin (a skill requiring considerable effort and experience). The plate is then inked all over, the surface is cleaned so that the ink remains only in the grooves, and when the plate is pressed against a piece of paper, it produces a print. In etching, the surface of the copper is covered with a thin wax coating, and the lines of the image are drawn through it to expose the metal underneath (a technique more easily acquired by an artist). The plate is then dipped in an acid bath, and the solution "bites" the lines into the copper, after which the wax is removed so that the plate can be inked and printed (see Jackson 1985; Gascoigne 1986).

4. Copies known to the author:

earliest version: Chelsea Physic Garden, London (facsimile in Meyers 1997: 17).

intermediate version: Natural History Museum, London; Hunt Institute for Botanical Documentation, Pittsburgh.

later version: Smithsonian Libraries, Washington, D.C.; Humanities Research Center, University of Texas, Austin (Brigham 1998: 110–111, note 22 dated the *Proposals* to 1729 based on this copy).

undetermined: Bodleian Library, Oxford; Peabody Institute, Baltimore, Maryland; Wagner Free Institute of Science, Philadelphia; Charleston Museum, Charleston, South Carolina.

5. Brigham 1998 discussed Catesby's subscribers.

6. Stevenson 1961: 146. That Catesby initially colored the plates himself is suggested by a letter dated 1 March 1729/30 to his niece Elizabeth Cocke Pratt Jones (Jones 1891: 219), which accompanied an uncolored copy of the first twenty plates and pages; the present location of this copy is unknown. Uncolored but incomplete copies are in the Natural History Museum, London (miscellaneous text and plates from volume I) and Chelsea Physic Garden, London (an incomplete but bound set: volume I and the first twenty pages and plates of volume II).

7. See note 6 above.

8. See Overstreet 2014. Frick & Stearns 1961: 37–42 presented a partial list of dates of the parts based on the Royal Society journal books, as did Meyers & Pritchard 1998: 15–16. Stearn 1958 relied solely on Mortimer's summaries of each part in the Society's *Philosophical transactions*, whose issue dates would seem to contradict the Society's own records in this regard, but the anomalies may be accounted for by delays in the journal's publication schedule.

9. Plomer et al. 1932: 232. I thank Charles Nelson for archival research to confirm this identification.

10. In the earlier typesetting, the phrase-name of the plant lacked a citation, and the English column was headed "Plum. Cat." This was corrected either by printed labels or, in the second typesetting, in type to the Latin name followed on the same line by "Plum. Cat." (referring to Plumier 1703) and the heading "Bignonia" for the English text.

11. Jones 1891: 219.

12. Thomas Knowlton (1691–1781) wrote to a colleague on 28 June 1733 that "Mr Catesby has published his fifth part with a preaface" (Henrey 1986: 135), but since Catesby in his "Note" stated

that the preface (as he defined it) was yet to come, it is possible that Knowlton was referring to the dedication leaf.

13. Smithsonian Libraries: call number fQH41 .C35 1731 c.1; available online from Biodiversity Heritage Library (www.biodiversitylibrary.org).

14. See Gaskell 1972.

15. See note 9 above.

16. Trinity College, Cambridge: shelf marks 5.17.67 and 5.17.68.

17. According to OCLC (WorldCat) (accessed 30 June 2013) more than seventy copies are in institutional collections. An unknown number of copies are in private collections.

18. Smithsonian Libraries: see note 14 above.

19. Philip Oldfield (University of Toronto Library), pers. comm., 16 November 2006.

20. Mortimer 1748: 173.

21. Stafleu and Cowan 1976–1988 asserted a date of [1748–] 1754 [–1756] but without evidence or explanation. Such a date range, beginning while Catesby was still alive, seems unlikely, as advertisements seeking subscribers have not been located in British newspapers prior to 1753; a prospectus for the edition (Houghton Library, Harvard University: call number pf NH 1557.54) titled "Proposals for re-publishing by subscriptions a natural history of Carolina . . ." is dated "London, Oct. 25, 1753." In addition, Collinson's inscription on the third front free endpaper in his personal copy (see note 3 above) noted that "His Widow subsisted on the Sale of it for about 2 years, afterwards [after Catesby's death]. Then the Work, Plates, &ct.—Sold for 400£. . . ."

22. Works applying Linnaean binomial nomenclature to Catesby's plants and animals began with Linnaeus himself (1753, 1758) and proliferated through the eighteenth century (see Ewan 1974; Reveal 2013; and pp. 331–350, this volume).

23. Papermakers James Whatman the elder and the younger produced high quality paper both handmade ("laid") and, later, made by machine ("wove"). The 1794 watermark appears in the copy at the Smithsonian Libraries (call number qQH41 .C35 1771), while the 1816 watermark appears in a privately owned copy.

CHAPTER 13. The plant collections of Mark Catesby in Oxford

ACKNOWLEDGMENTS

Software for the Catesby database was developed through the BRAHMS project, run by Denis Filer and partly funded by Plants for the 21st Century (University of Oxford). I thank Serena Marner (University of Oxford) and Anne Marie Catterall (University of Oxford) for their help in constructing the database, Henrietta McBurney for background information on Mark Catesby, and Charlie Jarvis (Natural History Museum, London) for the opportunity to study the Sloane Herbarium.

NOTES

1. Nicholls 2009.

2. Ogilvie 2006.

3. Gibson 1796.

4. Frick & Stearns 1961; Brigham 1998.

5. Frick & Stearns 1961: 114. William Sherard met More in 1718 and sent him to North America as a collector.

6. Forster 1772: xxvi.

7. Dandy 1958; Frick & Stearns 1961; Howard & Staples 1983; Wilbur 1990; McMillan et al. 2013.

8. Stearn 1957a: 121–122.

9. All of the specimens collected, or likely to have been collected, by Catesby have been found, photographed, and databased following an intensive search of the whole of the Oxford University Herbaria's early collections. Images, raw data, and label transcriptions of all OUH Catesby specimens are available in a publicly accessible, searchable database at http://herbaria.plants.ox.ac.uk/bol/catesby.

10. Frick & Stearns 1961.

11. RS CCLIII 63: MC to WS, 5 May 1722.

12. RS CCLIII 174: MC to WS, 16 January 1723/24.

13. RS CCLIII 171: MC to WS, 10 May 1723.

14. RS CCLIII 173: MC to [? G. Markham], 16 January 1723/24.

15. Petiver undated (c. 1695). This broadsheet is bound into some copies of *Musei Petiveriani. Centuria prima*, issued in London in 1695.

16. Petiver undated (c. 1695) (see note 15 above).

17. RS CCLIII 173: MC to [? G. Markham], 16 January 1723/24.

18. RS CCLIII 174: MC to WS, 16 January 1723/24.

19. OUH Sher-1249; a similar label is attached to NHM-HS 212, fol. 61.

20. OUH Sher-0604-12; NHM-HS 212, fol. 70.

21. Allen 1959.

22. For example, see RS CCLIII 179: MC to WS, 30 October 1724.

23. OUH 00087305N, 00087306O, 00087301J, Sher-1089, Sher-1089-2.

24. OUH Sher-1090-10.

25. RS CCLIII 179: MC to WS, 30 October 1724.

26. RS CCLIII 171: MC to WS, 10 May 1723.

27. OUH Sher-sn-ATH.

28. OUH 00087289Y, 00087293T.

29. Petiver undated (c. 1695) (see note 15 above).

30. OUH Sher-0020-5.

31. OUH Sher-1516-10.

32. OUH Sher-1078-3.

33. OUH Sher-1098-4; similar label is with NHM-HS 212, fol. 69.

34. OUH Sher-0125-b.

35. RS CCLIII 164: MC to WS, 20 June 1722.

36. OUH Sher-sn-ATH.

37. Petiver undated (c. 1695) (see note 15 above).

38. RS CCLIII 178: MC to WS, 16 August 1724.

39. Humboldt 1995: 8.

40. Dandy 1958; Allen 2008.

41. Bobart 1884; Clokie 1964.

42. Sherardian Library, Department of Plant Sciences, University of Oxford: MS Sherard 44 to MS Sherard 173: W. Sherard, "Sherardian Pinax."

43. Dandy 1958.

44. Seventy-one specimens bearing Catesby's name are unlocalized.

45. OUH Sher-1624.

46. OUH Sher-1504-10.

47. Pursh 1814: xviii.

48. W. Sherard, "Sherardian Pinax" (see note 42).

49. Druce 1897.

50. There is no evidence that Sherard's specimens were ever bound into books.

51. Sherardian Library, Department of Plant Sciences, University of Oxford: MS Sherard 22: W. Baxter, "Specimen of a catalogue of the Sherardian Herbarium in the Botanic Garden."

52. OUH Sher-0769-30.

53. "Entr." was used by John Ray to show he had entered a specimen into his *Historia Plantarum* (1704).

54. OUH Sher-0027-5.

55. OUH Sher-sn-AIF; a similar label is associated with NHM-HS 212, fol. 16.

56. Druce & Vines 1907.

57. Druce 1897: 7, although this was emended to eighty volumes in Druce 1928.

58. Druce 1928.

59. Pursh 1814.

60. Eleven specimens bearing Catesby's name are unlocalized.

61. RS CCLIII 184: MC to WS, 10 January 1724/25.

CHAPTER 14. Carl Linnaeus and the influence of Mark Catesby's botanical work

ACKNOWLEDGMENTS

I am most grateful for the help of a number of people in preparing this account. Gina Douglas and Elaine Charwat were of great assistance in tracing relevant records at the Linnean Society of London, while Charles Nelson, Eva Nyström (The Linnaean Correspondence project), Laila Österlund (Carolina Rediviva, Uppsala), Leslie Overstreet (Smithsonian Libraries), Karen Reeds, and Fiona Wild kindly provided valuable help and advice on literature and much else besides. I am particularly grateful to Bengt Jonsell for advice on the translation of a crucial part of one of De Geer's letters to Linnaeus.

NOTES

1. Bjurr later became a provincial physician in Västmanland, Sweden (Olsen 1997).

2. Linnaeus 1753b.

3. Stearn 1957a.

4. Stearn 1971.

5. Linnaeus 1753a.

6. Linnaeus 1758.

7. Linnaeus 1756: 12. *Vinca lutea ≡ Pentalinon luteum.*

8. Both names were published in Linnaeus 1753a: 209.

9. The application of scientific names is governed by an internationally accepted set of rules (in the case of plants, by McNeill et al. 2012). For information on the application of the type method to binomial names coined by Linnaeus, see Jarvis 2007: 13–60.

10. Blunt 1971: 95–108, 116–126.

11. Johannes Burman (1707–1779), director of the Botanic Garden in Amsterdam; Herman Boerhaave (1668–1738), the distinguished former director of the Hortus in Leiden; Albert Seba (1665–1736), a wealthy collector of natural history specimens; Adriaan van Royen (1704–1779), director of the Botanic Garden in Leiden; and Johan Gronovius (1686–1762), a physician and a keen botanist in Leiden.

12. Plumier 1693, 1703.

13. Merian 1705: Etheridge & Pieters pp. 39–56 (this volume).

14. Sloane 1707, 1725.

15. Plukenet 1691–1694, 1696, 1700, 1705.

16. Petiver 1695–1703, 1702–1709.

17. Linnaeus 1737 [1738].

18. Overstreet p. 158 (table 12-1) (this volume).

19. Overstreet p. 158 (table 12-1) (this volume).

20. The binomial names quoted here are those given to the corresponding species by Linnaeus in 1753 or later.

21. *NH* II: plates 43 *Andromeda paniculata*; 44 *Mimosa senegal*; 49 *Erythrina herbacea*; 50 *Laurus winterana*; 60 *Convolvulus batatas*.

22. Gronovius 1739.

23. Gronovius 1739: 104; *NH* II: plate 59.

24. Brigham 1998: 141–145.

25. Collinson's letters to Linnaeus can be viewed at http://linnaeus.c18.net. Selected letters have been transcribed by Armstrong 2002.

26. LS BL.151: James Edward Smith, "Bibliotheca Linnaeana Catalogus."

27. Schmidt 1965.

28. LS XVII 23: PC to CL, 14 May 1749.

29. Examples of annotated copies of comparable works in Linnaeus's library at the Linnean Society of London include Georg Rumphius's *Herbarium Amboinense* and Patrick Browne's *Civil and natural history of Jamaica*; see Savage (1940) for a comprehensive list. Leslie Overstreet (pers. comm., November 2012) reported that among the many copies of Catesby's book that she has studied, none appears to be annotated by Linnaeus.

30. Heller 2007: xlvii.

31. Stearn 1957b: 66.

32. This manuscript is at the Linnean Society of London and pre-dates Linnaeus's use of species epithets. The entries for the species that would later be named *Lonicera marilandica* (fol. 242) and *Magnolia virginiana* var. *tripetala* (fol. 677) both cite the corresponding Catesby plates from part 9 (*NH* II: plates 78 and 80, respectively). However, although many of the plates from parts 10 and 11 were cited in *Species plantarum* in 1753, not one of them appeared in the draft entry for the corresponding species in the 1746–1748 manuscript (see entries for *Sloanea emarginata* (fol. 635; *NH* II: plate 87), *Renealmia polystachya* (fol. 39; *NH* II: plate 89), *Cordia sebestena* (fol. 251; *NH* II: plate 91), *Plumeria rubra* (fol. 273; *NH* II: plate 92), *Passiflora cuprea* (fol. 977; *NH* II: plate 93), *Hamamelis virginiana* (fol. 199; *NH* II: Appendix, plate 2), *Theobroma cacao* (fol. 895; *NH* II: Appendix, plate 6), *Epidendrum vanilla* (fol. 975; *NH* II: Appendix, plate 7), and *Panax quinquefolius* (fol. 1101; *NH* II: Appendix, plate 16).

33. LS X 207v, 208: PC to CL, 14 October 1748.

34. Brigham 1998: 144. Queen Ulrika Eleonora died on 24 November 1741, probably before the list of "Encouragers" was published.

35. Identifications follow Reveal 2013.

36. Adler p. 262 (this volume).

37. For Plukenet, see, for example, *NH* II: plate 65, "Liquid-ambar Arbori, seu Styraciflua . . ." (subsequently Linnaeus's *Liquidambar styraciflua*); for Plumier, see *NH* I: plate 37, "Bignonia arbor pentaphylla: flore roseo, majore, siliquis planis" (Linnaeus's *Bignonia leucoxylon*); for Sloane, see *NH* I: plate 30, "Terebinthus major Betulae cortice, fructu trianguli" (Linnaeus's *Pistacia simaruba*).

38. For Ray, see *NH* I: plate 16 (Linnaeus's *Quercus phellos*); for Morison, see *NH* I: plate 70 (Linnaeus's *Gentiana saponaria*); for Parkinson, see *NH* I: plate 67 (Linnaeus's *Juglans nigra*); for Commelin, see *NH* II: plate 26 (Linnaeus's *Phyllanthus epiphyllanthus*); for Tournefort, see *NH* II: plate 92 (Linnaeus's *Plumeria rubra*).

39. John Clayton's specimens at NHM can be seen at http://www.nhm.ac.uk/research-curation/research/projects/clayton-herbarium/; specimens in the LS herbarium are at http://www.linnean-online.org/.

40. Frick & Stearns 1961: 92.

41. See *NH* II: Appendix, plates 2 (Linnaeus's *Hamamelis virginiana*), 12 (Linnaeus's *Amaryllis atamasca*), 15 (Linnaeus's *Magnolia acuminata*), and 16 (Linnaeus's *Panax quinquefolius*).

42. Frick & Stearns 1961: 92.

43. Many of Kalm's specimens can be found at LS (see Jarvis 2007: 215–217), and a second collection, heavily annotated by Kalm, is in Uppsala (see Lundqvist & Moberg 1993).

44. Harris pp. 173–188 (this volume).

45. MacGregor 1994; Walker et al. 2012.

46. Dandy 1958; Catesby's specimens are in NHM-HS 212 and 232, with the water tupelo and label in NHM-HS 212, fol. 67; Catesby's published account is in *NH* I: plate 60. The material of Carolina allspice is in NHM-HS 212, fol. 16. Images of these specimens, correlated with the published plates, can be seen at http://folio.furman.edu/projects/botanicacaroliniana/Parallels.html.

47. Dandy 1958: 112.

48. Linnaeus 1753a: 1055.

49. Linnaeus 1748: 94.

50. Gronovius 1739: 41.

51. See image at http://www.nhm.ac.uk/research-curation/research/projects/clayton-herbarium/search/detail.dsml?RowID=753&listPageURL=list%2edsml%3fsort%3dClaytonNo%26LinnaeanGenus%3dacer.

52. Hermann 1698: 1, t. 1.

53. Plukenet 1691: t. 2, fol. 2; 1696: 7.

54. For an image of the type specimen, see http://www.linnean-online.org/12342/.

55. Plukenet 1691: t. 160, fol. 7; 1696: 89.

56. Sloane 1707: 87, t. 191, fol. 2.

57. For example, LS sheet no. 80.55 (http://www.linnean-online.org/1289/).

58. Based on a report by Allen (1951: 468) of a second letter held in the Library of Congress, The Linnaean Correspondence Project (http://linnaeus.c18.net/) allocated to it the code L5288 (Eva Nyström, pers. comm., October 2012). All attempts to locate the letter have, however, proved unsuccessful, and it now seems likely that Allen's statement was erroneous (David Elliott, pers. comm., October 2013).

59. LS III 14–15 = letter L0612 (http://linnaeus.c18.net/), MC to CL, 6 April 1745.

The plants included *Taxodium distichum*, *Liriodendron tulipifera*, *Cornus florida*, *Populus balsamifera*, *Gelsemium sempervirens*, *Amorpha fruticosa*, *Wisteria frutescens*, *Lindera benzoin*, *Euonymus americanus*, *Aster linariifolius*, *Liquidambar styraciflua*, *Catalpa bignonioides*, *Aralia spinosa*, *Robinia pseudoacacia*, *Podophyllum peltatum*, and *Rubus occidentalis*.

60. LS X 207–208 = letter L0952 (http://linnaeus.c18.net/), J. Mitchell to CL, 1 October 1748.

61. Lucas 1892: 118–119.

62. LS XVII 25 = letter L5408 (http://linnaeus.c18.net/), PC to CL, 1 May 1750.

63. LS XVII 23–24 = L1034 (http://linnaeus.c18.net/), PC to CL, 6 March 1750.

64. LS XVII 27 = L5410 (http://linnaeus.c18.net/), PC to CL, 25 August 1749.

65. LS XVII 28 = L1592 (http://linnaeus.c18.net/), PC to CL, 8 May 1753.

66. LS XVII 37–38 = L1741 (http://linnaeus.c18.net/), PC to CL, 20 April 1754.

CHAPTER 15. The economic botany and ethnobotany of Mark Catesby

NOTES

1. Simpson pp. 71–84 (this volume).

2. Weidensaul 2007: 20.

3. Jarvis pp. 189–204 (this volume); Adler pp. 251–264 (this volume).

4. Reveal 2012a.

5. *NH* II: plate 100.

6. Cates 1996; Robertson pp. 127–140 (this volume).

7. Cates 1998; Robertson pp. 127–140 (this volume).

8. Northrop 1902, 1910.

9. Britton & Millspaugh 1920.

10. *NH* I: plate 11.

11. *NH* I: plate 17.

12. *NH* I: plate 69.

13. *NH* I: plate 26.

14. *NH* I: plate 55.

15. *NH* I: plate 52.

16. Lawson 1860.

17. *NH* II: plate 57.

18. OBL-S: MC to J. J. Dillenius, 10 December 1737, Sherard MS 202, fol. 15, transcribed by Serena Marner (Herbarium Manager, Oxford University Herbaria).

19. *NH* 1: plate 54.

20. NHM (Samuel Dale's herbarium); similar labels are on OUH Sher-1098 (see Harris p. 180, figure 13-4, this volume) and NHM-HS 212, fol. 69. *Tilia americana* var. *heterophylla* was not illustrated in *NH*.

21. NHM-HS 212, fol. 66; the medicinal uses were not mentioned in the text for *NH* I: plate 77.

22. NHM-HS 232, fol. 105 (*Aletris aurea*, golden colic-root); OUH Sher-1944-20 (*Rudbeckia* sp.): "Mr Catesby calls this Crux St. Andrew. . . . it hath a tuberous root used against y[e] bite of y[e] rattle snake."

23. *NH* I: plate 44; NHM-HS 212, fol. 013.

24. *NH* II: plate 95.

25. *NH* I: plate 30.

26. McClure & Eshbaugh 1983.

27. *NH* I: plate 40.

28. *NH* II: plate 81.

29. *NH* I: plate 98.

30. *NH* I: plate 85.

31. *NH* II: plate 99.

32. *NH* II: plate 46.

33. McCormack et al. 2011.

34. *NH* I: Account, xli.

35. *NH* I: Account, xli.

CHAPTER 16. "Of birds of passage": Mark Catesby and contemporary theories on bird migration and torpor

NOTES

1. Catesby 1747. On Catesby's North American and Caribbean birds, see Krech 2014.

2. Feduccia 1985: 5; Peterson 1985: xi.

3. Frick & Stearns 1961: 44, 55, 63.

4. Allen 1937: 359; Chaplin 1998: 64.

5. *NH* I: Preface, vii. In 1723 Catesby mentioned migration in correspondence with Sloane (Chaplin 1998: 61–62).

6. For "The Rice-Bird" and "Ortolan de la Caroline," Catesby's names for boblinks, see *NH* I: plate 14; Frick & Stearns 1961: 37; Mortimer 1729: 427.

7. *NH* I: Preface, vii, passim.

8. *NH* I: Account, xxxv–xxxvii.

9. Catesby 1748.

10. Catesby 1747: 435–436.

11. Allen 1937: 359.

12. A. B. 1748.

13. *NH* I: Account, xxxv–xxxvii; Derham 1708.

14. Frick & Stearns 1961: 44, 63–64.

15. Macaulay 1890: 1: book II: section 22; Thompson 1910: parts 3, 12, 16.

16. Bostock 1855; Bircham 2007: 126.

17. Ray 1678: 213.

18. *NH* I: Preface, v; Brigham 1998: 95–96; Frick 1960: 173; Frick and Stearns 1961: 3–10. See also Krech 2014. I wish to thank Charles Nelson for sharing his current thinking on the evidence for the relationship between Catesby and Ray.

19. Collinson 1760 (read to the Royal Society on 9 March 1758); Achard 1763; Cornelius 1994. On sand martins' arrivals, see White 1775, 1977: 109; and Sharpe & Wyatt 1885–1891.

20. Pennant 1776: 402, 414; Lyle 1978.

21. Barrington 1772; Lyle 1978.

22. White 1774, 1775.

23. Cornelius 1994: 231.

24. White 1977: 31, 39, 58, 63, 89, 91, 131, 138, 156, 164, 198–199, 242; see also Cornelius 1994. The strength of belief in hibernation is even greater in Cornish 1775.

25. Birkhead 2011: 154–155; Lewis 2005.

26. White 1977: 31; Collinson 1760: 462.

27. Jenner & Jenner 1824.

28. McKechnie & Lovegrove 2002.

29. Geiser & Ruf 1995; Barclay et al. 2001; Schleucher 2004.

30. McAtee 1947; McKechnie & Lovegrove 2002; Bartholomew et al. 1957: 145–155; Brigham 1992.

31. McAtee 1947: 193–194; McKechnie & Lovegrove 2002: 712; Alexander 1933; Newton 2007; Keskpaik 1972.

32. Prinzinger & Siedle 1988.

33. Brown & Brown 2000; Lasiewski & Thompson 1966; Serventy 1970.

34. Rowan 1968; Hobson et al. 2012.

CHAPTER 17. Catesby's animals (other than birds) in *The natural history of Carolina, Florida and the Bahama islands*

ACKNOWLEDGMENTS

I thank Leslie Overstreet (Smithsonian Libraries), David Elliott (Catesby Commemorative Trust), and Charles Nelson for their comments on the manuscript. Kraig Adler (Cornell University), Kevin de Queiroz (Smithsonian Institution), and James Reveal (Cornell University) kindly provided insights regarding the identity of Catesby's animals. This research was supported by the Gerald M. Lemole, M.D. Endowed Chair Funds at Villanova University.

NOTES

1. Frick & Stearns 1961.

2. For probable identities, see Reveal pp. 331–350 (this volume).

3. Tyler 1937.

4. Fish (*NH* II: plates 1–31), crabs (*NH* II: plates 32–37), sea turtles (*NH* II: plates 38–40), "snakes" (*NH* II: plates 41–60), "lizards" (*NH* II: plates 63–68), frogs and toads (*NH* II: plates 69–72), mammals (*NH* II: plates 62, 73–79), and insects (*NH* II: plates 83–84, 86, 88–91, 94–97, 100).

5. Mammals (*NH* II: Appendix, plates 18 and 20), fish (*NH* II: Appendix, plate 19), amphibian (*NH* II: Appendix, plate 10), centipede (*NH* II: Appendix, plate 2), and numerous insects (*NH* II: Appendix, plates 4, 5, 10, 11, 13, 15).

6. Ray's classification system was detailed in his major works on birds (1678, 1713), fish (1686, 1713), quadrupeds (1693), and insects (1710), all probably known to Catesby.

7. Ray 1693, 1713.

8. Ellis 1755.

9. *NH* I: Preface, x.

10. *NH* I: Account, xxxii–xxxv.

11. Frick & Stearns 1961.

12. *NH* II: plate 23; Raven 1964; Egerton 2006.

13. The other plates based on White's drawings are those of the remora (*NH* II: plate 26), globe fish (*NH* II: plate 28), land crab (*NH* II: plate 32), "Guana" (*NH* II: plate 64), and swallowtail butterfly (*NH* II: plate 97) (Feduccia 1985; McBurney 1997; Egerton 2006).

14. Frick & Stearns 1961.

15. Reveal 2009, 2012a, 2013, and pp. 331–350 (this volume).

16. Swift & Swift 1993.

17. *NH* I: Account, xxxiv.

18. Collins & Smith 1996.

19. Sloane 1707: 28–29.

20. Raven 1964.

21. Bartram 1791; Adler 2004.

22. Sloane 1707: lxxxviii.

23. Reveal 2009, 2012a, 2013; other authors (for example, McBurney 1997) have considered them to be unidentifiable.

24. Linnaeus 1758: 226 included Catesby's image in his synonymy of *Coluber* (now *Ahaetulla*) *mycterizans*, the Malayan green whipsnake.

25. Considered unidentified by McBurney 1997; see Reveal 2009, 2012a, 2013, and p. 338 (this volume).

26. Linnaeus 1766: 378 included it questionably under *Coluber fasciatus*, and it has been noted under the current name of that snake, *Nerodia fasciata* (southern watersnake), by several authors, including McBurney 1997. However, Reveal 2009: 330; 2012a: 11; 2013: 11; and p. 339 (this volume) gives it as the mudsnake, *Farancia abacura*.

27. Camp 1923; Schwenk 1988.

28. See, for example, McBurney 1997; Reveal 2009, 2012a.

29. Russell & Bauer 1991.

30. The original painting (RL 26018) depicts a green lizard and is doubtless the source of the engraved image in *The natural history of Carolina. . . .* This image is on the same sheet as a watercolor of a Carolina specimen of a spotted salamander (*Ambystoma maculatum*), and it may be that it represents the Carolina anole (*Anolis carolinensis*). If this is the case, then the model for the Jamaican anole on plate 66, here identified as *Anolis grahami aquarum*, would be this species. However, the two lizards are similar in appearance, and Catesby's renderings are not sufficiently detailed to distinguish between these options (although the body color of both the watercolor and the engraved image is more similar to that of the Carolina anole). Nonetheless, even if a Carolinian model was used, Catesby altered the image in the final etching, adding the extended dewlap mentioned in his text, and it appears that he intended the illustration to represent *Anolis grahami aquarum*, the only Jamaican species consistent with both the text description and the physical features depicted. Interestingly, the Royal Library image of the green lizard of Carolina has a dewlap, but the corresponding etching (*NH* II: plate 65) does not, suggesting that Catesby transposed this feature between the two anole species during the preparation of the plates for *The natural history of Carolina. . . .*

31. *NH* II: plate 68.

32. Rochefort 1658, 1666.

33. See Adler pp. 263–264 (this volume).

34. The actual limit today is slightly further north, near Albemarle Sound, but alligators have historically been more abundant south of Pamlico Sound, into which the Neuse flows (Palmer & Braswell 1995).

35. *NH* I: Account, xxiv–xxxi.

36. Harting 1880.

37. Llanover 1861–1862; Laird & Weisberg-Roberts 2009.

38. Winterrowd et al. 2005.

39. *NH* I: Account, ix.

40. Lister 1685–1692.

41. *NH* I: Preface, ix–x.

42. Darwin 1839.

43. *NH* I: Preface, x.

44. *NH* I: Account, v; Walley 2003: 1–13. If indeed Catesby saw and painted his "Green Snake" during his time in Virginia, this image would pre-date most of the others in *The natural history of Carolina*. . . .

45. Gibbons & Dorcas 2005.

46. Gibbons & Semlitsch 1991.

47. *NH* I: Preface, ix–x.

48. This species of catfish is endemic to northern South America.

49. Dandy 1958.

50. Clutton-Brock 1994; Johnston 2003.

51. Wilkins 1953.

52. Reveal 2012b, 2013.

53. Jordan 1884; Ewan 1974; Reveal 2009, 2012a, 2012b, 2013, and pp. 331–350 (this volume).

54. Linnaeus 1758, 1766 relied heavily on Catesby's illustrations as the basis of his descriptions (see Reveal 2012a; Adler pp. 251–264, this volume). Under the *International Code of Zoological Nomenclature* (ICZN 1999) new names published before 1931 may be based on previously published descriptions even if they were published before 1758 (Article 12.2.1), and their type specimens may include those represented by illustrations in such earlier works (Article 73.2.1).

CHAPTER 18. Catesby's fundamental contributions to Linnaeus's binomial catalog of North American animals

ACKNOWLEDGMENTS

I thank David Elliott for providing scans of Catesby's original plates and for many other courtesies. James Reveal and Richard Wahlgren kindly reviewed and commented on my manuscript. David M. Dennis generously provided his color photographs of living animals that correspond to Catesby's illustrations. I am indebted to Stig Söderlind for allowing me to use the map he drafted to show the routes traveled by Linnaeus's "apostles."

NOTES

1. Linnaeus 1737, 1753a.

2. Wahlgren 2012.

3. Adler 2012a; Broberg 2006.

4. Adler 2012a: 41.

5. Adler 2012b.

6. The animal class "Amphibia" of the eighteenth century is equivalent to two classes today: Amphibia (frogs, salamanders, and caecilians) and Reptilia (turtles, crocodiles, tuatara, lizards, and snakes).

7. Forster 1771 referenced Catesby in his lists of species using simply the letter "C." or as "Cat.," followed by the volume and plate number.

8. Adler 2004.

9. *NH* II: 47.

10. *NH* II: 59 (text misnumbered "56").

11. Linnaeus 1758: 194.

12. Linnaeus 1752.

13. Wahlgren 1999.

14. The scutes on the underside of snakes occur in two groups: (1) the abdominal or ventral scutes under the head and body, which Linnaeus called *sköldar* (*scutis* in Latin), and (2) the subcaudal scutes under the tail, which he called *stjert-fjäll* (*scutellis* in Latin).

15. Linnaeus 1766: 385.

16. Wahlgren 1999.

17. Reveal 2013: note 9.

18. One other species illustrated by Catesby was accepted as new by Linnaeus—the spotted salamander (*NH* II: Appendix, plate 10)—but absent-mindedly he gave it the name *Lacerta punctata*, which he had previously given to a species of skink from southern Asia now called *Eurprepis* (formerly *Lygosoma*) *punctatum*. The salamander was correctly named in 1802 and is known today as *Ambystoma maculatum* (see table 18-3).

19. There is also a Chinese species (*Alligator sinensis*).

20. McIlhenny 1935.

21. *NH* II: plate 63.

22. Formerly *Rana catesbeiana*.

23. They are:

Lithobates (formerly *Rana*) *catesbeianus* (Shaw 1802), a frog native to eastern North America.

Dipsas catesbyi (Seetzen 1796), a snake native to northern South America.

Uromacer catesbyi (Schlegel 1837), a snake native to Haiti.

Driophis catesbyi Schlegel 1837 (= *Oxybelis fulgidus* [Daudin 1803]), a snake native to Central and South America.

Heterodon catesbyi Günther 1858 (= *Heterodon simus* [Linnaeus 1766]), a snake native to the southeastern United States.

Crotalus catesbaei Fitzinger 1826 (= *Crotalus horridus* [Linnaeus 1758]), a snake native to the eastern United States.

Herpetologists have accorded Catesby more honor than have ornithologists, for I can find no birds that were named for him. I do notice one fish, a genus of eels from the Bahamas (*Catesbya*, named by Böhlke and Smith 1968). There are, of course, several plants named for him.

CHAPTER 19. Mark Catesby's plant introductions and English gardens of the eighteenth century

ACKNOWLEDGMENTS

I am very grateful to Charlie Jarvis, Joel Fry, and Jim Reveal for their advice.

NOTES

1. *NH* I: plate 49.

2. Coombs 1997.

3. Laird 2008.

4. Sir Thomas Robinson to Earl of Carlisle, 23 December 1734, original MS, J8/1/462, Castle Howard Archives; see Laird 2000.

5. Laird 2008; Riley 2011.

6. Riley 2011: 28.

7. Brigham 1998. While Brigham established biographical essentials for most of Catesby's subscribers, he stopped short of elaborating on gender, which deserves more attention. In addition to the ten women, including British and Swedish royalty, "Her Imperial Majesty of Russia" was represented by her special envoy, Semen Kirillovich Narischkin.

8. Laird 2002: 248.

9. Laird 2015.

10. Vickery 2009.

11. Riley 2011.

12. Anonymous 1730.

13. In the 1768 eighth edition, Miller wrote: "It is now become pretty common in some of the nurseries near London . . . but it is doubtful if they are different species."

14. *NH* I: plate 62.

15. Laird 1998: 199–201: on figure 41, X signifies the "Shrubbery" and V the "Wood."

16. Laird 1999: 86.

17. Mawe & Abercrombie 1778.

18. Anonymous 1730: 4.

19. Miller 1731 (under *Acer*). As a date of introduction, 1725 is given by Harvey 1988.

20. NHM-HS 212, fol. 14 has the heavily dissected leaves characteristic of *Acer saccharinum*, although this specimen was formerly identified as *Acer rubrum*. In OUH, there is a specimen of an unidentified *Acer* with leaves similar to NHM-HS 212, fol. 14, labeled "collectum in Carolina 1723 D. Catesby" (McMillan et al. 2013).

21. Trew 1772: plate 85.

22. Trew 1772: plate 86. The polynomial used by Ehret—"Acer Americanum, folio Majore, subtus argenteo, supra viridi splendente, floribus multis coccineis"—corresponds to that in the early editions of Miller's *Gardeners dictionary*, which Miller identified as "Sir Charles Wager's Maple." When, however, Ehret's gouache was published in Trew's *Plantae selectae*, the polynomial was changed to "Acer foliis triblobis, serratis, subtus glaucis; floribus pedunculatis simplicibus, confertim aggregatis," which no longer corresponds to any species in Miller's eighth edition (1768). As herbarium specimens of *Acer* species from Wager's garden are wanting, there remains a significant puzzle.

23. Catesby 1763: 32–34.

24. Berkeley & Berkeley 1992: 398.

25. Fry 2014.

26. Gray [c. 1740].

27. Baugh 2008.

28. Edwards 1743–1751: tab. 16 and tab. 86.

29. Armstrong 2002: 5.

30. Laird & Weisberg-Roberts 2009.

31. Baugh 2008.

32. Armstrong 2002: 256. "Isle of Iona" is possibly a lapse for Ionian Islands, and "Syrian Cedar" probably was a variant of cypress (*Cupressus horizontalis*).

33. Martyn 1728: plate 21; Dillenius 1732: 350.

34. Aiton 1789: 251. Under three varieties, "elliptica," "obovata," and "lanceolata," he wrote: "*Cult.* before 1737, by Sir John Colliton [*sic*]. *Catesb. carol.* 2. *p.* 61."

Sir John Colleton (1669–1754), third Baronet, inherited a proprietorship in Carolina in 1679 and lived in South Carolina for a short time after 1714. Thus he could have brought *Magnolia grandiflora* from North America himself and planted it in his garden at Exmouth.

35. *NH* II: plate 61.

36. Miller 1735.

37. Lamb 1995: 230, 244.

38. Lamb 1995: 230; Hulme & Barrow 1997: 408. Since 1979, the mean temperature of the winters of 1988–1989 at 6.5°C and of 1989–1990 at 6.2°C have been slightly warmer still.

39. A copy, "Delineated and Engraved by George, Dennis, Ehret," is in the Wellcome Library, London; see also Trew 1754: plate 33; Miller 1757: plate 172, for other versions. For a full discussion of all versions of Ehret's watercolor paintings, see Nelson 2014.

40. Tongiorgi-Tomasi 1997: 190–192.

41. Miller 1754: 2: see under "Magnolia."

42. Armstrong 2002: 84.

43. Miller 1754: 2: see under "Magnolia."

44. Glendinning 1835.

45. O'Neill & McLean 2008: 148–149.

46. Henrey 1986: 256.

47. Harvey 1981.

48. Mawe & Abercrombie 1778: see under "Magnolia."

49. Llanover 1862: ser. 2, vol. 1: 570.

50. Llanover 1862: ser. 2, vol. 2: 6.

51. Llanover 1862: ser. 2, vol. 2: 230.

52. Llanover 1862: ser. 2, vol. 2: 246.

CHAPTER 20. Following in the footsteps of Mark Catesby

NOTES

1. This theme runs through most of Humboldt's works; see Humboldt 1855: 1: introduction.

2. Wilson 1808–1814: 5: vi.

3. For example, Audubon 1844: 7: 274.

4. Smith 1821; Armstrong 2002: 198 (PC to CL, 12 May 1756).

5. Darlington 1849: 296 (PC to William Bartram, 16 February 1768).

6. Wilson 1808–1814: 5: x.

CHAPTER 21. Inspiration from *The natural history of Carolina . . .* by Mark Catesby

ACKNOWLEDGMENTS

I thank the Royal Botanic Gardens, Kew, for permission to reproduce the plates of Amazon plants by Margaret Mee, the Eden Project, W. Hardy Eshbaugh for getting me involved in the Catesby Symposium, and David Elliot for all his efforts for the symposium upon which this volume is based.

NOTES

1. Prance 1972; Prance & Sothers 2003a, 2003b.

2. *NH* I: plate 11.

3. Sloane 1696.

4. Vogel 1968, 1969. The technical term for these long pendulous inflorescences is *flagelliflory*.

5. Both the flowers and the fruits of the Lecythidaceae are fascinating and have been much

studied by my colleague Scott Mori and I over many years: see, for example, Prance & Mori 1979; Mori & Prance 1990.

6. Reveal 1997.

7. Dampier 1697; see also Preston and Preston pp. 57–70 (this volume).

8. Catesby (RL 24839) depicted a single cashew "pear" with a North American ground dove perched on it. These two species probably never met in reality, but many birds do eat the fleshy cashew pear and thereby distribute the seeds.

9. Reveal 2009.

10. The published illustration (*NH* II: plate 93) is the type for this name that was given to it by Linnaeus.

11. Granadilla is the name usually given to the South America edible species *Passiflora ligularis*, but it is sometimes applied to other species.

12. McDaniel 1971.

13. Prance & Arias 1975.

CHAPTER 22. Conclusions: Account, Appendix, *Hortus*, and other endings among Mark Catesby's work

ACKNOWLEDGMENTS

Too many people have made this book possible to acknowledge all by name. They include the directors and supporters of the Catesby Commemorative Trust (many are "Encouragers") and my colleagues involved in writing the book. At the risk of failing to include someone critical, I must mention Gina Douglas (Honorary Archivist, Linnean Society of London), Nancy Gwinn (Director, Smithsonian Libraries), and Elisabeth King Rudloff (Director Outdoor Operations, Kiawah Island Golf Resort)—all have been very helpful for over a decade. And very special thanks go to my wife, Sallie Lou Elliott, for her support as work on Catesby has absorbed so much of our lives.

NOTES

1. *NH* I: Preface, v.

2. *NH* I: Account, xxxiv.

3. *NH* I: Account, vii.

4. *NH* I: Account, vii–viii; see also Simpson pp. 71–84 (this volume).

5. *NH* I: Account, xvi.

6. Analysis by Eric Hollinger, Supervisory Archeologist, Smithsonian National Museum of Natural History, 2013.

7. *NH* II: Appendix, plate 6.

8. *NH* I: Account, xxxix.

9. *NH* I: Account, xxxv.

10. *NH* I: Account, vii.

11. *NH* I: Account, vii.

12. Cuvier 1796.

13. *NH* I: Account, iii.

14. *NH* I: Preface, xi.

15. *NH* I: Account, xxxiv.

16. U.S. Federal Register 32 FR 4001 (11 March 1967) and 63 FR 69,613 (17 December 1998).

17. RS CCLIII 174: MC to WS, 16 January 1723/24. Leslie Overstreet conducted a survey of members of the Society for the History of Natural History around 2005, but no responses conflicted with this statement.

18. RS CCLIII 174: MC to WS, 16 January 1723/24.

19. *NH* II: 100.

20. Bartram: *NH* II: Appendix, plates 1, 4, 5 ("from Pennsylvania"), 8 ("from Pennsylvania"), 11 ("from Pennsylvania"), 15, 16 ("from Pennsylvania"), and 17. While Bartram was the most likely source for items identified as "from Pennsylvania," he is not specifically identified.

21. *NH* II: Appendix, plate 20.

22. *NH* II: Appendix, plates 1, 8, 11, 13, 16, and 17.

23. *NH* II: Appendix, plates 2 and 13; Baigent 2004.

24. The chigoe should not be confused with the chigger (*Trombicula alfreddugesi*), a mite that is common to warmer and more humid parts of the United States and that, in comparison, is merely irritating.

25. *NH* II: Appendix, plate 10.

26. *NH* II: Appendix, plate 6. From a watercolor on vellum in the Royal Library (RL 26072) by an unknown artist, possibly a pupil or follower of Ehret (see McBurney 1997: 146).

27. From a watercolor on vellum in the Royal Library (RL 26072) by an unknown artist, possibly a pupil or follower of Ehret (see McBurney 1997: 148).

28. *NH* II: Appendix, plate 9.

29. *NH* II: Appendix, plate 8.

30. Nelson 2013.

31. White 1965.

32. Cannon 1847; Dalton 1912.

33. *NH* II: Appendix, plate 5. If Catesby painted this plant when he saw it, the original drawing has been lost; both this etching and the companion watercolor in the Royal Library (RL 26070) are composites, with subjects of different origins. While he may have seen a species of *Hymenocallis* in Carolina, the drawing is considered to be of a similar (and related) southern European plant, *Pancratium maritimum* (sea-daffodil). A logical explanation would be that Catesby saw it growing in London and, with the passage of twenty years, simply misidentified it.

34. *NH* II: Appendix, plate 2. The "Scolopendra" is shown with *Hamamelis virginiana* (witch-hazel), sent to Catesby by Clayton.

35. *NH* II: Appendix, plate 14.

36. *NH* II: Appendix, plate 16.

37. For current numbers see the American Birding Association and American Ornithological Union checklists.

38. *NH* I: Preface, vi.

39. *NH* II: Appendix, plate 19. The fish placed in Sloane's "Musæum" was *Acanthodoras cataphractus* (spiny catfish), which had been caught on the coast of New England.

40. *NH* II: Appendix, plate 1. Catesby's *Meadia* has had several more recent names, including *Dodecatheon*; the species he illustrated is now pride-of-Ohio (*Primula meadia*) (Mast & Reveal 2007).

41. Linnaeus 1741 (published 1746).

42. *NH* II: Appendix, plate 13; Schweizer 2004.

43. *NH* II: Appendix, plate 18. The identification of this animal as being an agouti native to the West Indies, including the Bahamas, was made by the early nineteenth century. It did not come from Java (Shaw 1801).

44. McCann 2004.

45. *NH* II: Appendix, plate 20.

46. See McBurney pp. 149–154 (this volume).

47. See Simpson pp. 71–84 (this volume).

48. Darlington 1849; Berkeley & Berkeley 1992.

49. Nelson 2014.

50. The full title was *Hortus britanno-americanus: or, a curious collection of trees and shrubs, the produce of the British colonies in North America; adapted to the soil and climate of England. With observations on their constitution, growth, and culture: and directions how they are to be collected, packed up, and secured during their passage.*

51. Vane-Wright & Hughes [2005]; Dance 2006.

52. On images of *Magnolia grandiflora*, see Nelson 2014.

53. Catesby 1763; 1767: plates 6 and 12.

54. Catesby 1763; 1767: Preface, ii; Henrey 1975: 3: 22. Comparison of copies of Catesby 1763 (Botany Library, Harvard University) and Catesby 1767 (Oak Spring Garden Library, Upperville, Virginia).

55. "Darwin, C. R. 'Books [read]' NOTEBOOK (1838–1858)," http://darwin-online.org.uk/converted/manuscripts/Darwin_C_R_CUL-DAR120-.html, accessed 11 July 2013; Romanes 1882. Jeremy Barlow (Darwin's great-great-grandson) provided this information.

56. eLife open source journal; European/U.S. team, 2013.

Appendix

NOTES

1. For more information on the development of this listing see Reveal 2013 (http://www.phytoneuron.net/2013Phytoneuron/06PhytoN-CatesbyIDs.pdf) and papers cited therein. Detailed notes about the use of Catesby's images and text by Linnaeus, and matters of typification, are in the notes attached to Reveal 2012a (http://www.phytoneuron.net/PhytoN-Catesby.pdf).

2. Edwards 1771.

3. Ewan 1974.

4. Howard & Staples 1983.

5. Wilbur 1990.

6. Reveal 2009, 2012a, 2013.

7. Feduccia 1985.

8. Olson & Reveal 2009.

9. Reveal et al. 2014.

A BIBLIOGRAPHY FOR MARK CATESBY

Achard, Mr., 1763. Remarks on swallows on the Rhine: in a letter . . . to Mr. Peter Collinson. . . . *Philosophical transactions* 53: 101–102.

Adler, K., 2004. America's first herpetological expedition: William Bartram's travels in southeastern United States (1773–1776). *Bonner Zoologische Beiträge* 52: 275–295.

Adler, K., 2012a. Herpetological exploration in the eighteenth century: spanning the globe with Linnaeus's students, pp. 39–52 in Bell, C. J. (editor), The herpetological legacy of Linnaeus: a celebration of the Linnaean tercentenary. *Bibliotheca herpetologica* 9.

Adler, K., 2012b. Herpetologists of the past, part 3, pp. 8–386 in Adler, K. (editor), Contributions to the history of herpetology, volume 3. *Society for the Study of Amphibians and Reptiles, Contributions to herpetology* 29.

Aiton, W., 1789. *Hortus Kewensis*. Volume 2. London.

Albin, E., 1720. *A natural history of English insects: illustrated with a hundred copper plates, curiously engraven from the life*. London.

Alexander, W. B., 1933. The swallow mortality in central Europe in September, 1931. *Journal of animal ecology* 2: 116–118.

Allen, D. E., 1959. The history of the vasculum. *Proceedings of the Botanical Society of the British Isles* 3: 135–150.

Allen, D. E., 1976. *The naturalist in Britain: a social history*. London.

Allen, D. E., 2008. Sherard, William (1659–1728). *Oxford dictionary of national biography*. Online edition (doi:10.1093/ref:odnb/25355). Oxford.

Allen, E. G., 1937. New light on Mark Catesby. *Auk* 54: 349–363.

Allen, E. G., 1951. The history of American ornithology before Audubon. *Transactions of the American Philosophical Society*, new series, 41 (3): 385–591.

Andrews, E. W. & Andrews, C. M. (editors), 1939. *Janet Schaw, journal of a lady of quality*. New Haven, Connecticut.

Anonymous, 1729. *The art of painting in miniature: teaching the speedy and perfect acquisition of that art without a master. Done from the original French*. London.

Anonymous, 1730. *Catalogus plantarum . . . A catalogue of trees, shrubs, plants, and flowers, both exotic and domestic*. London.

Anonymous, 1750. [Obituary of Mark Catesby]. *The gentleman's magazine* 20: 30.

Armstrong, A. W. (editor), 2002. *"Forget not mee and my garden . . ." Selected letters 1725–1768 of Peter Collinson, F.R.S.* Philadelphia.

Armytage, W. H. G., 1954. Letters on natural history of Carolina 1700–1705. *South Carolina historical magazine* 55 (2): 59–70.

Audubon, J. J., 1827–1838. *The birds of America*. London.

Audubon, J. J., 1844. *The birds of America*. New York. 7 volumes.

B., A., 1748. Some remarks on a printed paper concerning birds of passage. *The gentleman's magazine* 18: 445–447.

Bacon, F., 1620. *Novum organum*. London.

Baigent, E., 2004. Mitchell, John (1711–1768). *Oxford dictionary of national biography*. Online edition (doi:10.1093/ref:odnb/18842). Oxford.

Ballard, T., 1708. *Bibliotheca rayana: or, a catalogue of the library of Mr. John Ray*. London.

Barclay, R. M. R., Lausen, C. L. & Hollis, L., 2001. What's hot and what's not: defining torpor in free-ranging birds and mammals. *Canadian journal of zoology* 79: 1885–1890.

Barrett, P. H., Gautrey, P. J., Herbert, S., Kohn, D. & Smith, S. (editors), 1987. *Charles Darwin's notebooks, 1836–1844: geology, transmutation of species, metaphysical enquiries*. Cambridge.

Barrington, D., 1772. An essay on the periodical appearing and disappearing of certain birds, at different times of the year. In a letter . . . to William Watson, M.D. F.R.S. *Philosophical transactions* 62: 265–326.

Bartholomew, G. A., Howell, T. R. & Cade, T. J., 1957. Torpidity in the white-throated swift, Anna hummingbird, and poor-will. *The condor* 59: 145–155.

Bartram, W., 1791. *Travels through North and South Carolina, Georgia, East and West Florida*. Philadelphia.

Batho, G. R., 2000. Thomas Harriot and the Northumberland household, pp. 28–47 in Fox, R. (editor), *Thomas Harriot: an Elizabethan man of science*. Aldershot.

Baugh, D. A., 2008. Wager, Sir Charles (1666–1743). *Oxford dictionary of national biography*. Online edition (doi:10.1093/ref:odnb/28393). Oxford.

Beatty, R. C. & Mulloy, W. J. (editors), 1940. *William Byrd's Natural history of Virginia, or, the newley discovered Eden*. Richmond, Virginia.

Beck, H., 1987. *Alexander von Humboldt: life and work*. Ingelheim am Rhein.

Beer, W.-D. (editor), 1976. *Maria Sibylla Merian: Schmetterlinge, Käfer und andere Insekten. Leningrader Studienbuch*. 2 volumes. Leipzig & Lucerne.

Berkeley, E. & Berkeley, D. S. (editors), 1992. *The correspondence of John Bartram, 1734–1777*. Gainesville, Florida.

Beverley, R., 1705. *The history and present state of Virginia*. London.

Beverley, R., 1722. *The history and present state of Virginia*. Revised edition. London.

Bircham, P., 2007. *A history of ornithology*. London.

Birkhead, T., 2011. *The wisdom of birds: an illustrated history of ornithology*. London.

Bligh, W., 1790. *A narrative of the mutiny on board His Majesty's Ship Bounty*. London.

Blunt, W., 1971. *The compleat naturalist: a life of Linnaeus*. London.

Bobart, H. T., 1884. *A biographical sketch of Jacob Bobart of Oxford, together with an account of his two sons, Jacob and Tilleman*. Leicester.

Böhlke, J. E. & Chaplin, C. C. G., 1968. *Fishes of the Bahamas*. Wynnewood, Pennsylvania.

Böhlke, J. E. & Smith, D. G., 1968. A new xenocongrid eel from the Bahamas, with notes on other species in the family. *Proceedings of the Academy of Natural Sciences of Philadelphia* 120: 25–43.

Bostock, J. (translator), 1855. Pliny the Elder, *The natural history*. London.

Bott, G., 2001. *Ein Stück von allerlei Blumenwerk—ein Stück von Früchten—zwei Stück auf Tuch mit Hecht: Die Stillebenmaler Soreau, Binoit, Codino und Marrell in Hanau und Frankfurt 1600–1650*. Hanau.

Boulger, G. S., 1883. Samuel Dale (with portrait). *Journal of botany* 21: 193–197, 225–231.

Boulger, G. S., 1918. A seventeenth-century botanist friendship. *Journal of botany* 56: 197–202.

Boyle, A., Boston, C. & Witkin, A., 2005. *The archaeological experience at St Luke's Church, Old Street, Islington* (Archaeological Recording Action Report). Oxford.

Bradley, R., 1726. *A general treatise of husbandry and gardening. London.*

Briceland, A. V., 2000. John Lawson. *American national biography*. Online edition (http://www.anb.org.go.libproxy.wfubmc.edu/articles/01/01-00498.html).

Brickell, J., 1737. *The natural history of North-Carolina*. Dublin.

Bridgewater, W. & Kurtz, S. (editors), 1968. *The Columbia encyclopedia*. New York & London.

Brigham, D. R., 1998. Mark Catesby and the patronage of natural history, pp. 91–145 in Meyers, A. & Pritchard, M. B. (editors), *Empire's nature: Mark Catesby's New World vision*. Chapel Hill, North Carolina & London.

Brigham, R. M., 1992. Daily torpor in a free-ranging goatsucker, the common poor-will (*Phalaenoptilus nuttallii*). *Physiological zoology* 65: 457–472.

Britton, N. L. & Millspaugh, C. F., 1920. *The Bahama flora*. New York.
Broberg, G., 2006. *Carl Linnaeus*. Stockholm.
Brock, R. A. (editor), 1882. *The official letters of Alexander Spotswood, Lieutenant-Governor of the Colony of Virginia, 1710–1722. . . .* Richmond, Virginia.
Brown, C. R. & Brown, M. B., 2000. Weather-mediated natural selection on arrival time in cliff swallows (*Petrochelidon pyrrhonota*). *Behavioral ecology and sociobiology* 47: 339–345.
Browne, P., 1756. *The civil and natural history of Jamaica*. London.
Bürger, T. & Heilmeyer, M., 1999. *Maria Sibylla Merian. Neues Blumenbuch = New book of flowers*. Munich, London, & New York.
Calmann, G., 1977. *Ehret: flower painter extraordinary: an illustrated biography*. Oxford.
Camp, C. L., 1923. Classification of the lizards. *Bulletin of the American Museum of Natural History* 48: 289–307.
Cannon, R., 1847. *Historical record of the Tenth, or the North Lincolnshire, Regiment of Foot*. London.
Cates, D. L., 1996. Mark Catesby and *The natural history of Carolina, Florida and the Bahama islands*. *Journal of the Bahamas Historical Society* 18: 2–11.
Cates, D. L., 1998. Mark Catesby's Bahamian plants. *Bahamas journal of science* 5 (3): 16–21.
Catesby, M., [1729–]1731–1743[–1747]. *The natural history of Carolina, Florida and the Bahama islands*. 2 volumes. London.
Catesby, M., [1729?]. *Proposals, for printing an essay towards a natural history of Florida, Carolina and the Bahama islands*. London.
Catesby, M., 1747. Of birds of passage. *Philosophical transactions of the Royal Society of London* 44: 435–444.
Catesby, M., 1748. Extract from a paper . . . written by Mark Catesby, F.R.S. in *Phil. trans*. no. 483. *The gentleman's magazine* 18 (October): 447–448.
Catesby, M., 1750–[?]. *Piscium, serpentum, insectorum, aliorumque non nullorum animalum nec non plantarum . . . Die Abbildungen verschiedener Fische, Schlangen, Insecten, einiger andern Thiere, und Pflanzen*. Nuremberg.
Catesby, M., 1754. *The natural history of Carolina, Florida and the Bahama islands. Revis'd by Mr. Edwards, of the Royal College of Physicians, London*. Second edition. 2 volumes. London.
Catesby, M., 1763. *Hortus Britanno-Americanus: or, a curious collection of trees and shrubs, the produce of the British colonies in North America. . . .* London.
Catesby, M., 1767. *Hortus Europae Americanus: or, a collection of 85 curious trees and shrubs, the produce of North America. . . .* London.
Catesby, M., 1768–1776. *Recueil de divers oiseaux étrangers et peu communs*. Nuremburg.
Catesby, M., 1771. *The natural history of Carolina, Florida and the Bahama islands. Revis'd by Mr. Edwards, of the Royal College of Physicians, London*. Third edition. 2 volumes. London.
Chaplin, J., 1998. Mark Catesby, a skeptical Newtonian in America, pp. 34–90 in Meyers, A. & Pritchard, M. B. (editors), *Empire's nature: Mark Catesby's New World vision*. Chapel Hill, North Carolina & London.
Christakis, N. A. J. F., 2009. *Connected: the surprising power of our social networks and how they shape our lives*. New York.
Clokie, H. N., 1964. *An account of the herbaria of the Department of Botany in the University of Oxford*. Oxford.
Clutton-Brock, J., 1994. Vertebrate collections, pp. 77–92 in MacGregor, A. (editor), *Sir Hans Sloane, collector, scientist, antiquary, founding father of the British Museum*. London.
Coleridge, S. T., 1917. *The table talk and Omniana. . . .* Oxford.
Collins, M. R. & Smith, T. I. J., 1996. *Occurrence of sturgeons in South Carolina*. Charleston, South Carolina.
Collinson, P., 1760. A letter to the Honourable J. Th. Klein, Secretary to the City of Dantzick, . . . concerning the migration of swallows. *Philosophical transactions* 51: 459–464.

Commelin, J., 1697. *Horti medici Amstelodamensis rariorum*. Amsterdam.

Cook, H. J., 1993. The cutting edge of a revolution? Medicine and natural history near the shores of the North Sea, pp. 45–61 in Field, J. V. & James, F. A. J. L. (editors), *Renaissance and revolution: humanists, scholars, craftsmen and natural philosophers in Early Modern Europe*. Cambridge.

Cook, H. J., 2007. *Matters of exchange: commerce, medicine, and science in the Dutch Golden Age*. New Haven, Connecticut.

Cook, W. J., 2012. The correspondence of Thomas Dale (1700–1750): botany in the transatlantic republic of letters. *Studies in history and philosophy of biological and biomedical sciences* 43: 232–243.

Coombs, D., 1997. The garden at Carlton House of Frederick Prince of Wales and Augusta Princess Dowager of Wales. *Garden history* 25 (2): 153–177.

Cooper, T. (editor), 1838. *Statutes at Large of South Carolina*. Columbia, South Carolina.

Cornelius, P. F. S., 1994. Benjamin White (1725–1794), his older brother Gilbert, and notes on the hibernation of swallows. *Archives of natural history* 21: 231–236.

Cornish, J., 1775. Of the torpidity of swallows and martins . . . in sundry letters to the Honourable Daines Barrington, F.R.S. and Dr. Maty. . . . *Philosophical transactions* 65: 343–352.

Correll, D. S. & Correll, H. B., 1982. *Flora of the Bahama archipelago. . . .* Vaduz.

Costa, E. da, 1812. Notices and anecdotes of literati, collectors, &c. from a ms. by the late Mendes de Costa, and collected between 1747 and 1788. *The gentleman's magazine* 81: 205–207.

Coues, E., 1878. *Birds of the Colorado Valley*. Washington, D.C.

Crane, V., 1956. *The southern frontier, 1670–1732*. Ann Arbor, Michigan.

Crother, B. I. (editor), 2012. Scientific and standard English and French names of amphibians and reptiles of North America north of Mexico, with comments regarding confidence in our understanding. Seventh edition. *SSAR herpetological circular* 39.

Cuvier, G., 1796. Mémoire sur les épèces d'elephans tant vivantes que fossils, lu à la séance publique de l'Institut National le 15 germinal, an IV. *Magasin encyclopédique, 2E année*: 440–445.

Dale, S., 1693. *Pharmacologia, seu manuductio ad materiam medicam*. London.

Dalton, C., 1912. *George the First's army 1714–1727*. London.

Dampier, W., 1697. *A new voyage round the world*. London.

Dampier, W., 1699. *Voyages and descriptions. Volume 2. In three parts, viz. . . . 3. A discourse of trade-winds, breezes, storms, seasons of the year, tides and currents of the torrid zone throughout the world . . . &c.* London.

Dampier, W., 1703. *A voyage to New Holland, &c., in the year 1699: wherein are described the Canary-Islands. . . .* London.

Dampier, W., 1709. *A continuation of a voyage to New-Holland, &c. in the year 1699. Wherein are described, the islands Timor, Rotee and Anabao. . . .* London.

Dance, S. P., 2006. [Review of] *The Seymer legacy. . . . Archives of natural history* 33: 189–190.

Dandy, J. E., 1958. *The Sloane herbarium: an annotated list of the horti sicci composing it: with biographical accounts of the principal contributors*. London.

Darlington, W. (editor), 1849. *Memorials of John Bartram and Humpry Marshall with notices of their botanical contemporaries*. Philadelphia.

Darwin, C. R., 1839. *Narrative of the surveying voyages of His Majesty's Ships Adventure and Beagle between the years 1826 and 1836. . . . Journal and remarks. 1832–1836*. London.

Darwin, C. R., 1859. *On the origin of species by means of natural selection. . . .* London.

Day, D., 1989. *Vanished species*. Revised edition. New York.

Defoe, D., 1724. Tour through the eastern counties of England, letter 1 in volume 1, *A tour thro' the whole island of Great Britain*. London.

Deppermann, A., 2002. *Johann Jakob Schütz und die Anfänge des Pietismus*. Tübingen.

Derham, W., 1708. Part of a letter . . . concerning the migration of birds. *Philosophical transactions* 26: 123–124.

Dickinson, H. T., 1967. The poor Palatines and the parties. *English historical review* 82: 464–485.

Dillenius, J. J., 1732. *Hortus Elthamensis, seu, plantarum rariorum quas in horto suo Elthami in Cantio coluit . . . Jacobus Sherard*. London.

Dobson, A. (editor), 1906. *The diary of John Evelyn*. London.

Dobson, D., 1986. *Directory of Scots in the Carolinas 1680–1830*. Baltimore, Maryland.

Druce, G. C., 1897. *An account of the herbarium of the University of Oxford*. Oxford.

Druce, G. C., 1928. British plants contained in the Du Bois Herbarium in Oxford. *Botanical Exchange Club report* 1927: 463–493.

Druce, G. C. & Vines, S. H., 1907. *The Dillenian herbaria*. Oxford.

Edgar, W., 1998. *South Carolina: a history*. Columbia, South Carolina.

Edgar, W. (editor), 2006. *South Carolina encyclopedia*. Columbia, South Carolina.

Edgar, W. B. & Bailey, N. L., 1977. *Biographical directory of the South Carolina House of Representatives*. Columbia, South Carolina.

Edwards, G., 1743–1751. *A natural history of [uncommon] birds*. London. 4 volumes.

Edwards, G., 1771. A catalogue of the animals and plants represented in Catesby's *Natural history of Carolina* with the Linnaean names, pp. 1–2 in Catesby, M., *The natural history of Carolina, Florida and the Bahama islands: . . . revised by Mr. Edwards, of the Royal College of Physicians, London. To the whole is now added a Linnaean index of the animals and plants*. London.

Egerton, F. N., 2006. A history of the ecological sciences, part 22: early European naturalists in eastern North America. *Bulletin of the Ecological Society of America* 87: 341–356.

Ellis, J., 1755. *An essay towards a natural history of the Corallines and other marine productions found on the coasts of Great Britain and Ireland*. London.

Enright, B. J., 1960. An account of Charles Town in 1725. *South Carolina historical magazine* 61: 15–18.

Etheridge, K., 2010. Maria Sibylla Merian's frogs. *Bibliotheca herpetologica* 8: 20–27.

Etheridge, K., 2011. Maria Sibylla Merian: the first ecologist?, pp. 31–49 in Molinari, V. & Andreolle, D. (editors), *Women and science: pioneers, activists and protagonists*. Newcastle upon Tyne.

Ewan, J., 1974. Notes, pp. 89–100 in Catesby, M., *The natural history of Carolina, Florida and the Bahama islands*. . . . with an introduction by George Frick and notes by Joseph Ewan. Savannah, Georgia.

Ewan, J. & Ewan, N., 1970. *John Banister and his Natural history of Virginia, 1678–1692*. Urbana, Illinois.

Fairchild, T., 1722. *The city gardener*. London.

Feduccia, A. (editor), 1985. *Catesby's birds of colonial America*. Chapel Hill, North Carolina & London.

Findlen, P., 2008. Natural history, pp. 435–468 in Park, K. & Daston, L. (editors), *The Cambridge history of science*. Volume 3. *Early modern science*. Cambridge.

Forster, J. R., 1771. *A catalogue of the animals of North America*. . . . London.

Forster, J. R., 1772. *Voyage around the world. Performed by order of his most Christian Majesty, in the years 1766, 1767, 1768, and 1769, by Lewis de Bougainville*. London.

French, H. R., 2000. "Ingenious & learned gentlemen"—social perceptions and self-fashioning among parish elites in Essex, 1680–1740. *Social history* 25: 44–66.

Frick, G. F., 1960. Mark Catesby: the discovery of a naturalist. *Papers of the Bibliographical Society of America* 54: 163–175.

Frick, G. F. & Stearns, R. P., 1961. *Mark Catesby: the colonial Audubon*. Urbana, Illinois.

Frick, G. F., 1974 . . . in Catesby, M., *The natural history of Carolina, Florida and the Bahama islands*. . . . with an introduction by George Frick and notes by Joseph Ewan. Savannah, Georgia.

Fry, J. T., 2014. Inside the box: John Bartram and the science and commerce of the transatlantic plant trade, pp. 194–200 in Smith, P. H., Meyers, A. R. W. & Cook, H. J. (editors), *Ways of making and knowing: the material culture of empirical knowledge*. Ann Arbor, Michigan.

Gascoigne, B., 1986. *How to identify prints: a complete guide to manual and mechanical processes from woodcut to ink-jet*. London.

Gascoigne, J., 1994. *Joseph Banks and the English Enlightenment: useful knowledge and polite culture*. Cambridge & New York.

Gaskell, P., 1972. *A new introduction to bibliography*. New York & Oxford.

Geiser, F. & Ruf, T., 1995. Hibernation versus daily torpor in mammals and birds: physiological variables and classification of torpor patterns. *Physiological zoology* 68: 935–966.

Gerard, J., 1598. *The herball, or, generall historie of plantes. . . .* London.

Gerard, J., 1633. *The herball, or, generall historie of plantes, . . . very much enlarged and amended by Thomas Johnson citizen and apothecarye of London*. London.

Gerard, J., 1636. *The herball, or, generall historie of plantes, . . . very much enlarged and amended by Thomas Johnson citizen and apothecarye of London*. London.

Gibbons, J. W. & Dorcas, M. E., 2005. *Snakes of the Southeast*. Athens, Georgia.

Gibbons, J. W. & Semlitsch, R. D., 1991. *Guide to the reptiles and amphibians of the Savannah River Site*. Athens, Georgia.

Gibson, J., 1796. A short account of several gardens near London, with remarks on some particulars wherein they excel, or are deficient, upon a view of them in December 1691 . . . *Archaeologia* 12: 181–192.

Glendinning, R., 1835. The history of the original plant of the Exmouth magnolia, in the garden of Sir John Colliton, in Exmouth. *The gardener's magazine*, new series, 1: 70–71.

Goedaert, J., 1662–1669. *Metamorphosis naturalis, ofte historische beschryvinghe. . . .* 3 volumes. Middelburgh.

Gräffin, M. S. *See* Merian, M. S.

Gray, C., [c. 1740]. *A catalogue of American trees and shrubs that will endure the climate of England*. Fulham.

Greene, J. P. (editor), 1965. *The diary of Colonel Landon Carter of Sabine Hall, 1752–1778*. Charlottesville, Virginia.

Grimwood, C. G. & Kay, S. A., 1952. *History of Sudbury Suffolk*. Sudbury.

Gronovius, J. F., 1739. *Flora Virginica*, part 1. Leiden.

Gronovius, J. F., 1743. *Flora Virginica*, part 2. Leiden.

Gunther, R. T., 1922. *Early British botanists and their gardens, based on unpublished writings of Goodyer, Tradescant, and others*. Oxford.

Hall, R. L., 1991. Savoring Africa in the New World, pp. 161–169 in Viola, H. J. & Margolis, C. (editors), *Seeds of change*. Washington, D.C.

Harkness, D. E., 2007. *The jewel house: Elizabethan London and the scientific revolution*. New Haven, Connecticut.

Harkness, D. E., 2009. Elizabethan London's naturalists and the work of John White, pp. 44–50 in Sloan, K. (editor), *European visions: American voices*. London.

Harriot [Hariot], T., 1588. *A brief and true report of the new found land of Virginia*. London.

Harriot [Hariot], T., 1590. *A briefe and true report of the new found land of Virginia: of the commodities and of the nature and manners of the naturall inhabitants. . . .* Frankfurt am Main.

Harris, J., 2004. London, George (d. 1714). *Oxford dictionary of national biography*. Online edition (doi:10.1093/ref:odnb/37686). Oxford.

Harris, S. J., 2008. Networks of travel, correspondence and exchange, pp. 347–362 in Park, K. & Daston, L. (editors), *The Cambridge history of science*. Volume 3. *Early Modern science*. Cambridge.

Harting, J. E., 1880. *British animals extinct within historic times; with some account of British wild white cattle*. London.

Harvey, J. H., 1981. A Scottish botanist in London in 1766. *Garden history* 9 (1): 40–75.

Harvey, J. H., 1988. *The availability of hardy plants of the late eighteenth century*. Garden History Society.

Hayes, K. J., 1997. *The library of William Byrd of Westover*. Madison, Wisconsin & Philadelphia.

Heitzler, M., 2010. Boochawee: Plantation land and legacy in Goose Creek. *South Carolina historical magazine* 111 (1–2): 34–70.

Heller, J. L., 1976. Linnaeus on sumptuous books. *Taxon* 25 (1): 33–52.

Heller, J. L., 2007. *Index of the books and authors cited in the zoological works of Linnaeus (Index librorum et auctorum in operibus zoologicis Linnaei citatorum). Edited by J. H. Penhallurick.* London.

Henrey, B., 1975. *British botanical and horticultural literature before 1800.* 3 volumes. Oxford.

Henrey, B., 1986. *No ordinary gardener: Thomas Knowlton, 1691–1781. . . . Edited by A. O. Chater.* London.

Hermann, P., 1698. *Paradisus batavus . . . cui accessit catalogus plantarum. . . .* Leiden.

Hobson, K. A., Van Wilgenburg, S. L., Piersma, T. & Wassenaar, L. I., 2012. Solving a migration riddle using isoscapes: house martins from a Dutch village winter over West Africa. *PLOS ONE* 7 (9) e45005: 1–7 (www.plosone.org).

Houttuyn, M. (translator), 1772–1781. *Verzamerling van uitlandsche en zeldzaame vogelen . . . door G. Edwards en M. Catesby. . . .* Amsterdam.

Howard, R. A. & Staples, G. W., 1983. The modern names for Catesby's plants. *Journal of the Arnold Arboretum* 64: 511–546.

Hulme, M. & Barrow, E. (editors), 1997. *Climates of the British Isles: present, past and future.* London & New York.

Humboldt, A. von, 1855. *Cosmos: sketch of a physical description of the universe translated . . .* [by Mrs. Sabine]. New edition. London.

Humboldt, A. von, 1865. *De la physionomie des plantes.* Paris.

Humboldt, A. von, 1995. *Personal narrative of a journey to the equinoctial regions of the new continent. . . . abridged and translated with an introduction by Jason Wilson.* London.

Huth, G. L., 1749–1770. *Sammlung verschiedener auslandischer und seltener Vogel.* Nuremberg.

ICZN (International Commission on Zoological Nomenclature), 1999. *International code of zoological nomenclature.* Fourth edition. London.

Ivers, L. E., 1970. *Colonial forts of South Carolina 1670–1775.* Columbia, South Carolina.

Ivers, L. E., 1972. Scouting the Inland Passage, 1685–1737. *South Carolina historical magazine* 73: 117–129.

Jackson, B. D., 1876. *A catalogue of plants cultivated in the garden of John Gerard, in the years 1596–1599 . . . and a life of the author.* London.

Jackson, C. E., 1985. *Bird etchings: the illustrators and their books, 1655–1855.* Ithaca, New York & London.

James, K. A., 2004 "Humbly dedicated": Petiver and the audience for natural history in early eighteenth-century Britain *Archives of natural history* 31 (2): 318–329.

Jarvis, C., 2007. *Order out of chaos: Linnaean plant names and their types.* London.

Jenner, E. & Jenner, G. C., 1824. Some observations on the migration of birds. *Philosophical transactions* 114: 11–44.

Johns, A., 2008. Coffee houses, pp. 320–340 in Park, K. & Daston, L. (editors), *The Cambridge history of science.* Volume 3. *Early modern science.* Cambridge.

Johnston, D. W., 2003. *The history of ornithology in Virginia.* Charlottesville, Virginia.

Jones, H., 1724. *The present state of Virginia: giving a particular and short account of the Indian, English and Negro inhabitants of that colony.* London.

Jones, L. H., 1891. *Captain Roger Jones of London and Virginia.* Albany, New York.

Jonston, J., 1653. *Historiae naturalis de insectis libri III.* Frankfurt am Main.

Jonston, J., 1660. *Jonstons naeukeurige beschryving van de natuur der vier-voetige dieren, vissen en bloedlooze waterdieren, vogelen, kronkel-dieren, slangen en draken, 4. . . .* Amsterdam.

Jordan, D. S., 1884. An identification of the figures of fishes in Catesby's *Natural history of Carolina, Florida, and the Bahama islands. Proceedings of the United States National Museum* 7: 190–199.

Jorink, E., 2010. *Reading the book of nature in the Dutch Golden Age, 1575–1715*. Leiden.

Kalm, P., 1770. *Travels into North America . . . translated into English by John Reinhold Forster*. London.

Kalm, P., 1892. *See* Lucas 1892.

Keskpaik, J., 1972. Summary: temporary hypothermy in sand-martins (*Riparia r. riparia*) in natural conditions. *Communications of the Baltic Commission for the Study of Bird Migration* 7: 182–183.

Keynes, R. D. (editor), 1988. *Charles Darwin's Beagle diary*. Cambridge.

Kornwolf, J. D., 1993. "Doing good to posterity": Francis Nicholson, first patron of architecture, landscape design, and town planning in Virginia, Maryland, and South Carolina, 1688–1725. *Virginia magazine of history and biography* 101 (3): 333–374.

Kovacik, C. F. & Winberry, J. J., 1987. *South Carolina: the making of a landscape*. Columbia, South Carolina.

Krech, S., III, 2014. The birds in Catesby's Natural history: collected, imaged, commodified, in Galdy, A. (editor), *Collecting nature*. Newcastle Upon Tyne.

Laird, M., 1998. From *Callicarpa* to *Catalpa*: the impact of Mark Catesby's plant introductions on English gardens of the eighteenth century, pp. 184–227 in Meyers, A. R. W. & Pritchard, M. B. (editors), *Empire's nature: Mark Catesby's New World vision*. Chapel Hill, North Carolina & London.

Laird, M., 1999. *The flowering of the landscape garden: English pleasure grounds, 1720–1800*. Philadelphia.

Laird, M., 2000. Exotics and botanical illustration, pp. 93–113 in Ridgway, C. & Williams, R. (editors), *John Vanbrugh and landscape architecture in Baroque England, 1690–1730*. Stroud.

Laird, M., 2002. The culture of horticulture: class, consumption, and gender in the English landscape garden, p. 248 in Conan, M. (editor), *Bourgeois and aristocratic cultural encounters in garden art, 1550–1850*. Washington, D.C.

Laird, M., 2008. The congenial climate of coffee-house horticulture: the *Historia plantarum rariorum* and the *Catalogus plantarum*, pp. 226–259 in O'Malley, T. & Meyers, A. R. W. (editors), *The art of natural history: illustrated treatises and botanical paintings, 1400–1850*. New Haven, Connecticut & London.

Laird, M., 2015. *A natural history of English gardening: 1650–1800*. New Haven, Connecticut & London.

Laird, M. & Weisberg-Roberts, A. (editors), 2009. *Mrs. Delany and her circle*. New Haven, Connecticut & London.

Lamb, H. H., 1995. *Climate, history and the modern world*. Second edition. London & New York.

Lasiewski, R. C. & Thompson, H. J., 1966. Field observations of torpidity in the violet-green swallow. *Condor* 68: 102–103.

Latham, J., 1781. *A general synopsis of birds*. London.

Lawson, J., 1709. *A new voyage to Carolina; containing the exact description and natural history of that country*. London.

Lawson, J., 1714. *The history of Carolina; containing the exact description and natural history of that country*. London.

Lawson, J., 1718. *The history of Carolina*. London.

Lawson, J., 1860. *The history of Carolina, containing the exact description and natural history of that country. . . .* London & Raleigh.

Leapman, M., 2000. *The ingenious Mr Fairchild: the forgotten father of the flower garden*. London.

Lebedeva, I., 1976. On the history of the "Leningrad book of notes and studies," pp. 43-47 in Beer, W.-D. (editor), *Maria Sibylla Merian: Schmetterlinge, Käfer und andere Insekten. Leningrader Studienbuch*. Volume 2. Leipzig & Lucerne.

Lefler, H. T. (editor), 1967a. *A new voyage to Carolina by John Lawson*. Chapel Hill, North Carolina.

Lefler, H. T., 1967b. Promotional literature of the southern colonies. *Journal of southern history* 33 (1): 3–25.

Lefler, H. T. & Newsome, A. R., 1979. *North Carolina: the history of a southern state*. Chapel Hill, North Carolina.

Leith-Ross, P., 2006. *The John Tradescants: gardeners to the Rose and Lily Queen*. Revised edition. London.

Le Rougetel, H., 1971. Gardener extraordinary: Philip Miller of Chelsea (1691–1771). *Journal of the Royal Horticultural Society* 96: 556–563.

Lesser, C. H., 1995. *South Carolina begins: the records of a Proprietary Colony, 1663–1721*. Columbia, South Carolina.

Lewis, A. J., 2005. A democracy of facts, an empire of reason: swallow submersion and natural history in the early American republic. *William and Mary quarterly* 62: 663–696.

Linder, S. C., 2000. *Anglican churches in Colonial South Carolina: their history and architecture*. Charleston, South Carolina.

Linder, S. C., 2006. Periagua, p. 712 in Edgar, W. (editor), *South Carolina encyclopedia*. Columbia, South Carolina.

Linnaeus, C., 1735. *Systema naturæ, sive regna tria naturæ systematice proposita per Classes, Ordines, Genera, et Species*. Leiden.

Linnaeus, C., 1737. *Genera plantarum*. Leiden.

Linnaeus, C., 1737 [1738]. *Hortus Cliffortianus*. Amsterdam.

Linnaeus, C., 1741 [1746]. Decem plantarum genera (no. 1025; pp. 79–80, tab. II). *Societas Regia Scientiarum Upsaliensis. Acta*.

Linnaeus, C., 1748. *Hortus Upsaliensis*. Stockholm.

Linnaeus, C., 1752. Anmärkning om ormarnas skilje-märken. *Kongliga Svenska vetenskaps-academiens handlingar* 1752: 206–207.

Linnaeus, C., 1753a. *Species plantarum*. Stockholm. 2 volumes

Linnaeus, C., 1753b. *Incrementa botanices proxime praeterlapsi semiseculi*. Stockholm.

Linnaeus, C., 1756. *Centuria II plantarum*. Stockholm.

Linnaeus, C., 1758. *Systema naturæ per regna tria naturæ, secundum Classes, Ordines, Genera, Species, cum characteribus, differentiis, synonymis, locis*. Tomus I. Tenth edition. Stockholm.

Linnaeus, C., 1766. *Systema naturæ per regna tria naturæ, secundum Classes, Ordines, Genera, Species, cum characteribus, differentiis, synonymis, locis*. Tomus I. Twelfth edition. Stockholm.

Lister, M., 1685–1692. *Historiae sive synopsis methodicae conchyliorum*. . . . 2 volumes. London.

Llanover, A. (editor), 1861–1862. *The autobiography and correspondence of Mary Granville, Mrs Delany: with interesting reminiscences of King George the Third and Queen Charlotte*. 6 volumes. London.

Lounsbury, C., 2000. Ornaments of civic aspiration: the public buildings of Williamsburg, pp. 25–38 in Maccubbin, R. P. & Hamilton-Phillips, M. (editors), *Williamsburg, Virginia: a city before the State, 1699–1999*. Williamsburg, Virginia.

Lucas, J. (translator), 1892. *Kalm's account of his visit to England on his way to America in 1748*. London & New York.

Lundqvist, S. & Moberg, R., 1993. The Pehr Kalm herbarium in UPS, a collection of North American plants. *Thunbergia* 19: 1–62.

Lyle, I. F., 1978. John Hunter, Gilbert White, and the migration of swallows. *Annals of the Royal College of Surgeons of England* 60: 485–491.

Lysons, D., 1811. *The environs of London*. Volume 2 (part 2). *County of Middlesex. Hornsey–Wilsdon*. Second edition. London.

McAtee, W. L., 1947. Torpidity in birds. *American Midland naturalist* 38: 191–206.

McAtee, W. L., 1955–1956. The birds in Lawson's *New voyage to Carolina. Chat* 19: 74–77; 20: 23–27.

Macaulay, G. C. (translator), 1890. *The history of Herodotus*. Volume I. Book II. Section 22. London.

McBurney, H., 1997. *Mark Catesby's natural history of America: the watercolors from the Royal Library, Windsor Castle*. Houston & London.

McCann, T., 2004. Lennox, Charles, second Duke of Richmond . . . (1701–1750). *Oxford dictionary of national biography*. Online edition (doi:10.1093/ref:odnb/1645). Oxford.

McClure, S. A. & Eshbaugh, W. H., 1983. Love potions of Andros Island, Bahamas. *Journal of ethnobiology* 3 (2): 149–156.

McCord, D. J. (editor), 1839. *Statutes at Large of South Carolina*. Volume 6. Columbia, South Carolina.

McCormack, J. H., Maier, K. & Wallens, P. B., 2011. *Bush medicine of the Bahamas—a cross-cultural perspective from San Salvador Island, including pharmacology and oral histories*. Charlottesville, Virginia.

McDaniel, S., 1971. The genus *Sarracenia* (Sarraceniaceae). *Bulletin of Tall Timbers Research Station* 9: 1–36.

McDowell, W. L., Jr. (editor), 1992. *Journals of the commissioners of the Indian Trade*. Columbia, South Carolina.

MacGregor, A. (editor), 1994. *Sir Hans Sloane: collector, scientist, antiquary—founding father of the British Museum*. London.

McIlhenny, E. A., 1935. *The alligator's life history*. Boston.

McKechnie, A. E. & Lovegrove, B. G., 2002. Avian facultative responses: a review. *Condor* 104: 705–724.

McMillan, P. D., Blackwell, A. H., Blackwell, C. & Spencer, M., 2013. Catesby plants in the Sloane Herbarium. *Phytoneuron* 2013-7: 1–37.

McNeill, J. et al. (editors), 2012. International code of nomenclature for algae, fungi and plants (Melbourne code). *Regnum vegetabile* 154.

Marambaud, P., 1971. *William Byrd of Westover, 1674–1744*. Charlottesville, Virginia.

Martyn, J., 1728[–1737]. *Historia plantarum rariorum*. London.

Mast, A. R. & Reveal, J. L., 2007. Transfer of *Dodecatheon* to *Primula* (Primulaceae). *Brittonia* 59 (1): 79–82.

Mawe, T. & Abercrombie, J., 1778. *The universal gardener and botanist*. London.

Meacham, S. H., 2009. *Every home a distillery: alcohol, gender, and science in the Colonial Chesapeake*. Baltimore, Maryland.

[Merian, M. S.] Gräffin, M. S., 1679. *Der Raupen wunderbare Verwandelung und sonderbare Blumen-nahrung*. Nuremberg.

[Merian, M. S.] Gräffin, M. S., 1680. *Neues Blumenbuch allen kunstverständigen Liebhabern zu Lust Nutz und Dienst mit Fleisz verfertiget*. Nuremberg.

[Merian, M. S.] Gräffin, M. S., 1683. *Der Raupen wunderbare Verwandlung und sonderbare Blumen-nahrung . . . anderer Theil*. Frankfurt am Main.

Merian, M. S., [1705]. *Metamorphosis insectorum surinamensium. . . .* Amsterdam.

Merian, M. S., 1717. *Der Rupsen begin, voedzel en wonderbaare verandering . . . Derde en laatste Deel*. Amsterdam.

Merian, M. S., 1718. *Erucarum ortus alimentum et paradoxa metamorphosis, in qua origo, pabulum, transformatio . . . exhibentur*. Amsterdam.

[Merian], M. S., 1719. *Over de voortteeling en wonderbaerlyke veranderingen der Surinaemsche insecten. . . .* Amsterdam.

Merian, M. S., 1726. *Dissertatio de generatione et metamorphosibus insectorum surinamensium . . . Accedit appendix transformationum piscium in ranas et ranarum in pisces*. The Hague.

Merian, M. S., 1976. *Schmetterlinge, Käfer und andere Insekten. Leningrader Studienbuch*. 2 volumes. Leipzig & Lucerne.

Merrens, H. R., 1969. The physical environment of Early America. *Geographical review* 59 (4): 528–556.

Meyers, A. R. W., 1997. "The perfecting of natural history": Mark Catesby's drawings of American flora and fauna in the Royal Library, Windsor Castle, pp. 11–27 in McBurney, H. (editor), *Mark Catesby's natural history of America: the watercolors from the Royal Library, Windsor Castle*. Houston & London.

Meyers, A. R. W. & Pritchard, M. B., 1998. Introduction: toward an understanding of Catesby, pp. 1–33 in Meyers, A. R. W. & Pritchard, M. B. (editors), *Empire's nature: Mark Catesby's New World vision*. Chapel Hill, North Carolina & London.

Miller, J., 1997. *The Glorious Revolution*. London.

Miller, P., 1731. *The gardeners dictionary*. London.

Miller, P., 1735. *Appendix to the gardeners dictionary*. London.

Miller, P., 1754. *The gardeners dictionary*. Fourth edition. 3 volumes. London.

Miller, P., 1757. *Figures of the most beautiful, useful, and uncommon plants. . . .* London.

Miller, P., 1768. *The gardeners dictionary*. Eighth edition. London.

Minter, S., 2000. *The Apothecaries' Garden*. London.

Monardes, N., 1577. *Ioyfull nevves out of the newe founde worlde written in Spanish . . . and Englished by John Frampton*. London.

Monardes, N., 1593. *Simplicivm medicamentorvm: ex novo orbe delatorvm, qvorvm in medicina vsvs est, historia; Hispanico sermone duobus libris descripta . . . Latio deinde donata, et in vnum volumen contracta, insuper annotationibus, iconibusque affabre depictis illustrata à Carolo Clvsio*. Antwerp.

Monardes, N., 1605. *Caroli Clvsii Atrebatis . . . Exoticorvm libri decem: quibus animalium, plantarum, aromatum, aliorum que peregrinorum fructuum historiae describuntur item Petri Bellonii observatione; eodem Carolo Clusio interprete*. Antwerp.

Mori, S. A. & Prance, G. T., 1990. Lecythidaceae—Part II. The zygomorphic-flowered New World Lecythidaceae. *Flora neotropica. Monograph* 21 (I).

Morison, R., 1699. *Plantarum historiae universalis Oxoniensis pars tertia*. Oxford.

Morison, S. E., 1942. *Admiral of the ocean sea: a life of Christopher Columbus*. Boston.

Morris, A. D., 1974. Samuel Dale (1659–1739), physician and geologist. *Proceedings of the Royal Society of Medicine* 67: 4–8.

Mortimer, C., 1729. An account of Mr. Mark Catesby's essay towards the natural history of Carolina and the Bahama Islands, with some extracts out of the first three sets. *Philosophical transactions* 36: 425–434.

Mortimer, C., 1748. A continuation of an account of an essay towards a natural history of Carolina, and the Bahama islands; by Mark Catesby. . . . with some extracts out of the appendix. *Philosophical transactions* 45: 157–173.

Morton, C., 1703. *An essay towards the probable solution of this question. Whence come the stork and the turtle, the crane and the swallow, when they know and observe the appointed time of their coming. Or where those birds do probably make their recess and abode, which are absent from our climate at some certain times and seasons of the year*. London.

Nelson, E. C., 2013. The Catesby brothers and the early eighteenth-century natural history of Gibraltar. *Archives of natural history* 40 (2): 357–360.

Nelson, E. C., 2014. Georg Dionysius Ehret, Mark Catesby and Sir Charles Wager's *Magnolia grandiflora*: an early eighteenth-century picture puzzle resolved. *Rhododendrons, camellias and magnolias* 65: 36–51.

Newton, I., 2007. Weather-related mass-mortality events in migrants. *Ibis* 149: 453–467.

Nicholls, S., 2009. *Paradise found. Nature in America at the time of discovery*. Chicago & London.

Northrop, A. R., 1902. Flora of New Providence and Andros, with an enumeration of the plants collected by John I. Northrop and Alice R. Northrop, in 1890. *Memoirs of the Torrey Botanical Club* 12 (1): 1–98, plates 1–19.

Northrop, A. R., 1910. Flora of New Providence and Andros, pp. 118–211 in Osborne, H. F. (editor), *A naturalist in the Bahamas: John I. Northrop, October 12, 1861–June 25, 1891.* New York.

Oberg, B. B. (editor), 2002. *The papers of Thomas Jefferson.* Volume 29. Princeton, New Jersey.

O'Connell, S., 2008. Goupy, Joseph (1689–1769). *Oxford dictionary of national biography.* Online edition (doi:10.1093/ref:odnb/11159). Oxford.

Ogilvie, B. W., 2006. *The science of describing. Natural history in Renaissance Europe.* Chicago & London.

Olsen, S.-E. S., 1997. *Bibliographia discipuli Linnaei. Bibliographies of the 331 pupils of Linnaeus.* Copenhagen.

Olson, S. L. & Reveal, J. L., 2009. Nomenclatural history and a new name for the blue-winged warbler (Aves: Parulidae). *Wilson journal of ornithology* 121: 618–620.

O'Malley, T., 1998. Mark Catesby and the culture of gardens, pp. 147–183 in Meyers, A. R. W. & Pritchard, M. B. (editors), *Empire's nature: Mark Catesby's New World vision.* Chapel Hill, North Carolina & London.

O'Neill, J. & McLean, E. P., 2008. *Peter Collinson and the eighteenth-century natural history exchange.* Philadelphia.

Onuf, P. S. (editor), 2010. *Thomas Jefferson, Notes on the State of Virginia.* New York.

Overstreet, L., 2014. The dates of the parts of Catesby's *The natural history of Carolina, Florida, and the Bahama islands* (London, 1731–1743). *Archives of natural history* 41 (2).

Palmer, W. M. & Braswell, A. L., 1995. *Reptiles of North Carolina.* Chapel Hill, North Carolina.

Parkinson, A., 2007. *Nature's alchemist: John Parkinson, herbalist to Charles I.* London.

Parkinson, J., 1629. *Paradisi in sole paradisus terrestris, or, A garden of all sorts of pleasant flowers which our English ayre will permitt to be noursed vp. . . .* London.

Parkinson, J., 1640. *Theatrum botanicum: the theater of plants.* London.

Pavord, A., 1999. *The tulip.* London.

Peltonen, M., 2004. Bacon, Francis, Viscount St Alban (1561–1626). *Oxford dictionary of national biography.* Online edition (doi:10.1093/ref:odnb/990). Oxford.

Pennant, T., [1761–]1766. *The British zoology.* London.

Pennant, T., 1776. *British zoology.* Fourth edition. London.

Peterson, R. W., 1985. Foreword, p. x in Feduccia, A., *Catesby's birds of Colonial America.* Chapel Hill, North Carolina & London.

Petiver, J., 1695–1703. *Musei Petiveriani centuria. . . .* London.

Petiver, J., not dated [c. 1695]. *Brief directions for the easie making, and preserving collections of all natural curiosities.* London.

Petiver, J., 1702–1709. *Gazophylacii naturae et artis. . . .* London.

Petiver, J., 1707. Herbarium Virginianum Banisteri. *Monthly miscellany or memoirs for the curious,* decas no. 7, December 1707. (Reprinted in Ewan & Ewan 1970: 265–271.)

Petiver, J., 1714. Botanicum hortense IV. *Philosophical transactions* 29: 353–364.

Pieters, F. F. J. M., 1988. The first 150 years of "Artis" and the Artis Library. *Bijdragen tot de Dierkunde* 58 (1): 1–6.

Pieters, F. F. J. M. & Winthagen, D., 1999. Maria Sibylla Merian, naturalist and artist (1647–1717): a commemoration on the occasion of the 350th anniversary of her birth. *Archives of natural history* 26: 1–18.

Plomer, H. R., Bushnell, G. H. & Dix, E. R. McC., 1932. *A dictionary of the printers and booksellers who were at work in England, Scotland and Ireland from 1726 to 1775.* Oxford.

Plukenet, L., 1691–1694. *Phytographia sive stirpium . . . icones.* London.

Plukenet, L., 1696. *Almagestum botanicum sive Phytographiae Pluknetianae onomasticon.* London.

Plukenet, L., 1696–1725. *Opera omnia botanica, in sex tomos divisa*. London.

Plukenet, L., 1700. *Almagesti botanici mantissa . . . cum inice totius operis*. London.

Plukenet, L., 1705. *Amaltheum botanicum*. London.

Plukenett, L., 1942. Leonard Plukenett to William Byrd I. *William and Mary quarterly*, series 2, 4: 244–247.

Plumier, C., 1693. *Description de plantes de l'Amérique*. Paris.

Plumier, C., 1703. *Nova plantarum Americanum genera*. Paris.

Potter, J., 2006. *Strange blooms: the curious lives and adventures of the John Tradescants*. London.

Prance, G. T., 1972. Chrysobalanaceae. *Flora neotropica. Monograph* 9.

Prance, G. T. & Arias, J. R., 1975. A study of the floral biology of *Victoria amazonica* (Poepp.) Sowerby (Nymphaeaceae). *Acta amazonica* 5 (2): 5–35.

Prance, G. T. & Mori, S. A., 1979. Lecythidaceae—part I. The actinomorphic-flowered New World Lecythidaceae. *Flora neotropica. Monograph* 21 (II): 1–270.

Prance, G. T. & Sothers, C. A., 2003a. Chrysobalanaceae 1: *Chrysobalanus* to *Parinari*. *Species plantarum: Flora of the World* part 9: 1–319.

Prance, G. T. & Sothers, C. A., 2003b. Chrysobalanaceae 2: *Acioa* to *Magnistipula*. *Species plantarum: Flora of the World* part 10: 1–268.

Preston, D. & Preston, M., 2004. *A pirate of exquisite mind: the life of William Dampier*. London.

Prinzinger, R. & Siedle, K., 1988. Ontogeny of metabolism, thermoregulation and torpor in the house martin *Delichon u. urbica* (L.) and its ecological significance. *Oecologia* 76: 307–312.

Pursh, F., 1814. *Flora Americae septentrionalis; or, a systematic arrangement of the plants of North America*. London.

Quinn, D. B. (editor), 1952–1955. *The Roanoke voyages 1584–1590: documents to illustrate the English voyages to North America under the patent granted to Walter Raleigh in 1584*. London.

Randolph, L. R., 1994. An ethnobiological investigation of Andros Island, Bahamas. Doctoral thesis, Miami University, Oxford, Ohio.

Ransome, D. R., 2004. Lawson, John (d. 1711). *Oxford dictionary of national biography*. Online edition (doi:10.1093/ref:odnb/16203). Oxford.

Raphael, S., 2004. Gray, Christopher (1694–1764). *Oxford dictionary of national biography*. Online edition (doi:10.1093/ref:odnb/37479). Oxford.

Raven, C. E., 1947. *English naturalists from Neckam to Ray: a study in the making of the modern world*. Cambridge.

Raven, C. E., 1964. John White's significance for natural history, pp. 47–52 in Hulton, P. & Quinn, D. B. (editors), *The American drawings of John White, 1577–1590*. London.

Ray, J. (editor), 1678. *The ornithology of Francis Willughby. . . .* London.

Ray, J. (editor), 1686a. *Francisci Willughbeii de historia piscium libri quatuor. . . .* Oxford.

Ray, J., 1686b. *Historia plantarum*. Volume 1. London.

Ray, J., 1688. *Historia plantarum*. Volume 2. London.

Ray, J., 1691. *A collection of English words not generally used*. Second edition. London.

Ray, J., 1693. *Synopsis methodica animalium quadrupedum et serpenti generis. . . .* London.

Ray, J., 1698. An account of a book. Museo de Piante rare della Sicilia, Malta, Corsica, Italia, Piemonte e Germania, &c. di Don Paolo Boccone, &c. with additional remarks. *Philosophical transactions* 20: 462–468.

Ray, J., 1704. *Historia plantarum*. Volume 3. London.

Ray, J., 1710. *Historia insectorum: opus posthumum. . . .* London.

Ray, J., 1713. *Synopsis methodica avium et piscium: opus posthumum. . . .* London.

Rediker, M., 1987. *Between the devil and the deep blue sea: merchant seamen, pirates and the Anglo-American maritime world, 1700–1750*. Cambridge

Reeds, K., 2009. "Don't eat, don't touch": Roanoke colonists, natural knowledge, and dangerous plants of North America, pp. 51–57 in Sloan, K. (editor), *European visions: American voices*. London.

Reitsma, E. & Ulenberg, S., 2008. *Maria Sibylla Merian and daughters. Women of art and science*. Amsterdam, Los Angeles & Zwolle.

Reveal, J. L., 1997. *Erythrina herbacea*, p. 469 in Jarvis, C. & Turland, N. J. (editors), Typification of Linnaean specific and varietal names in the Leguminosae (Fabaceae). *Taxon* 46: 457–485.

Reveal, J. L., 2009. Identification of the plant and associated animal images in Catesby's *Natural history*, with nomenclatural notes and comments. *Rhodora* 111: 273–388.

Reveal, J. L., 2012a. A nomenclature summary of the plant and animal names based on images in Mark Catesby's *Natural history* (1729–1747)). *Phytoneuron* 2012-11: 1–32.

Reveal, J. L., 2012b. The shell of Catesby's hermit-crab. *Phytoneuron* 2012-18: 1–2.

Reveal, J. L., 2013. Identification of the plants and animals illustrated by Mark Catesby for his *Natural history of Carolina, Florida and the Bahama islands*. *Phytoneuron* 2013-6: 1–55.

Reveal, J. L., Gandhi, K. N. & Jarvis, C. E., 2014. Epitypification of *Laurus borbonia* L. (*Lauraceae*). *Taxon* 63 (5): 918–920.

Richey-Abbey, L., 2012. Bush medicine in the Family Islands: the medical ethnobotany of Cat Island and Long Island, Bahamas. Doctoral thesis, Miami University, Oxford, Ohio.

Rickman, M. L., 2011. Making the "Herball": John Gerard and the fashioning of an Elizabethan herbarist. Doctoral thesis, University of Oklahoma, Norman, Oklahoma.

Riley, M., 2006. The club at the Temple Coffee House revisited. *Archives of natural history* 33: 90–100.

Riley, M., 2011. "Procurers of plants and encouragers of gardening": William and James Sherard, and Charles du Bois, case studies in late seventeenth- and early eighteenth-century botanical and horticultural patronage. Doctoral thesis, University of Buckingham.

Robertson, R., 2005. Large conchs (*Strombus*) are endangered herbivores having many predators and needing dense populations of adults to reproduce successfully. *American conchologist* 33 (3): 3–7.

Robertson, R., 2011a. Catesby's gallery: a trailblazing naturalist in the New World. *Natural history* 119 (4): 32–37.

Robertson, R., 2011b. Cracking a queen conch (*Strombus gigas*), vanishing uses, and rare abnormalities. *American conchologist* 39 (3): 21–24.

Robinson, F. J. G. & Wallis, P. J., 1975. *Book subscription lists: a revised guide*. Newcastle-upon-Tyne.

Rochefort, C. de, 1658. *Histoire naturelle et morale des iles Antilles de l'Amerique. Enrichie de plusieurs belles figures des raretez les plus considerables qui y sont d'écrites. Avec vn vocabulaire Caraïbe*. Rotterdam.

Rochefort, C. de, 1666. *The history of the Caribby-Islands, viz. Barbados, St Christophers, St Vincents, Martinico, Dominico, Barbouthos, Monserrat, Mevis* [*sic*], *Antego, &c. in all XXVIII. In two books. The first containing the natural; the second, the moral history of those islands*. London.

Romanes, G. J., 1882. *Animal intelligence*. London.

Rountree, H., 1990. *Pocahontas's people: the Powhatan Indians of Virginia through four centuries*. Norman, Oklahoma.

Rowan, M. K., 1968. The origins of European swallows "wintering" in South Africa. *The ostrich* 39: 76–84.

Royle, G., 1995. Family links between George London and John Rose: new light on the "pineapple paintings." *Garden history* 23 (2): 246–249.

Rumphius, G. E., 1741–1750. *Herbarium Amboinense*. . . . Amsterdam.

Russell, A. P. & Bauer, A. M., 1991. *Anolis grahami*. *Catalogue of American amphibians and reptiles* 514: 1–4.

Russo, J. B. & Russo, J. E., 2012. *Planting an empire: the early Chesapeake in British North America*. Baltimore, Maryland.

Savage, S., 1940. *Synopsis of the annotations by Linnaeus and his contemporaries in his library of printed books*. London.

Schaw, J., 1921. *Journal of a lady of quality; being the narrative of a journey from Scotland to the West Indies, North Carolina, and Portugal, in the years 1774 to 1776*. New Haven, Connecticut.

Schleucher, E., 2004. Torpor in birds: taxonomy, energetics, and ecology. *Physiological and biochemical zoology* 77: 942–949.

Schmidt, H., 1965. Der "Hortus Elthamensis" aus der Bibliothek Carl von Linnés. *Feddes repertorium* 70: 69–108, tab. xvi–xxxii.

Schweizer, K. W., 2004. Stuart, John, third Earl of Bute (1713–1792). *Oxford dictionary of national biography*. Online edition (doi:10.1093/ref:odnb/26716). Oxford.

Schwenk, K., 1988. Comparative morphology of the lepidosaur tongue and its relevance to squamate phylogeny, pp. 569–598 in Estes, R. & Pregill, G. (editors), *Phylogenetic relationships of the lizard families: essays commemorating Charles L. Camp*. Stanford, California.

Sealey, N. E., 1994. *Bahamian landscapes: an introduction to the geography of the Bahamas*. Second edition. Nassau.

Secord, A., 1994a. Science in the pub: artisan botanists in early nineteenth-century Lancashire. *History of science* 32: 269–315.

Secord, A., 1994b. Corresponding interests: artisans and gentlemen in nineteenth-century natural history. *British journal for the history of science* 27: 383–408.

Segal, S., 2012. *Belief in nature. Flowers with a message*. Amsterdam.

Serventy, D. L., 1970. Torpidity in the white-backed swallow. *The emu* 70: 27–28.

Shammas, C., 1980. The domestic environment in Early Modern England and America. *Journal of social history* 14 (1): 3–24.

Sharpe, R. B. & Wyatt, C. W., 1885–1891. *A monograph of the Hirundinidae: or family of swallows*, Volume 1. London.

Shaw, G., 1801. *General zoology or systematic natural history*. London.

Simpson, M. B., Jr. & Simpson, S. W., 2008. John Lawson's *A new voyage to Carolina*: notes on the publication history of the London (1709) edition. *Archives of natural history* 35: 223–242.

Simpson, M. B., Jr., Simpson, S. W. & Johnston, D. W., 2010. Zoological material for John Lawson's "Compleat History" of Carolina: specimens recorded in Hans Sloane's catalogues. *Archives of natural history* 37: 333–345.

Sloan, K., 2007. *A new world: England's first view of America*. Chapel Hill, North Carolina.

Sloan, K. (editor), 2009. *European visions: American voices*. (*British Museum research publication* 172.) London.

Sloan, K., 2012. Sloane's "Pictures and drawings in frames" and "Books of miniature & painting, designs, &tc.," pp. 168-189 in Hunter, M., Walker, A. & MacGregor, A. (editors), *From books to bezoars*. London.

Sloane, H., 1696. *Catalogus plantarum quae in insula Jamaica sponte proveniunt*. London.

Sloane, H., 1707. *A voyage to the islands Madera, Barbados, Nieves, S. Christophers and Jamaica. . . .* Volume 1. London.

Sloane, H., 1725. *A voyage to the islands Madera, Barbados, Nieves, S. Christophers and Jamaica. . . .* Volume 2. London.

Sloane, H., 1727. An account of elephants teeth and bones found under ground. *Philosophical transactions* 35: 457–471, 497–514.

Smith, B. S., 1986. Hannah English Williams: America's first woman natural history collector. *South Carolina historical magazine* 87 (2): 83–92.

Smith, G., 1957. *A history of England*. New York.

Smith, H. A. M., 1905. The town of Dorchester in South Carolina—a sketch of its history. *South Carolina historical magazine* 6: 62–95.

Smith, H. A. M., 1918. Charleston and Charleston Neck. The original grantees and the settlements along the Ashley and Cooper rivers. *South Carolina historical magazine* 19 (1): 3–76.

Smith, J., 1624. *Generall historie of Virginia, New-England, and the Summer Isles: with the names of the Adventurers, Planters, and Governours from their first beginning, Ano: 1584. to this present 1624.* London.

Smith, J. E., 1821. *A selection of the correspondence of Linnaeus and other naturalists.* Volume 1. London.

Smith, P. H. & Findlen, P. (editors), 2002. *Merchants and marvels: commerce, science, and art in early modern Europe.* New York.

Snapp, J. R., 2000. Christoph von Graffenried. *American national biography.* Online edition (http://www.anb.org.go.libproxy.wfubmc.edu/articles/01/01-00336.html).

Spary, E. C., 2000. *Utopia's garden: French natural history from Old Regime to Revolution.* Chicago.

Sperling, C. F. D., 1895a. Thomas Jekyll. *The Essex review* 4: 254–261.

Sperling, C. F. D., 1895b. William Holman. *The Essex review* 4: 261–266.

Spotswood, A., 1882. *The official letters of Alexander Spotswood, Lieutenant-Governor of the Colony of Virginia, 1710–1722.* 2 volumes. Richmond, Virginia.

Sprat, T., 1667. *The history of the Royal Society of London, for the Improving of Natural Knowledge.* London.

Stafleu, F. A. & Cowan, R. S., 1976–1988. *Taxonomic literature.* Second edition. Utrecht.

Stearn, W. T., 1957a. An introduction to the *Species plantarum* and cognate botanical works of Carl Linnaeus, volume 1, pp. 1–157 in Linnaeus, C., *Species plantarum* (Ray Society facsimile). London.

Stearn, W. T., 1957b. Linnaeus's sexual system of classification, volume 1, pp. 24–35 in Linnaeus, C., *Species plantarum* (Ray Society facsimile). London.

Stearn, W. T., 1957c. The preparation of the *Species plantarum* and the introduction of binomial nomenclature, volume 1, pp. 65–74 in Linnaeus, C., *Species plantarum* (Ray Society facsimile). London.

Stearn, W. T., 1958. Publication of Catesby's *Natural history of Carolina. Journal of the Society for the Bibliography of Natural History* 3 (6): 328.

Stearn, W. T., 1971. Linnaean classification, nomenclature, and method, pp. 242–249 (appendix) in Blunt, W., *The compleat naturalist: a life of Linnaeus.* Second edition. London.

Stearns, R. P., 1952. James Petiver, promoter of natural science, c. 1663–1718. *Proceedings of the American Antiquarian Society* 62: 241–365.

Stearns, R. P., 1970. *Science in the British colonies of America.* Urbana, Illinois.

Stephenson, M. A., 1961. (Cocke-Jones Lots) Block 31. Colonial Williamsburg Foundation Library research report series—1614 (http://research.history.org/DigitalLibrary/View/index.cfm?doc=ResearchReports%5CRR1614.xml).

Stevens, J. (editor), 1708–1710. *A new collection of voyages and travels.* London.

Stevenson, A., 1961. *Catalogue of botanical books in the collection of Rachel McMasters Miller Hunt.* Pittsburgh, Pennsylvania.

Stiverson, G. A. & Butler, P. H., III (editors), 1977. Virginia in 1732: the travel journal of William Hugh Grove. *Virginia magazine of history and biography* 85 (1): 18–44.

Swanton, J. R., 1922. The distribution of Indian tribes in the Southeast about the year 1715. *Bureau of American Ethnology, bulletin* 73.

Swem, E. G., 1949. *Brothers of the spade: correspondence of Peter Collinson, of London, and of John Custis, of Williamsburg, Virginia, 1734–1746.* Worcester, Massachusetts.

Swift, A. E. B. & Swift, T. R., 1993. Ciguatera. *Clinical toxicology* 31: 1–29.

Switzer, S., 1718. *Ichnographia rustica.* Volume 1. London.

Taylor, W. B., 1913. *Catalogue of the library of the Society of Apothecaries. Edited and with an introduction by J. E. Harting.* London.

Thompson, D'A. W. (translator), 1910. Aristotle, *Historia animalium*. Oxford.

Tinling, M. (editor), 1977. *The correspondence of the three William Byrds of Westover, Virginia 1684–1776*. Charlottesville, Virginia.

Tjaden, W. L., 1977. William and James Sherard and John James Dillenius: some errors in the biographies. *Journal of the Society for the Bibliography of Natural History* 8: 143–147.

Tongiorgi-Tomasi, L., 1997. *An Oak Spring flora: flower illustration from the fifteenth century to the present time*. Upperville, Virginia.

Tournefort, J. P. de, 1700. *Institutiones rei herbariae; editio altera; tomus primus*. Paris.

Tradescant, J. [the Elder], 1634. *Plantarum in horto Iohannem Tradescanti nascentium catalogus. . . .* [no place].

Tradescant, J. [the Younger], 1656. *Musaeum Tradescantium: or, a collection of rarities preserved at South-Lambeth neer London*. London.

Trepp, A.-Ch., 2009. *Von der Glückseligkeit alles zu wissen: die Erforschung der Natur als religiöse Praxis in der frühen Neuzeit*. Frankfurt & New York.

Trevor-Roper, H. R., 2006. *Europe's physician: the life of Sir Theodore de Mayerne, 1573–1655*. New Haven, Connecticut.

Trew, C. J., 1754. *Plantae selectae quarum imagines ad exemplaria naturalis Londini in hortis curiosorum nutrita manu arteficiosa. . . .* Decuria 4. Nuremberg.

Trew, C. J., 1772. *Plantae selectae quarum imagines ad exemplaria naturalis Londini in hortis curiosorum nutrita manu arteficiosa. . . .* Decuria 9. Nuremberg.

[Turner, D.], 1835. *Extracts from the literary and scientific correspondence of Richard Richardson, M.D., F.R.S., of Brierly, Yorkshire. . . .* Yarmouth.

Tyler, W. M., 1937. Turkey vulture (*Cathartes aura*). *United States National Museum bulletin* 167 (1): 12–28.

Vane-Wright, R. I. & Hughes, H. W. D., [2005]. *The Seymer legacy: Henry Seymer and Henry Seymer Jnr of Dorset and their entomological paintings, with a catalogue of butterflies and plants (1755–1783)*. Cardigan.

Vickery, A., 2009. *Behind closed doors: at home in Georgian England*. New Haven, Connecticut & London.

Vines, S. H. & Druce, G. C., 1914. *An account of the Morisonian herbarium . . . with biographical and critical sketches of Morison and the two Bobarts and their works and the early history of the Physic Garden, 1619–1720*. Oxford.

Vogel, S., 1968. Chiropterophilie in der neotropischen Flora. *Flora* (Abt. B) 157: 562, 602.

Vogel, S., 1969. Chiropterophilie in der neotropischen Flora. *Flora* (Abt. B) 158: 289–323.

Wahlgren, R., 1999. Herpetology in the *Transactions of the Royal Swedish Academy of Sciences*. A listing of titles 1739–1825, translated into English, with annotations and unabridged translations of selected contributions and a brief history of the Academy. *International Society for the History and Bibliography of Herpetology* 1 (2): 7–26.

Wahlgren, R., 2012. Carl Linnaeus and the Amphibia, pp. 5–38 in Bell, C. J. (editor), The herpetological legacy of Linnaeus: a celebration of the Linnaean tercentenary. *Bibliotheca herpetologica* 9.

Walker, A., MacGregor, A. & Hunter, M. (editors), 2012. *From books to bezoars*. London.

Walley, H. D., 2003. *Liochlorophis, L. vernalis. Catalogue of American amphibians and reptiles* 776. Salt Lake City.

Weidensaul, S., 2007. *Of a feather. A brief history of American birding*. Orlando, Florida.

Weir, R. M., 1997. *Colonial South Carolina: a history*. Columbia, South Carolina.

Wells, C., 1993. The planter's prospect: houses, outbuildings, and rural landscapes in eighteenth-century Virginia. *Winterthur portfolio* 28 (1).

Wenger, M., 1981. Westover: William Byrd's mansion reconsidered. Master of Arts thesis, University of Virginia.

Wettengl, K., 1998. Maria Sibylla Merian. Artist and naturalist between Frankfurt and Suriname, pp. 12–36 in Wettengl, K. (editor), *Maria Sibylla Merian 1647–1717 artist and naturalist*. Berlin.

White, A. S., 1965. *Bibliography of regimental histories of the British army*. London.

White, G., 1774. Account of the house-martin, or martlet. In a letter . . . to the Hon. Daines Barrington. *Philosophical transactions* 64: 196–201.

White, G., 1775. Of the house-swallow, swift, and sand-martin . . . in three letters to the Hon. Daines Barrington. . . . *Philosophical transactions* 65: 258–276.

White, G., 1977. *The natural history of Selborne*. London.

Wilbur, R. L., 1990. Identification of the plants illustrated and described in Catesby's *Natural history of the Carolinas, Florida and the Bahamas*. *Sida* 14: 29–48.

Wiles, R. M., 1957. *Serial publication in England before 1750*. Cambridge.

Wilkins, G. L., 1953. A catalogue and historical account of the Sloane shell collection. *Bulletin of the British Museum (Natural History), historical series* 1: 1–48, plates 1–12.

Wilkinson, C., 1929 *William Dampier. London.*

Willson, E. J., 1982. *West London nursery gardens: the nursery gardens of Chelsea, Fulham . . . and part of Westminster, founded before 1900*. Fulham & Hammersmith Historical Society.

Wilson, A., 1808–1814. *American ornithology*. 7 volumes. Philadelphia.

Winnington-Ingram, A. F., 1908. *The early English colonies: a summary of the lecture with additional notes and illustrations. . . .* London.

Winterrowd, M. F., Gergits, W. F., Laves, K. S. & Weigl, P. D., 2005. Relatedness within nest groups of the southern flying squirrel using microsatellite and discriminant function analyses. *Journal of mammalogy* 86: 841–846.

Wolf, E., II, 1958. The dispersal of the library of William Byrd of Westover. *Proceedings of the American Antiquarian Society* 68: 19–106.

Wolf, E., II, 1978. More books from the library of the Byrds of Westover. *Proceedings of the American Antiquarian Society* 88: 51–82.

Woodfin, M. H. (editor) & Tinling, M. (translator), 1942. *Another secret diary of William Byrd of Westover, 1739–1741; with letters and literary exercises 1696–1726*. Richmond, Virginia.

Wright, L. B. & Tinling, M. (editors), 1941. *The secret diary of William Byrd of Westover, 1709–1712*. Richmond, Virginia.

Wright, L. B. & Tinling, M. (editors), 1958. *The London diary (1717–1721) and other writings*. New York.

Wulf, A., 2009. *The brother gardeners: botany, empire and the birth of an obsession*. New York.

Zierden, M., Linder, H. & Anthony R., 1999. *Willtown: An archeological and historical perspective*. Charleston, South Carolina.

CONTRIBUTORS

KRAIG ADLER is professor of biology at Cornell University and formerly was chairman of the Department of Neurobiology and Behavior and vice provost for Life Sciences. His research has been on amphibians and reptiles, and he has conducted fieldwork in China, Taiwan, Colombia, Mexico, and the United States. Professor Adler has had a long interest in the history of natural history; his books include *Herpetology of China* (1993), *Encyclopedia of reptiles and amphibians* (2002), and the three-volume *Contributions to the history of herpetology* (1989–2012), which contains biographies of nearly eight hundred naturalists, including Mark Catesby, Alexander Garden, and Carl Linnaeus.

AARON BAUER is the Gerald M. Lemole, M.D., Professor of Integrative Biology at Villanova University, Pennsylvania. A native of New York, he obtained degrees in zoology and history from Michigan State University in 1982 and a doctorate in zoology from the University of California at Berkeley in 1986. His research interests are in the systematics, morphology, and biogeography of reptiles, and he works extensively in Africa, India, and the South Pacific. He has a long-standing interest in the history and bibliography of herpetology. Professor Bauer has written several books and more than 525 papers in his areas of interest.

JANET BROWNE is a professor of the history of science at Harvard University. Until 2006, she lived and worked in England. She has specialized in reassessing Charles Darwin's work, first as an associate editor of *The correspondence of Charles Darwin* and more recently as the author of a two-volume biographical study that integrates Darwin's science with his life and times. This biography was awarded the James Tait Black Prize for nonfiction and the Pfizer Prize from the History of Science Society in 2004. At Harvard, Professor Browne teaches an introductory history of Darwinism from Darwin's day to now.

DAVID J. ELLIOTT has been executive director of the Catesby Commemorative Trust since 2002 and was executive producer of *The Curious Mister Catesby* documentary. He was educated at Dartington Hall School in England, where he developed his interest in natural history, and Trinity College in Connecticut. He retired from a thirty-seven-year career at Procter & Gamble as director of international trade and has served as deputy assistant secretary at the U.S. Department of Commerce. He was founder and first chairman of the Kiawah Conservancy, which is devoted to the protection of wildlife habitat.

W. HARDY ESHBAUGH is professor emeritus of botany, Miami University, Oxford, Ohio. In 2005 he received the Audubon Great Egret Award in conservation. He is president of the EWH Charitable Foundation and a fellow of the American Association for the Advancement of Science, of the Ohio Academy of Science, and of the Explorers Club, and he serves on the boards of the Atlantic Salmon Federation, the American Botanical Council, and the Catesby Commemorative Trust. He is a past president of the American Institute of Biological Sciences, the Botanical Society of America, the American Society of Plant Taxonomists, and the Society for Economic Botany.

KAY ETHERIDGE is professor of biology at Gettysburg College. Her current scholarship centers on the integration of natural history images and the history of biology, with a focus on Maria Sibylla Merian in recent publications. Etheridge also is investigating Renaissance natural history art and illustration in her ongoing work. Her publications in physiology and ecology include work on tropical bats, manatees, hibernation in lizards, aestivation in aquatic salamanders, and hormonal control of metabolism. In addition to biology courses, she teaches a seminar on creativity in art and science and a course on Renaissance "curiosity cabinets."

STEPHEN HARRIS is the Druce Curator of the Oxford University Herbaria. His research concentrates on the use of molecular markers in evolutionary and conservation biology, especially hybridization, polyploidy, the evolutionary consequences of human-mediated plant movement, and conservation genetics. He is also interested in the problems of using herbarium specimens as a source of DNA for evolutionary studies and in the history of botany.

VALERIE HERBERT lives in Sudbury, Suffolk, where Catesby spent his boyhood, and is a member of the Sudbury Museum Trust. Since her retirement as a daily newspaper journalist, she has indulged her passion for history by researching, writing, and lecturing. She has collaborated as writer or editor in the publication of three books: *No glorious dead*, the effects of twentieth-century wars on Sudbury; a history of the Saxon town's street names; and, in 2013, a celebration of the century since a princess unveiled in Sudbury the national monument to the artist Thomas Gainsborough, who was born in the town.

SUZANNE LINDER HURLEY graduated from Converse College; completed her master's degree at Wake Forest University; and gained her doctorate from the University of South Carolina, where she was a research fellow in the Institute for Southern Studies. In 2005, Converse College awarded her an honorary degree of doctor of humane letters. She is the author or co-author of ten books on the history of the Carolinas, including *Anglican churches in Colonial South Carolina*, *A river in time: the Yadkin–Pee Dee River system*, and *From the Highlands to high finance: the Carolina McColls.*

CHARLES E. JARVIS is in the Department of Life Sciences at the Natural History Museum, London, with particular interests in seventeenth- and eighteenth-century plant collections, especially those assembled by Sir Hans Sloane. He has published extensively on the scientific names published by Carl Linnaeus and is the author of *Order out of chaos: Linnaean plant names and their types* (2007).

SHEPARD KRECH III is professor emeritus of anthropology at Brown University and a research associate in the Department of Anthropology, National Museum of Natural History, Smithsonian Institution. A trustee of the National Humanities Center and recipient of numerous fellowships and grants, he is the author or editor of more than 150 books, essays, and reviews, including *The ecological Indian: myth and history*, *Encyclopedia of world environmental history* (edited with John McNeill and Carolyn Merchant), *Spirits of the air: birds and American Indians in the South*, and *Indigenous knowledge and the environment in Africa and North America* (edited with David Gordon).

MARK LAIRD is adjunct professor in the history of landscape architecture, Harvard University. Based in Toronto as a consultant in historic landscape preservation, he advises on sites in Europe and North America. His research on eighteenth-century planting developed from his practice and is presented in *The flowering of the landscape garden* (1999). Educated at the Universities of Oxford, Edinburgh, and York, he was research fellow at Chelsea Physic Garden in London, has been twice a fellow at Dumbarton Oaks, and has taught at the University of Toronto. He has recently completed *A natural history of English gardening: 1650–1800.*

HENRIETTA MCBURNEY was keeper of fine and decorative art at Eton College (2004–2013), previously having served as deputy curator of prints and drawings in the Royal Library, Windsor Castle. She was editor of the series *The natural history drawings at Windsor Castle* and of the Royal Collection's series *The paper museum of Cassiano dal Pozzo: catalogue raisonné,* to which she is contributing a volume on the ornithological drawings collected by Cassiano. Her publications include *Mark Catesby's natural history of America* (1997), and she is currently writing a book titled *The art and science of Mark Catesby*.

JUDITH MAGEE is collections development manager in the Library of the Natural History Museum, London. She manages the historical collections of published literature, manuscripts, and artwork of the library. She has been at the Natural History Museum since 1994, working primarily with the botanical and zoological library collections. Her main interest lies in the collectors, artists, and naturalists of the eighteenth and nineteenth centuries. She is a member of the Society for the History of Natural History and a patron of the Florilegium Society at Sheffield Botanic Gardens.

SARAH HAND MEACHAM, a graduate of Smith College, of Vanderbilt University, and of the University of Virginia, is an associate professor of history at Virginia Commonwealth University. Professor Meacham is the author of *Every home a distillery: alcohol, gender and technology in the early Chesapeake*. She has published articles demonstrating that it was women's work to make cider and manage taverns in early Virginia, explaining why Colonial Virginians living in the woods kept pets, and examining the first women's benevolent association in Virginia. She is currently working on a book about how Colonial Americans felt and the history of emotions.

CYNTHIA P. NEAL is the producer and director of *The Curious Mister Catesby* documentary film. She is a veteran of the motion picture production and broadcast industries. Her productions have won Emmy and other major awards, including at the New York Film Festival. She has produced a significant body of work devoted to wildlife and natural habitat conservation across the North and South American continents. She brings to her productions a background in fine arts and performance. She was educated at West Virginia University and Carnegie Institute in theater and stage as well as broadcast communications.

CHARLES NELSON was educated at Portora Royal School, Enniskillen, Northern Ireland, and University College of Wales, Aberystwyth. After completing his doctoral research at the Australian National University, Canberra, he was appointed horticultural taxonomist in the National Botanic Gardens, Glasnevin, Dublin. Now a freelance author and editor, he was honorary editor of *Archives of natural history* from 1999 until 2012 and was awarded the Founders' Medal of the Society for the History of Natural History in 2013. Two of his books have received the accolade of Reference Book of the Year from the Garden Media Guild.

LESLIE K. OVERSTREET is the Curator of Natural-History Rare Books and heads the Smithsonian Libraries' Joseph F. Cullman 3rd Library of Natural History. She has co-authored papers on early books in the natural sciences and on the history and technologies of printing as they affect scientific publications. For many years her primary research has focused on the printing of Mark Catesby's *The natural history of Carolina, Florida and the Bahama islands*. She has given talks on the work and served as a consultant to the documentary film *The Curious Mister Catesby*.

FLORENCE F. J. M. PIETERS studied biology and philosophy of science at Radboud University Nijmegen. In 1969, after a period of college teaching, she was appointed as scientific librarian of the Artis Library, University of Amsterdam, where she later became conservator. At the Artis Library, which has an outstanding collection of natural history works, she became fascinated with the work of Maria Sibylla Merian. Since her retirement she has continued her investigations into the history of early Dutch collections and collectors of natural history, working as guest researcher at the Artis Library and at the Naturalis Biodiversity Center, Leiden.

GHILLEAN PRANCE was educated at Malvern College and Keble College, Oxford. He has been senior vice president for science at the New York Botanical Garden and director of the Royal Botanic Gardens, Kew. He is currently McBryde Senior Fellow at the National Tropical Botanical Garden, Hawaii, as well as scientific director and a trustee of the Eden Project in Cornwall and visiting professor at Reading University. He was elected a fellow of the Royal Society in 1993 and was knighted in July 1995. Sir Ghillean Prance continues to be active with research in plant systematics and in conservation of the tropical rain forest.

DIANA PRESTON is an historian, writer, and broadcaster educated at Oxford University. Her husband, MICHAEL PRESTON, also is an Oxford graduate. Now based in London, they are joint authors of several books, including a biography of William Dampier, *A pirate of exquisite mind.* Their book *Before the fall-out: from Marie Curie to Hiroshima* won the Los Angeles Times Prize for Science and Technology, and their latest book, *The dark defile*, deals with Britain's first disastrous foray into Afghanistan in 1839–1842.

KAREN REEDS is an independent scholar and curator living in Princeton. Dr. Reeds is a visiting scholar in the history and sociology of science at the University of Pennsylvania and a member of the Princeton Research Forum. She served as the science medicine editor at Rutgers University Press and the University of California Press. Her four books deal with the history of biology and medicine from the Middle Ages to our own time. For the 2007 Linnaeus tercentennial she curated the exhibition *Come into a New World: Linnaeus and America* at the American Swedish Historical Museum and New Jersey State Museum.

JAMES L. REVEAL is an adjunct professor at Cornell University, an honorary curator at the New York Botanical Garden, and professor emeritus at the University of Maryland. A vascular plant taxonomist and botanical historian, he is the author of more than five hundred books and scientific articles, ranging from floristics in North America, monographic studies mainly of wild buckwheat (*Eriogonum*) and relatives, vascular plant classification, supergeneric nomenclature, and botanical explorations and discoveries in North America. He received his undergraduate and master's degrees from Utah State University (1963, 1965) and his doctorate in systematic botany from Brigham Young University (1969).

ROBERT ROBERTSON is emeritus curator of malacology at the Academy of Natural Sciences of Drexel University, Philadelphia. He traveled widely with his parents before going to college. Among other places, he lived in the Bahamas, hence his subsequent interest in Catesby. Later, he went to Stanford University and received a doctorate from Harvard. He went on to spend his entire career studying mollusks at the academy, where he also studied its copy of the first edition of Catesby's *The natural history of Carolina, Florida and the Bahama islands.* He has published nearly one hundred papers.

MARCUS SIMPSON was educated at Davidson College and the University of North Carolina, with postdoctoral training at Yale University and Johns Hopkins University. After a thirty-five-year career in academic medicine, he retired as director of clinical laboratories and vice-chairman of pathology at Wake Forest University School of Medicine in 2011. His historical writings include biographies in the *Oxford dictionary of national biography* and *American national biography* and papers in *Archives of natural history* and the *North Carolina historical review.* His nonmedical research interests involve ornithology, natural history art and illustration, biography, and publication history.

INDEX TO SCIENTIFIC NAMES OF ANIMALS AND PLANTS

INDEX TO COMMON NAMES OF ANIMALS AND PLANTS

GENERAL INDEX

Mark Catesby's name is abbreviated as "MC" throughout.